Quaternary Period in Saudi Arabia

Volume 2

Sedimentological, Hydrogeological, Hydrochemical, Geomorphological, Geochronological and Climatological Investigations in Western Saudi Arabia

A Cooperative Research Project
of the University of Petroleum and Minerals, Dhahran,
and the Austrian Academy of Sciences, Vienna

Edited by

Abdul Raof Jado
and Josef G. Zötl

Springer-Verlag Wien NewYork

Dr. ABDUL RAOF JADO
Associate Professor, Earth Sciences Department, University of Petroleum and Minerals, Dhahran, Saudi Arabia

Prof. Dr. JOSEF GEORG ZÖTL
Institute for Geothermics and Hydrogeology, Research Centre, Graz, Austria and Quaternary Commission, Austrian Academy of Sciences, Vienna, Austria

Part of German contributions translated by EUGENIA LAMONT

Printed in Austria

With 124 Figures and 6 Folding Plates

Library of Congress Cataloging in Publication Data. (Revised for volume 2) Main entry under title: Quaternary period in Saudi Arabia. Bibliography: p. v. 1, p. [312]–330. Includes index. CONTENTS: 1. Sedimentological, hydrogeological, hydrochemical, geomorphological and climatological investigations in central and eastern Saudi Arabia. – 2. Sedimentological, hydrogeological, hydrochemical, geochronological, geomorphological, and climatological investigations in western Saudi Arabia. 1. Geology, Stratigraphic-Quaternary. 2. Geology – Saudi Arabia. I. Al-Sayari, Saad S. II. Zötl, Josef. III. Jāmi'at al-Bitrūl wa-al-Ma'ādin. IV. Österreichische Akademie der Wissenschaften. QE696.Q34 551.7'9'09538 78-1708

ISBN 3-211-81749-2 Springer-Verlag Wien–New York
ISBN 0-387-81749-2 Springer Verlag New York–Wien

Foreword

The publication of this second volume of "Quaternary Period in Saudi Arabia" is another fruitful result of international scientific cooperation. It is the outcome of the second phase of continued and expanded field visits, laboratory analysis, map and section drawings, and scientific interpretations. The cooperative research work has been carried out by the University of Petroleum and Minerals and the Austrian Academy of Science; in addition, the West German University of Karlsruhe has been strongly involved. The expansion is evidenced by the larger number of participants from the different institutions, including graduate students, more frequent and longer field excursions, and the coverage of new and large areas. In fact, the two volumes now cover most of the Kingdom.

As in the first volume, the scientists, administrators, and officials who contributed to this volume are too numerous to be mentioned in this brief foreword. However, the continued supervision and contributions of Professor Dr. ZÖTL are specially appreciated. Outstanding overall efforts and enthusiasm are undeniably credited to Professor Dr. HÖTZL. Field arrangement and responsibilities in Saudi Arabia were kindly handled by Dr. JADO in addition to his scientific participation and contributions.

It is a pleasure to acknowledge the introduction of this second volume on the Quaternary Geology of the Kingdom of Saudi Arabia. We hope the scientific contributions presented here in the form of original research will benefit students of earth sciences as well as government and industry organizations. Above all we hope this humble effort will contribute to the progress and welfare of this country.

Dhahran, March 1984 BAKR ABDULLAH BAKR

Contents

List of Authors

AL-SAYARI, SAAD S., Department of Earth Sciences, University of Petroleum and Minerals, Dhahran, Saudi Arabia; new address: Saudi Vitrified Clay Pipe Co. Ltd., P. O. Box 6415, Riyadh, Saudi Arabia.

ANTON, DANILO, Research Institute, University of Petroleum and Minerals, Dhahran, Saudi Arabia.

BAYER, HANS-JOACHIM, Institut für Geologie, Universität Karlsruhe, Kaiserstraße 12, D-7500 Karlsruhe, Federal Republic of Germany.

BLÜMEL, WOLF-DIETER, Professor, Geographisches Institut, Universität Karlsruhe, Kaiserstraße 12, D-7500 Karlsruhe, Federal Republic of Germany.

BRIEM, ELMAR, Akad. Rat, Geographisches Institut, Universität Karlsruhe, Kaiserstraße 12, D-7500 Karlsruhe, Federal Republic of Germany.

DABBAGH, ABDALLAH E., Director, Research Institute, University of Petroleum and Minerals, Dhahran, Saudi Arabia.

DULLO, WOLF-CHRISTIAN, Institut für Paläontologie, Universität Erlangen, Loewenichstraße 28, D-8520 Erlangen, Federal Republic of Germany.

EMMERMANN, ROLF, Professor, Institut für Petrographie und Geochemie, Universität Karlsruhe, Kaiserstraße 12, D-7500 Karlsruhe; new address: Institut für Mineralogie und Petrologie, Universität Gießen, Landgraf-Philipp-Platz 4–6, D-6300 Gießen, Federal Republic of Germany.

FREDRICH, KLAUS, Fa. Gitec, Augustastraße, D-4000 Düsseldorf, Federal Republic of Germany.

HACKER, PETER, Bundesversuchs- und Forschungsanstalt, Geotechnisches Institut, Arsenal, A-1030 Wien, Austria.

HÖTZL, HEINZ, Professor, Institut für Geologie, Universität Karlsruhe, Kaiserstraße 12, D-7500 Karlsruhe, Federal Republic of Germany.

JADO, ABDUL RAOF, Professor, Department of Earth Sciences, University of Petroleum and Minerals, Dhahran, Saudi Arabia.

KLINGE, HANS, Hambacher Tal, D-6148 Heppenheim 3, Federal Republic of Germany.

KOLLMANN, WALTER, Geologische Bundesanstalt (GBA), Rasumofskygasse 23, A-1031 Wien, Austria.

LIPPOLT, HANS-JOACHIM, Professor, Laboratorium für Geochronologie, Universität Heidelberg, Im Neuenheimer Feld 234, D-6900 Heidelberg, Federal Republic of Germany.

MOSER, HERIBERT, Professor, Head, GSF-Institut für Radiohydrometrie, Ingolstädter Landstraße 1, D-8042 Neuherberg, Federal Republic of Germany.

MÜLLER, ERHART, Agrar- und Hydrotechnik, Huyssenallee 66–68, D-4300 Essen, Federal Republic of Germany.

PUCHELT, HARALD, Professor, Institut für Petrographie und Geochemie, Universität Karlsruhe, Kaiserstraße 12, D-7500 Karlsruhe, Federal Republic of Germany.

QUIEL, FRIEDRICH, Institut für Photogrammetrie und Topographie, Universität Karlsruhe, Kaiserstraße 12, D-7500 Karlsruhe, Federal Republic of Germany.

RAUERT, WERNER, GSF-Institut für Radiohydrometrie, Ingolstädter Landstraße 1, D-8042 Neuherberg, Federal Republic of Germany.

RONNER, FELIX, Professor, Late Director, Geologische Bundesanstalt (GBA), Rasumofskygasse 23, A-1031 Wien, Austria.

SCHNIER, HAJO, Institut für Geologie, Universität Karlsruhe, Kaiserstraße 12, D-7500 Karlsruhe; new address: Bundesanstalt für Geowissenschaften und Rohstoffe, Stilleweg 2, D-3000 Hannover 51, Federal Republic of Germany.

STICHLER, WILLIBALD, GSF-Institut für Radiohydrometrie, Ingolstädter Landstraße 1, D-8042 Neuherberg, Federal Republic of Germany.

WOLF, MANFRED, GSF-Institut für Radiohydrometrie, Ingolstädter Landstraße 1, D-8042 Neuherberg, Federal Republic of Germany.

ZÖTL, JOSEF G., Professor, Head, FGJ-Institut für Geothermie und Hydrogeologie am Forschungszentrum Graz, Elisabethstraße 16/I, A-8010 Graz, Austria.

Introduction

The first volume on the **Quaternary Period in Saudi Arabia** was published in 1978 and covered the results of field work on parts of the Shelf, in the eastern part of the Arabian Peninsula.

In view of the size of the Kingdom of Saudi Arabia, the expeditions had from the outset to be limited to chosen areas; notwithstanding careful study of maps, aerial photographs and consulting reports, a certain amount of luck was necessary to find typical outcrops for the evaluation of sedimentological, hydrogeological, geomorphological and climatological processes during the Quaternary period.

Traces of these processes could most likely be expected to be found in the tremendous former river systems of Wadi Ar Rimah, Wadi Birk and Wadi Ad Dawasir, as well as in the cuesta region of the Tuwayq Mountains, the As Sulb Plateau, the Rub' Al Khali and Al Hasa, Al Qatif and parts of the Arabian Gulf coast. The choice of scientific methods used depended on the type of locality involved.

Within this context, we should like to reply briefly to some critical remarks in reviews of Volume 1. In general, the echo that this work found was pleasingly positive. Only the absence of palynological studies received negative comment. It is quite understandable that specialists tend to see their own field as being of utmost importance. In our case, the large excavations involved in the massive renewal of the irrigation system at the Al Hasa oasis had already been completed when we began our field work, and further excavations were not possible; this prevented us from including this particular field in this study.

The large number of people involved in the project was often noted with astonishment. Here it must be noted that the list of authors also includes personnel from the various laboratories involved in the individual chapters, without regard to their academic degrees. We are aware that this has not always been the usual practice, either in older or newer publications. Had we eliminated the names of these colleagues, the number of authors would be decreased by one third. We arc, however, of the opinion that laboratory work is equal in value to field work. Decades of cooperation with the most various institutions has shown that this is the only basis for a good long-term relationship. The colleague in the laboratory is no longer willing to be degraded to a helper when a large part of the information obtained is based on the data he has provided. The resulting discussions with the various teams also proved fruitful.

The names of collaborators for the same chapter are listed in alphabetic order.

This present second volume is also based on selected field reconnaissance, this time, however, in the area of the Red Sea coast, the Shield and part of the western Shelf platform.

The final chapters present a summary of the hydrogeological situation, and information on geochronology and climate during the Quaternary period in Saudi Arabia as compared to climatic events in the Middle East, Europe and North Africa.

As in Volume 1, we have not here attempted to cover all the problems involved in their entirety. We do, however, believe that we have provided a generous survey, and plentiful suggestions for further specialized work. We feel that we are justified in making this assumption, as a thorough literature search on the state of research in southern Asia, the Middle East and North Africa showed areas of neglect for the Arabian area; these matters have now been given the attention they deserve.

Acknowledgements

For the publication of the second volume of the **Quaternary Period in Saudi Arabia,** the co-editors are primarily indebted to His Excellency Dr. BAKR ABDULLAH BAKR, Rector of the University of Petroleum and Minerals in Dhahran. As he had, during all the years of field work, held a protective hand over the joint research project undertaken by the University and the Austrian Academy of Sciences, which is now completed.

We owe special thanks to the following officers and members of the Austrian Academy of Sciences: the President, Professor Dr. ERWIN PLÖCKINGER, for his personal attention to the new contract agreement with the publisher; Professor E. W. PETRASCHECK, Chairman of the Commission for Basic Research in Raw Minerals, and Professor Dr. H. ZAPFE, Chairman of the Commission on Quaternary Problems. As was the case with Volume 1, Prof. Dr. L. SCHMETTERER was most helpful in coordination with the publisher.

The Austrian Fund for the Promotion of Scientific Research supported this project with grants to cover the costs of air transport and laboratory work done in Austria. The University of Petroleum and Minerals in Dhahran generously covered the very considerable expenses for equipment, field work, inland flights and living expenses for the entire team in Saudi Arabia. We are also most obliged to Dr. ABDUL AZIZ GWAIZ, the former Vice Rector of the University, for his further support and advice.

We owe a debt of gratitude to Dr. ABDALLAH E. DABBAGH, Director of the Research Institute at the same University, for his great personal involvement both in the scientific work and in organization matters; the same applies to Dr. SAAD S. AL-SAYARI, co-editor of Volume 1, who is no longer attached to the University.

We also received generous support for our work in the Jizan area from MUSTAFA NOORY, Director General of the Department of Geology, Ministry of Agriculture and Water, Riyadh; and, for the Jeddah area, from Dr. AHMED AL-SHANTI, the former Director of the Institute for Applied Geology, King Abdul Aziz University, Jeddah.

The Deputy Ministry of Mineral Resources (Jeddah) supporting us with maps and aerial photos as well as the U. S. Geological Survey (Jeddah), the French Bureau de Recherches Géologiques et Minières, and the Arabian-American Oil Company (Dhahran) were most cooperative.

This project could not have been completed on time without the selfless help of our colleague and friend, Professor Dr. HEINZ HÖTZL, University of Karlsruhe, Federal Republic of Germany. At the same time we appreciate the financial support he and his group received by the 'Special Research Branch on Stress and Strain in the Lithosphere' sponsored by the German Research Foundation (DFG).

The co-operation with the „GSF-Institut für Radiohydrometrie" in Munich-Neuherberg (Prof. Dr. H. MOSER, W. RAUERT, W. STICHLER), Federal Republic of Germany, was very successful.

The completion of this project justifies a recollection of its beginning. It was first suggested by DDr. ELMAR WALTER, Austria, Federal Ministry of Industry and Trade, Vienna, when the Institute for Applied Geology in Jeddah was established. He deserves our thanks for his encouragement and help in getting the project started, and his personal contribution to the development of an atmosphere of confidence between the partners. We hope with this volume to show all our sponsors proof of successful scientific cooperation between Saudi Arabia and Austria.

We should like to express our sincere thanks to the Springer-Verlag for their excellent co-operation and the fine quality of the printed book.

1. General Geology of Western Saudi Arabia

1.1. The Arabian Shield

(H.-J. BAYER)

1.1.1. General Introduction

1.1.1.1. Tectonic position

The Arabian Shield is an extensive occurrence of Precambrian crystalline and metamorphic rocks in the western part of the Arabian Peninsula. It covers a total area of 610,000 km^2 and is bounded in the west by the Red Sea graben. In the north and east, sediments from Cambrian to Quaternary time cover the Arabian Shield; these are thickest in the area of the Arabian Gulf (sediments exceeding 6,000 meters).

The Arabian Shield is part of the Arabian-Nubian Shield belonging to the African Shield complex. Only in the early Tertiary was this Arabian-Nubian Shield divided by beginning of the rifting into separate Nubian and Arabian Shields (Fig. 1); the morphological uplift of the Arabian Shield was somewhat greater due to upheaval of the graben shoulders to heights up to 3,000 meters above m. s. l. (Fig. 2).

1.1.1.2. Geological exploration

A systematic geological mapping survey of the entire Arabian Shield only began in 1950. Maps with scales of 1 : 2,000,000 and 1 : 500,000 were completed in 1963 (Kingdom of Saudi Arabia, Ministry of Petroleum and Mineral Resources, published by the U. S. Geological Survey). First publications on the stratigraphy and tectonics of the Precambrian include R. KARPOFF (1957, 1960), G. F. BROWN and R. O. JACKSON (1960), and parts of R. G. BOGNE (1953). In the meanwhile, a geological-petrographical and structural mapping of the entire Arabian Shield with a scale of 1 : 100 000 is in progress. Basic works have also been published on the geology of the Arabian Shield by L. T. ALDRICH et al. (1978), J. C. BAUBRON et al. (1976), G. F. BROWN and R. G. COLEMAN (1972), R. G. COLEMAN (1973), J. DELFOUR (1970), R. J. FLECK et al. (1976), W. R. GREENWOOD et al. (1975, 1980), D. G. HADLEY (1979), and D. L. SCHMIDT et al. (1973, 1979).

These works use chemistry, the degree of metamorphosis, structure and measured absolute age to classify and when possible to reconstruct the events that took place in the formation of the Shield.

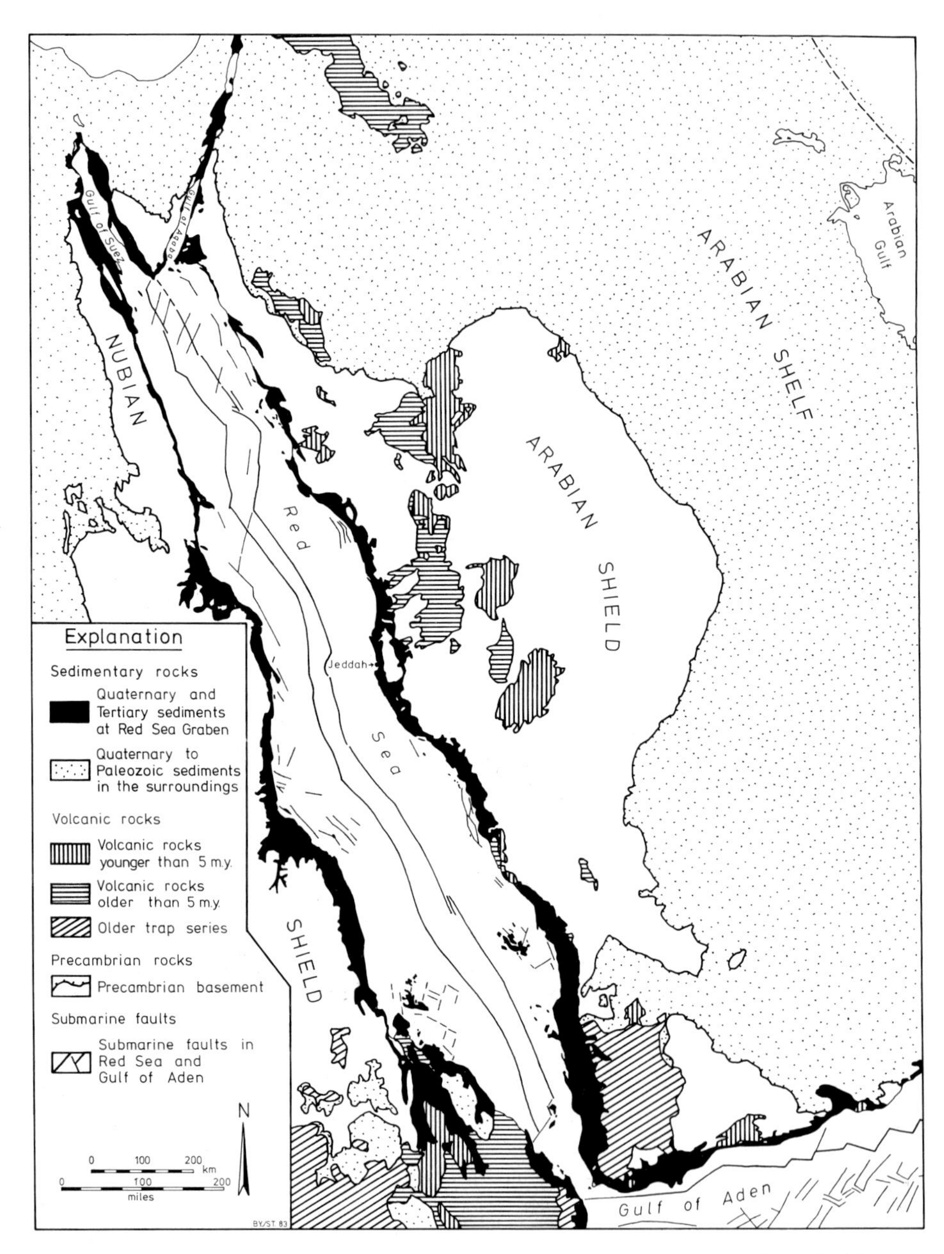

Fig. 1. Tectonic situation of the Red Sea with the adjacent Arabian and Nubian Shield.

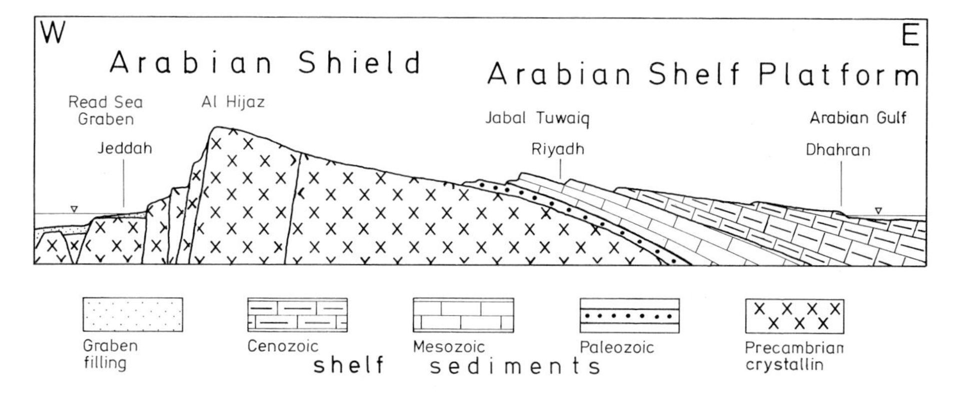

Fig. 2. Geologic cross section of the Arabian Peninsula from the Red Sea to the Arabian Gulf showing the tilted position with the uplifted western part where the Precambrian basement is outcropping.

1.1.1.3. Geological evolution

In its early development, the Arabian Shield was apparently formed by a number of more or less intraoceanic island arcs with basaltic to dacitic volcanism, accompanying sedimentation and gabbroid to granitic magmatism. The island arcs very likely were located upon a subduction zone dipping to the northeast that was in the forefield of already partially cratonized Africa. The Arabian-Nubian Shield is thus to be seen as a frontal collision structure (Arabian-Nubian neo-craton) on older cratonic elements of Africa. Absolute age dating showed 1,170 million years for the oldest elements of the Arabian Shield, and 510 million years for the youngest; the latter would correspond purely chronologically, but not developmentally to Cambrian.

The first rocks in the island arcs were submarine, then there was a phase of submarine and near-surface volcanic rocks, sediments and magmatites. As the island arcs grew together and the collision on the west side occurred, the events were removed almost completely from the submarine area. Seen chronologically, this led first to extrusions of oceanic basaltic series, then the calc-alkali series, and alkalic and peralkalic rocks were formed. In the central and eastern part of the Arabian Shield there are a number of ophiolite belts bearing witness to previous plate collisions.

The regional distribution of identical outcropping petrographic structures makes possible a reconstruction based on the volcanic rocks of the earlier series of island arcs that today overlap. Going from southwest to northeast, there are the following rock belts:

a) an older belt of tholeiitic basalt and basaltic andesite (Baish group); b) a somewhat younger belt of mainly tholeiitic basalt, andesite and dacite (Jeddah group); c) a younger belt with calc-alkalis and tholeiitic andesite and rhyolite (Halaban group), and d) a youngest belt with sedimentary rocks containing some andesite and rhyolite (Murdana group). Smaller taphrogeosynclinal deposits related to the Najd fault system are the youngest rock elements of the Shield.

As a result of the earlier island arcs, repeated phases of metamorphosis, folding and fault structures created a structural pattern on the Shield running north and northeast. The oldest rock structures thus show the most metamorphosis, while younger elements (Shammar group, Jubaylah group) did not undergo metamorphosis at all. The orogenic folding phases were thus followed by a purely fault-tectonic stress upon the Arabian Shield, mainly by structures trending NW–SE and N–S. It is remarkable that since the youngest Precambrian, the Najd fault zone (NW–SE trending) has been causing a left-lateral strike slip. The fault structures of the Arabian Shield are equivalent to those in the area of the North African Shield and must be seen in relation to them.

1.1.2. Geological-petrographical Formations

The stratigraphic units of the Shield are not to be seen as a sequence of layers, but rather as a petrographic contour sequence with parallel discharges differing in development; D. L. SCHMIDT et al. (1973) evaluated and narrowed them down according to lithostratigraphic assemblages (see also Table 1, Summary of Precambrian rock units). Going from older to younger, the following main units may be seen:

1. An assemblage of biotite schist, amphibolite and orthogneiss (Khamis Mushayt gneiss, Hali group; after W. R. GREENWOOD et al., 1980),
2. A metabasalt-graywacke-chert assemblage (Baish and Bahah groups),
3. A meta-andesitic assemblage (Jeddah, Ablah, Halaban and Murdama groups), and
4. A nonmetamorphosed assemblage of felsic volcanic rocks (Shammar group) and taphrogeosynclinal volcanic and sedimentary rocks (Jubaylah group).

Highly metamorphic areas of the Baish-to-Ablah group as well as parts of the younger groups are usually also classified in the biotite-schist-amphibolite-orthogneiss assemblage, which is mainly to be seen as a highly metamorphic parallel development to the metabasalt-graywacke-chert assemblage.

1.1.2.1. Biotite-schist-amphibolite-orthogneiss assemblage

a) Khamis Mushayt gneiss

These are migmatitic and quartz-feldspar gneisses that sometimes are older than the Hali group, and sometimes represent recrystallized plutonic rocks intruded into the Jeddah and Ablah groups.

b) Hali group

The rocks of the Hali group include repeatedly metamorphosed garnet-biotite-schist and paragneisses, amphibolite and amphibolite schist, quartzites bearing garnet biotite, marble and conglomerate schist. They belong to the higher almandine-amphibolite facies of the Barrow type and show slight to moderate constrictive conditions during repeated metamorphoses. In addition to various types of folded structures, the strongly developed foliation, usually in the form of double foliation, is especially characteristic.

Table 1. *Precambrian rock units, tectonism, plutonism and volcanism in the Arabian Shield* (ex W. R. GREENWOOD et al., 1980, completed after D. L. SCHMIDT et al., 1973)

Assemblages	Units	Major rock types	Tectonism episodes		Plutonism	Radiometric ages (m.y.)
Nonmetamorphosed rocks (included with meta-andesitic assemblage)	Jubaylah Group	Clastic rocks, minor volcanic rocks, and marine limestone	Najd faulting: Northwest-trending left-lateral wrench faulting		Granite to granodiorite	~530
	Shammar Group	Rhyolite, Trachyte, and minor clastic	–?–		Granite to granodiorite	570 (?)
Meta-andesitic	Murdama Group	Conglomerate and graywacke, minor andesite and rhyolite, locally thick marble	Hijaz Tectonic Cycle	Bishah orogeny: Folds and faults, north trends; greenschist metamorphism	Quartz monzonite, subordinate diorite, gabbro, granite	550–570
	Halaban Group	Andesite to rhyolite volcanic and pyroclastic rocks and volcaniclastic rocks		Yafikh orogeny: Folds and faults, north trends: greenschist metamorphism	Quartz monzonite, subordinate diorite, gabbro, granite	600–650
	Ablah Group	Basalt to rhyodacite, volcanic and pyroclastic rocks and volcaniclastic rocks, locally thick marble		Ranyah orogeny: Folds and faults, north and northeast trends; greenschist metamorphism; late gneiss doming and amphibolite to granulite facies metamorphism	Granodiorite gneiss Second diorite series; gabbro gabbro to trondjhemite, mainly diorite and quartz diorite	~750 ~835
	Jiddah Group	Basalt to dacite volcanic, pyroclastic, and volcaniclastic rocks, minor rhyodacite		Aqiq orogeny: Folds and faults, north and northeast trends; greenschist metamorphism	First dioritic series; gabbro to trondjhemite, mainly diorite and quartz diorite	890
Metabasalt-graywacke-chert	Bahah Group	Graywacke, chert, minor marble, and tuff, all locally graphitic, and minor basalt				
	Baish Group	Mafic volcanic rocks, mafic tuff, graywacke, and minor chert and marble			Quartz porphyry orthoschist, diabase	1165
Biotite-schist-amphibolite-orthogneiss	Hali Group Khamis Mushayt	Gneiss	Asir tectonism Basement tectonism			

1.1.2.2. Metabasalt-graywacke-chert assemblage

Rocks belonging to this sequence appear almost exclusively in the southern and middle parts of the Shield; between the two formative series (Baish group and Bahah group) interfingering can occur.

a) Baish group

This group is characterized by volcanoclastic flow layers, breccias and tuffs with basaltic to andesitic chemistry; there are also tuffaceous sediment layers, chert and marble. The volcanoclastics have generally metamorphosed to higher green-schist facies, include pyritic, silicate-to-calcareous and in places rather large graphitic components, are distinctly foliated, especially in lateral shift zones, and show different forms of folding and a subordinate internal lineation pattern. In the Baish group, dikes and sills of quartz porphyry appear as plutonic rocks. The mafic, tholeiitic volcanic rocks and the intermediate graphitic chert and marble are typical for the situation of a submarine island chain above a subduction zone.

b) Bahah group

The Bahah group is mainly made up of chlorite-sericite-quartz-albite schists that come from clayey to sandy graywacke and clayey chert. There are smaller amounts of dark marble, arkoses, graphite-rich chert and phyllites. Fine chert and mudstone breccias are actually common in the Bahah group. Occasional preserved sedimentary structures show, in addition to typical noncontinental graywacke sedimentation, current turbidity (mudflow) deposits, as can occur on the flanks of submerged island arcs.

1.1.2.3. Meta-andesitic assemblage

This sequence mainly consists of basaltic to andesitic flows, flow breccias with a considerable amount of more silicate volcanic rocks and intercalated volcanoclastics. The younger part shows rhyolites and polymitic conglomerates. Each of the stratigraphic groups in the meta-andesitic assemblage is to be placed in a metamorphic superimpositional phase resulting from the Hijaz tectonic cycle.

a) Jeddah group

This group consists of basaltic, andesitic to dacitic volcanics and pyroclastics with conglomerate sandstones, schists, graphitic phyllites and chert. Petrologically, the Jeddah-group rocks show the characteristics found today in the volcanic island chains on the edge of the Pacific basin. Jeddah-group rocks usually have metamorphosed to higher green-schist and amphibolite facies. Unlike the Baish group, the Jeddah-group volcanics already show some calc-alkali flows.

b) Ablah group

This includes a basal clastic unit, a middle unit made of andesitic to dacitic volcanics and pyroclastics, and an upper clastic unit. The clastic sediments consist mainly of graywacke, conglomerates and sometimes clastic or stromatolitic carbonate rocks. The rocks in this group have generally been metamorphosed to green-schist facies and thus for the first time contain larger sedimentary cycles.

c) Halaban group

This group is also made up of three units. The lower one contains conglomerates and finely granular clastics with basaltic to andesitic agglomerates, breccias and tuffs. The middle series is taken up by andesitic volcanic flows, and the upper unit consists of rhyolitic and trachytic volcanic flows and pyroclastics; ash layers are common. The Halaban rocks show major facial changes vertically and horizontally, have also metamorphosed to green-schist facies, and reach thicknesses up to 10,000 meters. The presence of ophiolites in the series indicates initial plate collisions.

d) Murdama group

This sequence is up to 2,700 meters thick, has metamorphosed to lower green-schist facies and is to be found above all in the southern part of the Arabian Shield. It contains basal conglomerates and flows of acid-to-intermediate magmatites and volcanics (Hishbi formation), a middle limestone and marble sequence (Farida formation) and an upper sandstone-siltstone sequence (Hadiyah formation). The Murdama sediments are characteristic for a period of uplift and erosion. The uplifting is primarily due to an intensive granitic magma ascent that had its main phase during the Murdama deposit period.

While the Baish-to-Murdama groups were being laid down, the Arabian Shield consolidated and cratonized. During the Hijaz tectonic cycle there were folding cycles and plutonic magma ascents covering large areas and sometimes very intensive, so that the sedimentation cycles kept being interrupted.

1.1.2.4. Nonmetamorphosed assemblages

a) Shammar group

The Shammar group is to be found only in the north of the Shield. It is made up of rhyolitic flows and ashflow tuffs, along with other alkalic and peralkalic volcanics that sometimes are associated with gently folded sedimentary layers. K-Ar datings of granite plutons contemporary to the rhyolite volcanism showed age values between 572 and 555 millions of years.

b) Jubaylah group

This sequence appears only in the northern part of the Shield near the Najd fault zone and is made up mainly of interlayers of clastics, siliceous limestones and thin andesitic and basaltic lava flows. Age datings of the andesites showed ages between 515 and 528 million years; the Jubaylah limestones of the same age, however, carry only stromatolitic mats and no Cambrian fossils. Fault-tectonic events, such as the left-sided shifts in the Najd fault zone dominated, and there were even gabbro intrusions in the Najd fault zone.

1.1.3. Tectonism and Plutonism in the Shield

Plutonic rocks cover more than 45% of the Arabian Shield, and if orthogneiss surface outcrops are added to this, then it is more than half of the exposed surface of the Shield. There was plutonism in phases of pronounced folding stress (orogenic cycles), four of which occurred in the Hijaz tectonic cycle (the Aqiq, Ranyah, Yafikh and Bishah orogenies). The rocks from these tectonic cycles reflect the corresponding petrological and chemical developmental trends. Volcanics of basaltic to calcareous composition, the deposition of "immature" sedimentary rocks and the intension of granitic to dioritic magmas are products of the major Hijaz tectonic cycles, even if the initial crust material under the Shield is not uniform (basaltic material under the southern part of the Shield, and ultrabasic material under the middle part and in the north).

1.1.3.1. Aqiq orogeny

This folding phase caused the first green-schist metamorphosis with a north-northeast dipping fold structure, allowed intrusions of gabbroid-to-quartz dioritic magma and ended 890 million years ago, as shown by Rb/Sr age dating. The Aqiq orogeny is equivalent in time to the Jeddah group. The placing of these rather extensive batholites in this phase was mainly tectonic (in folding arcs, etc.); at times, these plutonites also underwent a deformative overprint in the same phase.

1.1.3.2. Ranyah orogeny

During this phase, the rocks of the Ablah group and older units were folded with a N-NE vergency, metamorphosed to green-schist facies, and there was intrusion of gabbroid-to-dioritic and quartz-dioritic magmas. Age dating of the intrusions produced values between 835 and 785 million years. The placing of the plutonites was sometimes almost diapiric, and older gneiss components were even metamorphosed to granulite facies. Broad, flat shear zones form the southern part of the Shield, and gneiss rocks were sometimes intercalated there.

1.1.3.3. Yafikh orogeny

The Halaban rocks were narrowly folded 650 to 600 million years ago, metamorphosed to green-schist facies and invaded by gabbroid-to-granitic plutons. Ring-shaped intrusions are typical, and sometimes even ring-shaped dikes developed.

1.1.3.4. Bishah orogeny

This folding phase some 570 to 550 millions of years ago was similar to the one proceeding it, only the intrusives extended from granitic to dioritic as well as gabbroid magmatism and the green-schist metamorphosis was less intensive. At the time of the Bishah orogeny the crust was already distinctly thickened, there was again local melting and numerous new and renewed magmatic intrusions.

After these folding phases, mainly fault tectonics were at work on the Arabian

Shield, especially at the time of the Jubaylah group, in which the Najd fault system was formed. This fault system divides the northeast of the Shield from its central and southern parts. On four main fault lines with distinct vertical displacement there were also left-lateral movements which, according to G. F. BROWN (1972) together created a strike slip of 240 kilometers. In this Najd fault system, metamorphosis effects still occurred as a result of shear movements; here there were gabbro intrusions as well, and very extensive hydrothermal activity.

As the Arabian and Nubian Shields were connected before the Tertiary, attempts were made to correlate sequences found on both sides of the Red Sea; these showed very considerable similarity. The remaining differences can be explained by the regional structure of the Arabian-Nubian Shield.

1.2. The Red Sea

(H. HÖTZL)

1.2.1. Geological Context

In the last two decades, the geology of the Red Sea has been the focal point of a variety of research programs in the geosciences. This was not only the result of economic interests (petroleum and ore sludge), but also because of the tectonic position that makes the Red Sea a prime example of the formation of ocean systems according to the modern concept of "new global tectonics". In the course of this research work, numerous reports were published with new results. It is not possible to go into all of them here. The reader is directed especially to the following collections of articles and proceedings of meetings: N. L. FALCON et al. (1972); E. T. DEGENS and D. A. ROSS (1969); R. B. WHITEMARSH et al. (1974); L. S. HILPERT (1974). Newer works by individual authors include X. LE PICHON and J. FRANCHETEAU (1978); P. STYLES and S. A. HALL (1980), and J. R. COCHRAN (1983). For the Gulf of Aqaba there is the work of R. FREUND and Z. GARFUNKEL (1981), and for the Afar Depression from A. PILGER and A. ROESLER (1976).

1.2.1.1. Tectonic position

The Red Sea is a narrow oceanic trough some 2,000 kilometers long. It begins in the area of the Strait of Bab al Mandeb (Fig. 4), which connects this sea with the Indian Ocean via the Gulf of Aden. It extends from 13° to 28° N, and at its northern end it divides into the two arms of the Gulf of Suez and the Gulf of Aqaba (Fig. 3). In the north the Red Sea is 180 kilometers wide. To the south, it widens to 350 kilometers (between 15° and 18° N), and then narrows down at the Strait of Bab el Mandeb to 30 kilometers. Bathymetry shows a distinct axial trough. This is roughly limited by the 500 meters isobath. It attains depths around 1,000 meters, and in isolated basins, nearly 2,000 meters. Broad, flat shelf areas accompany this central trough.

Fig. 3. The Northern part of Red Sea with the Gulf of Suez and the Gulf of Aqaba. The darkgrey surfaces represent the outcropping crystalline shield; light grey colors are mainly for the Paleozoic to Cenozoic sediments; light colors for recent sand accumulations. Landsat mosaic (constructed for NASA and U.S. Geological Survey, prepared by General Electrics, Space Division).

The elongated form with the clearly defined, relatively straight edges underscores, along with the morphology of the framework, the tectonic structure of the sea. The Red Sea is a tectonic graben that was formed when the originally connected African and Arabian land masses broke apart. At first it was a continental rift, then, as Arabia drifted away, developed into an intercontinental system that today separates the independent Arabian plate from the African plate.

The Red Sea graben is part of an extensive global system of faults running approximately north to south (Fig. 5). It begins in southern Africa and extends over the East African graben system and the Afar Depression, a triple junction,

Fig. 4. The Southern Part of the Red Sea with the Strait of Bab El Mandeb and the Gulf of Aden. In the upper part the southwestern corner of the Arabian Peninsula with the wide coastal plain and the Yemen mountains, in the lower part, left Afar Depression with parts of Ethiopia and Somalia. Landsat mosaic (constructed for NASA and U.S. Geological Survey, prepared by General electrics, Space Division).

from which both the Gulf of Aden and the Red Sea proceed. At the northern end of the Red Sea this graben system again splits into the Gulf of Suez and the Gulf of Aqaba, two units with completely different structures. While the first is a tensional rift system, the latter is part of an extensive shear system that may be followed to the Dead Sea and Jordan, and from there to the alpidic chains of the Taurus in southern Turkey. In all, this enormous meridional fault system may be understood to be part of the fault lines on which the old Gondwana continent falls into separate plates.

Within the entire meridional system, the Red Sea is the most advanced section, as far as the breaking-apart is concerned. The continental crust is broken

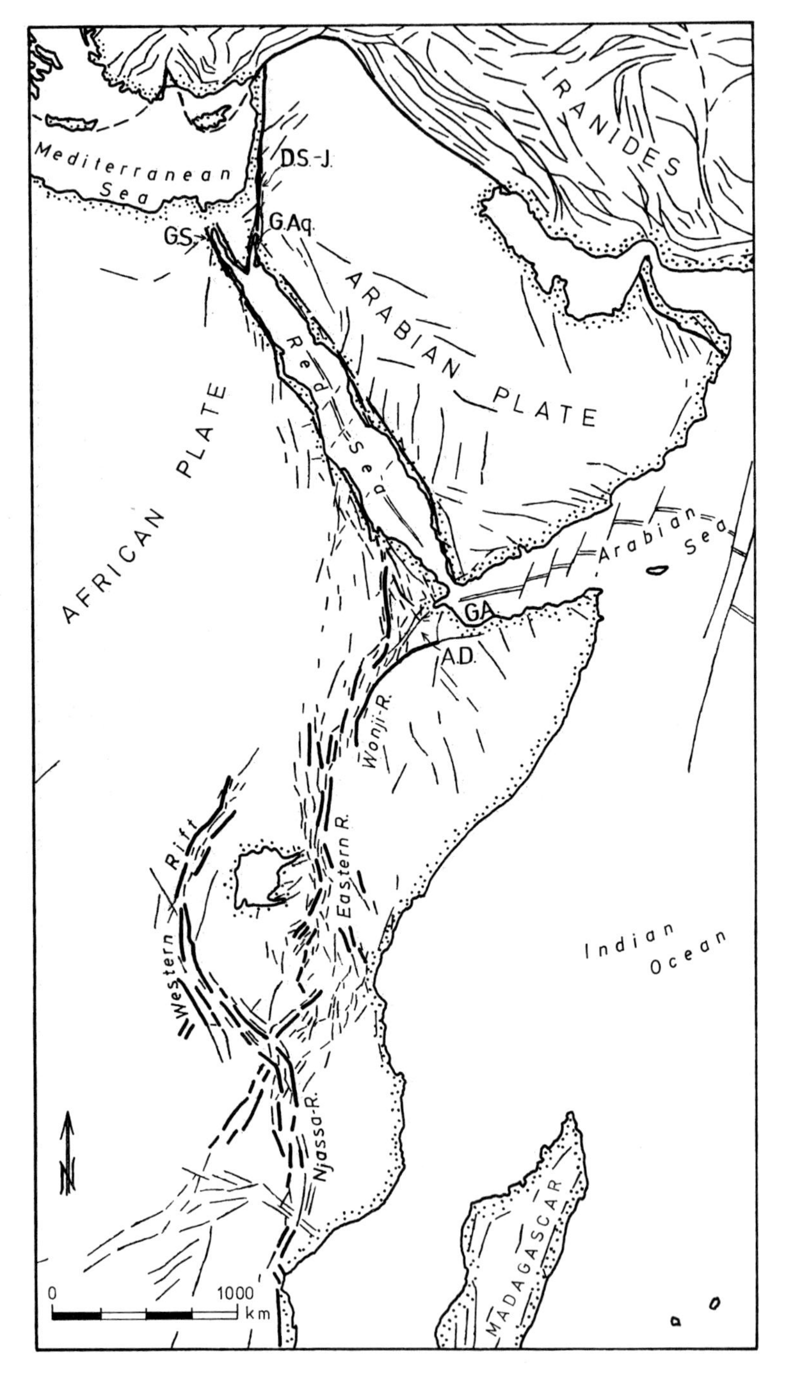

Fig. 5. The African-Arabian rift system (*D. S.-J.* Dead Sea-Jordan rift; *G. Aq.* Gulf of Aqaba; *G. S.* Gulf of Suez; *G. A.* Gulf of Aden; *A. D.* Afar Depression).

open along almost its entire length. Lava rising from the earth's mantle forms a new oceanic crust in the area of the axial trough. This process of active ocean spreading is responsible for the ever-increasing drift of the Arabian plate. Its side facing the Red Sea continues to develop from an originally seismically and tectonically active graben rim to a passive continental margin. The decreasing tectonic activity in the marginal areas may be an expression of the gradual establishment of a state of equilibrium on the new edge of the continent.

These tectonics, if only in a weakened form, were in many ways nonetheless determinative for the Quaternary geology of the western edge of the Arabian Peninsula. Very young vertical shifts, such as the elevation of occasional blocks in the north or the extensive sinking of the coastal plain in the south have, together with young volcanism, influenced erosional and sedimentational events.

1.2.1.2. Magmatism

Today, the graben event is seen in relation to thermic and magmatogenic processes in the asthenosphere and lithosphere (H. Illies, 1970; J. D. Lowell, G. J. Genik, 1972). These processes led to an arching under the crust, in the course of which the earth's crust is stretched, so that the crust finally breaks open at the top of the dome. The magma uses these faults and fissures to rise farther up. The graben event is thus often accompanied by magmatic, or especially volcanogenic phenomena. R. G. Coleman et al. (1977) presented a first model of the magmatic processes in the Red Sea (Fig. 6).

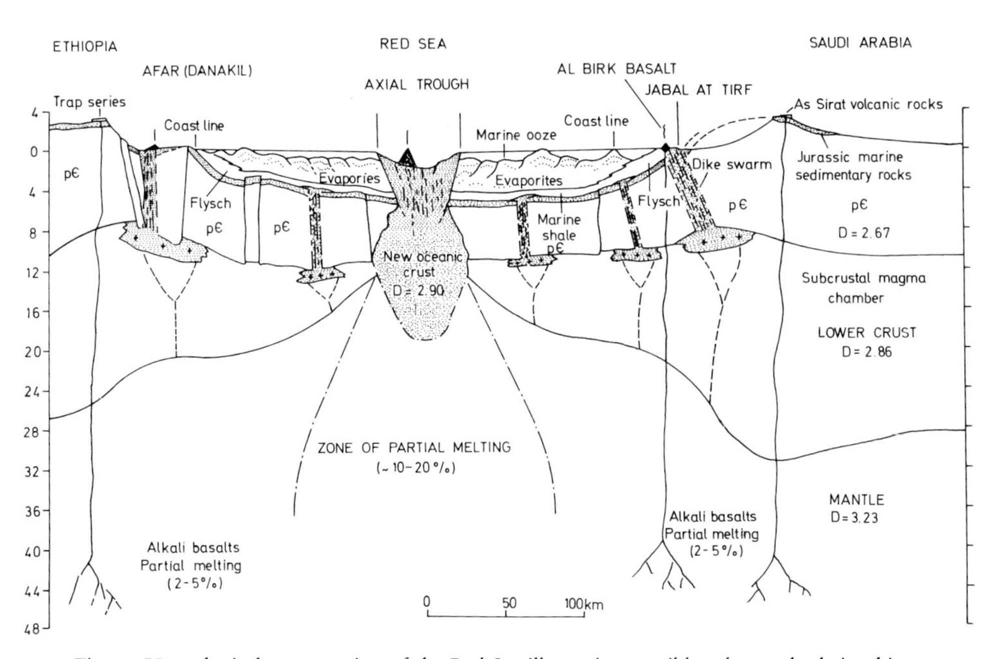

Fig. 6. Hypothetical cross section of the Red Sea illustrating possible subcrustal relationships. (Ex: R. G. Coleman et al., 1977.)

The first volcanogenic indications of the beginning graben event are seen in the thick Eocene to Oligocene trap basalts at the southern end of the Red Sea (Yemen, Ethiopia). They are part of the actual doming. With the opening of the earth's crust and the formation of the tectonic graben structure in the Oligocene, and during its continuation in the Miocene, continental tholeiites intruded and rose along faults parallel to the rift.

As the graben continued to open up in the Middle and Upper Miocene, a chemical differentiation became increasingly apparent between the magmas of the central trough (oceanic tholeiites) and the volcanic rocks of the rest of the edge of the rift (alkali basalts). These volcanic rocks sometimes extend far back into the hinterland of the Arabian Shield, and there they indicate not only faults parallel to the graben, but also older, youthfully activated fault systems. Particularly characteristic is the Hail arch, trending north–south at an acute angle to the Red Sea; since the Miocene it has continued to show new fissure eruptions and occasional volcanic cones. A slight decrease may be assumed for the general magmatic activity in the Middle to Upper Miocene.

Increased magma production has led to an intensification or recommencement of the spreading processes in the area of the axial rift since the Pliocene. In the course of these processes the Arabian plate is drifting ever-farther to the north-east. Continuing magmatic activity is to be found on the edge of the rift, both in the area of the downwarped faulted blocks and in the hinterland in the elevated graben shoulders; this has been occurring up through recorded history.

1.2.1.3. Erosion and sedimentation

Before graben tectonics began in the Tertiary, the Arabian-Nubian Shield showed a generally even relief. As a result of long-lasting erosion processes throughout the Paleozoic and Mesozoic eras, the erosion and sedimentation processes in the area of the fully developed peneplain had more or less come to a stop. Soil-formation processes, sometimes with extensive laterite formation, dominated.

The beginning graben event with its doming and rifting reactivated not only the erosion and sedimentation processes, but also determined their nature and course. The sinking trench created a new sedimentation trough. Tremendous amounts of debris fell into this new sedimentation space from the steeply rising flanks of the trench.

The penetration of the sea into the graben changed the course of sedimentation, especially in the central trough, where carbonates and evaporites were deposited, along with fine-clastic sediments.

In the marginal areas, coarse-clastic intercalations remained dominant. Their chronological distribution also shows the sudden cessation of the vertical graben tectonics. The flanks and shoulders of the graben were increasingly eroded in the process, and the originally uniform escarpment was dissected and moved backward.

There has not as yet been much study of the graben sediments in the Red Sea area. This is because most of them are under water, and only the narrow coastal strips are directly accessible. Then there is the fact that as a result of the continued sinking and sedimentation, usually only the youngest deposits are exposed. Older series are only exposed on the surface in places where marginal crustal blocks have been involved in the uplift. Various exploratory oil borings have provided us with new information on the marginal area (M. GILLMANN, 1968; J. F. MASON and Q. M. MOORE, 1970; P. SKIPWITH, 1973). Results of research borings are available for the central part of the trough (R. B. WHITEMARSH et al., 1974; P. STOFFERS and D. A. ROSS, 1974).

Table 2 shows the sequence of graben sediments for various areas based on published data. The processing varies considerably. The comparison of the individual profiles suffers especially from insufficient stratigraphic arrangement of the individual series, which leads to difficulties in correlation, especially with the facial differentiation common on the edges, and the dependence on local influences. The best studied sequences are from the Gulf of Suez; here, earlier general profiles (R. SAID, 1962) have been supplemented by newer studies (M. KERDANY, 1968; R. SAID, 1969).

For the Arabian Red Sea coast there is newer material for the area from the Midyan Peninsula in the north, where the Tertiary sediments are more widely distributed (M. M. A. BOKHARI, 1982; E. MOTTI, 1982, C. DULLO et al., 1983). New classifications were made for the section from Al Wajh to Yanbu in the course of geological surveying of various map sheets (M. BIGOT and B. ALABOUVETTE, 1976; B. ALABOUVETTE, 1977; C. PELLATON, 1979, 1982). The Tertiary sediments in the sections south of Jeddah are comparatively little known, as they are scarcely exposed (G. F. BROWN and R. O. JACKSON, 1959; P. SKIPWITH, 1973, G. M. FAIRER, 1982, 1983).

In spite of facial differentiation and sometimes more pronounced regional deviation, all in all there is a relatively uniform picture for the sedimentation processes in the entire Red Sea. First there is graben sedimentation with clastic continental erosion products, found at the edges as relatively coarse conglomerates and becoming finer down into the trough as sandstones, arkoses and siltstones. As yet, it was nowhere possible to make a chronological classification of these series. They lie discordantly upon basement, or in places upon Cretaceous or Eocene series. Most authors thus place them in the Oligocene (P. DADET et al., 1970; C. PELLATON, 1982; C. DULLO et al., 1983). There is a special series with intercalations of hematitic iron oolites in the Jeddah area (A. M. S. AL-SHANTI, 1966).

These continental red series are overlaid by a marine clastic-to-carbonate sequence. Recent studies by C. DULLO et al. (1983) showed that the marine transgression had already taken place in the uppermost Oligocene (Chatt), and is thus an important indication that rifting began before the Miocene. During this first marine ingression with a shallow covering of water, major facial differentiations naturally became apparent over a small area. Thus, on Midyan marine sandstones containing sea urchins could be seen along with reef inclusions and evaporitic inclusions as well.

The Lower Miocene (Aquitan) carbonate sediments are increasingly covered by alternating clastic accretions on the edges. These go both laterally, in the direction of the basin, and vertically toward stratigraphically younger sediments,

Table 2. *Sedimentary sequences of the Red Sea graben*

Period	Epoch		Stage	Gulf of Suez after R. SAID (1962, 1969)		Gulf of Aqaba (Midyan) after C. DULLO et al. (1893), R. VAZQUEZ-LOPEZ (1981), M. M. A. BOKHARI (1981)				Northern Red Sea Coast (Dhuba to Yanbu al Bahr) after C. PELLATON (1979, 1982a, b) M. BIGOT, B. ALABOUVETTE (1977)	
Quaternary	H.				reef and gravel terraces				reef and gravel terraces		reef limestones gravel terraces
	Pleistocene										
Tertiary	Pliocene	Upp.	Ast.		continental clastics corals-, nulliporen-limestone		Ifal-Fm.		clastic continental sequence with few gypsum	Buwanah-F.	gypsiferious sand and clays (evaporites ?)
		Low.	Piac.								
	Miocene	Upper	Mess.	Zeit-Fm.	shales, dolomites partly evaporites	Raghama Group	Al Bad-Fm.		evaporites (lowerpart dominant marls and shales)	Raghama-F.	evaporites
			Tort.								
		Middle	Serr.	South Gharib-Fm.	evaporites						
			Lang.	Belayim-Fm.	Globigerina marls						
		Lower	Burd.	Karem-Fm. Rudeis-Fm.			Jabal Tayran-F.	Wadi Telah-Mem.	sandstone conglomerate with reef intercalation		conglomerates sandstone
			Aqui.	Nukhul-Fm.	limestones						reef limestone
	Oligocene		Chatt	Abu Zenima-Fm.	red clastic sequence			Wadi Al Kils-M.	limestone and sandstone	Dhaylan-F.	red clastic sequence (partly limestone)
			Rupel					Wadi Al Hamd-M.	red clastic sequence		

Period	Epoch	Stage	Middle Red Sea Coast (Jeddah area) after P. SKIPWITH (1973)		Southern Red Sea Coast (Jizan area) after M. GILLMAN (1968), P. SKIPWITH (1973), D. G. HADLEY, R. J. FLECK (1980)			Red Sea Graben after P. STOFFERS, D. A. ROSS (1974)	
Quaternary	H. / Pleistocene			reef limestones gravel terraces		reef limestones gravel terraces		Unit I	silty clay and chalk
Tertiary	Pliocene	Ast. Upp.		unconsolitated clastic rocks	? Upper Continental Mem. (Bathan-F.):	?		Unit II	calcareous clay and silt
		Piac. Low.			Middle Continental Mem.	clastic sequence shales		Unit III	dolomite siltstone
	Miocene	Mess. / Tort. Upper	Raghama-F.		Lower Continental Mem.	clastic sequence	? BAYD-F.: shales and tuffs?		
		Serr. / Lang. Middle		evaporites	? Evaporite Mem.	evaporites (salt with anhydrite)		Unit IV	evaporites (salt with anhydrite)
		Burd. / Aqui. Lower		consolitated clastic rocks limestones	Infra evaporitic-Mem.	shales sandstone conglomerate		Upper Glo- bigerina Group	marls
	Oligocene	Chatt / Rupel	Shumaysi-F.	red clastic sequence with intercalation of iron oolite		shales and volcanic rocks			

becoming finer grained and more marly. Globigerina marls are deposited in the graben itself. They indicate increasing water depth.

From the Middle Miocene on, evaporitic intercalations become increasingly common; they take over the basin and can reach thicknesses in excess of 3,000 meters (M. GILLMAN, 1968; P. STOFFERS and D. A. ROSS, 1974). Particularly on the edges there are mainly anhydrite and gypsum rocks that get thicker going toward the basin, and tend to pass into halite deposits.

C. DULLO et al. (1983) could demonstrate Tortonian for the higher part of these evaporites in the Midyan area. In the basin the uppermost parts may be even younger. As the evaporites sometimes reach as far as the escarpment, a certain phase of stability can be assumed for the erosion process.

Postevaporitic there were extensive changes in graben sedimentation. This is apparent not only in a facies change – here to marine deposits from an initially quite enclosed and later open sea – but also in a distinct unconformity (A. D. ROSS, 1974; M. M. A. BOKHARI, 1982). The edge of the evaporite sequence is usually covered by a very thick clastic series, indicating pronounced erosion. The few borings from near the edge of the graben report thicknesses as great as 2,000 meters (e. g. well Barquan – J. F. MASON and Q. M. MOORE, 1970, or well Mansiyah I – M. GILLMAN, 1968).

Besides purely clastic series there are shallow-water limestones with oyster banks and echinoid shells. There has not yet been a precise stratigraphic classification of this sediment sequence, but work on it is proceeding. The general position suggests essentially a Pliocene age. This classification is supported by the observation that in the present-day coastal area these sediments in the covering layers pass into reef formations, which in turn are covered by Pleistocene reefs, or on which younger Pleistocene reefs are deposited.

For the Quaternary, the present coastal plain is remarkable for the interfingering of clastic accumulation terraces with the marine beach and reef terraces, and the relative constancy of the coastline. This series of terraces shows both tectonic effects and the influence of eustatic changes in sea level and varying climatic effects. In the area of the Red Sea basin, mainly calcareous, silty and clayey sediments with nannoplankton ooze and chalk have been deposited since the Pliocene.

1.2.2. Tectonic Development of the Red Sea Graben

1.2.2.1. Pre-rifting phase

The uncovered and denuded Arabian-Nubian Shield naturally offers few clues as to when the graben event began. Unlike the areas covered by sediment series, where updoming as a preliminary phase to the actual rifting is seen in increasing erosion, in the crystalline area this is difficult to differentiate from the earlier uplift.

Particular attention is thus given to sedimentation processes that indicate the beginning of a depression of this sort in the axial area of the updoming. Only after a long erosion phase in the Paleozoic and early Mesozoic the sedimentation reached the Shield in the vicinity of the Red Sea in the Upper Jurassic (R. W. POWERS et al., 1966).

Quarzitic sandstones of the Nubian type were deposited. They are gathered under the term Khums Formation and were apparently deposited in a narrow, tongue-shaped basin extending from the south over the mountains in what today is Yemen to 18° N latitude. There is no more exact information on the original form and orientation of the sedimentation area. These sediments today are limited to Yemen, as well as a narrow zone located on the edge of the graben (G. F. BROWN and R. O. JACKSON, 1959). These sandstones are conformably overlaid by gray-blue, occasionally silicified limestones from the Hanifa Formation. They are, however, extensively distributed only in the south. This is even more true of the Lower Cretaceous sandstones and shales of the Taouillah Formation, which are only to be found in Yemen.

In the Upper Cretaceous there was a marine intrusion from the Mediterranean toward the south. Large areas in Egypt and Libya are still covered by Cretaceous and Paleocene sand- and limestones. This is also true of the Clysmic Gulf area and, apparently, of narrower small troughs to the immediate south of the present-day Red Sea area. The Cretaceous-Early Tertiary sequences in the Azlam trough, running at an acute angle to what later became the Red Sea graben, should be mentioned here (B. ALABOUVETTE et al., 1979; F. B. DAVIES, 1980). There are also the Upper Cretaceous-to-Paleocene sediments in the Usfan Formation in the Jeddah area (G. F. BROWN et al., 1973; P. SKIPWITH, 1973). Sedimentation continued into the Eocene, at least in the area of the Clysmic Gulf (R. SAID, 1962).

Some authors believe that these marine advances mark the first signs of graben development. Disregarding the fact that this is somewhat contrary to the modern tectonic concept of the development of a rift with doming in the initial phase, the Upper Jurassic and Cretaceous marine advances are by no means structurally related to the Red Sea system. They are these very deviating structures, such as the more N-W trending Azlam trough, that show that there was a development independent of the later Red Sea graben.

Indications of the actual beginning of the more pronounced updoming as a preliminary phase of graben formation are much more likely to be found in the activation of magmatic processes. The oldest related volcanic activity is that of the trap basalts found on both sides of the southern end of the Red Sea. They are always located on top of the old peneplain, and their alkaline tholeiitic chemistry is that typical for doming magmas. They probably belong to the beginning activity of the Afar hot spots (R. G. COLEMAN, 1977).

In general, the updoming in the Red Sea area seems to have begun from this mantle plum in the area of the present-day triple junction at the southern end of the Red Sea. This is also indicated by the gradual retreat of the marine sedimentation in the north. So there are the Paleocene sediments mentioned above in the Jeddah area and the Eocene sediments at the northern end of the Red Sea, both of which were deposited before the general rearrangement caused by the graben.

1.2.2.2. Early rifting in the Oligocene and Miocene

There are various presentations of the chronological beginning of the rifting process to be found in the literature. The usual stratigraphic classification of the graben sediments was usually seen as indication that rifting began in the Early Miocene (A. D. ROSS and J. SCHLEE, 1973).

Datings from basalts in the area of the rift shoulder as well as from basic dikes seem to confirm this. There is to be no doubt that the Early Miocene was a pronounced rifting phase, and this will be dealt with in more details below. But this would require advanced thinning of the crust, or graben formation. It is from this pushing-up of dikes that R. G. COLEMAN et al. (1977) conclude that there must have already been a sort of proto-Red Sea.

On the basis of the classification as Oligocene of the clastic continental series subjacent to the fossil-dated Miocene sediments, it must be assumed that there was a depression of corresponding age at the apex of this updoming. The recent literature thus usually indicate of an Oligocene to Early Miocene beginning of the graben. C. DULLO et al. (1983) could, however, show that some of the basal limestones previously classified as Miocene belong to the Upper Oligocene, and therefore that the marine ingression in the graben trough thus took place before the Miocene.

The break up of a huge continental plate has to be caused by a supraregional stress field. It thus would not be surprising if the doming and rifting were essentially involved in the Mid-Tertiary event (uppermost Eocene – lowest Oligocene, 40–36 million years B. P.). This event was responsible for a number of movement processes on the edges of the plate fragments, and most particularly in the Mediterranean area.

The starting mechanism of graben depression is discussed differently. While I. G. GASS (1970) or J. D. LOWELL and G. J. GEMIK (1972) here speak first of a graben collapse at the apex of the Arabian-Nubian updoming zone, R. G. COLEMAN et al. (1977) suggest that there was first a broad, flexure-like downwarping as a result of the crustal extension. The letter is supported by certain geophysical findings, as well as by the gradually steepening dip of the Jurassic-Cretaceous strata series in the southern section of the graben.

Only after further tensional stress was there a complete break of the crust with intensive block faulting from the Oligocene or early Miocene. As a result of the related further subsidence there was then the marine ingression from the Mediterranean mentioned above and the subsequent deepening of the trough. Blockfaulting in the graben area and the simultaneous opening of fissures distinctly intensified volcanic activity. Fissures parallel to the rift spilled thick alkali basalts down the shoulders of the graben. At the same time, there was extensive magmatic activity in the trough itself. Special mention should be given to the numerous dikes on the edge of the southern Tihama near Jizan, where in addition to the dikes, the layered gabbro body of the Jabal at Tirf has also penetrated Tertiary sediments (M. GILLMAN, 1968; R. G. COLEMAN et al., 1977).

We still do not know whether in this first phase of rifting there was also formation of an oceanic crust via spreading processes in the axial trough. R. W. GIRDLER and P. STYLES (1974), as well as S. A. HALL et al. (1977) think there is a process of this sort in the lamellar anomalies in the magnetic field of the Red Sea. Based on magnetic measurements in the southern Red Sea (17°–20° N lat), they postulate the existence of Neogenic oceanic crust for the entire trough area. It is supposed to have developed during two spreading phases: the older one between 40–34 million years B. P., and a younger one (in the axial trough area) in the last 5 million years. With further measurements in the adjacent section to the north, the period for the older spreading phase was corrected to 20–30 million years B. P. (P. STYLES and S. A. HALL, 1980).

J. R. COCHRAN (1983) doubted the existence of an oceanic crust in the broad marginal area of the Shelf and used the result of borings and rock formation on individual islands to suggest a continental crust divided into horsts and special grabens. Blockfaulting parallel to the rift also could be responsible for the zonal arrangement of the magnetic field. Further study will, however, be required to provide definite information on the extent of the oceanic crust in the Red Sea.

The sedimentation process also conforms with the development mentioned above of an initially flattish depression followed by breakage into blocks. The subsidence of the graben depression led to marine ingression in the Upper Oligocene and the continuation of this process finally brought about the transition from shallow-water to basin facies. The blockfaulting in the trough is also seen in the numerous coarse-clastic intercalations during the Aquitan and Burdigal. A decrease in tectonic activity toward the end of this first rifting phase is to be seen in the gradual transition from coarse- to fine-clastic sediments in the hanging layer, as well as ultimately in the transition to evaporitic sedimentation.

1.2.2.3. Spreading and shearing in the Plio- and Pleistocene

There is a reactivation of tectonic activity in the upper-most Miocene or lowest Pliocene. The evaporites are followed by coarse-to-fine clastic series, particularly in marginal areas and sometimes with distinct discordance. They attain thicknesses in excess of 1,000 meters. They reflect the recurring vertical tectonics that led to involvement, blockfaulting and tilting of the Miocene sediment sequence. At the same time, a breakthrough to the south developed that connected the Red Sea rift with the open ocean via the Gulf of Aden. The basin, which previously had been quite isolated, was again flooded. In the course of this movement there was also increased uplift of the graben shoulders, e. g. Sinai and the Asir highland, upon which the currently existant morphological opposite to the graben basin rests.

What is decisive for the present-day tectonic picture is the recurrence or renewal of the spreading process in the trough axis. Further extension there opens up the sediment filling, creating a graben in the graben. There, tholeiitic magma in the form of a mid-ocean ridge is brought up from the mantle and forms a new oceanic crust. Paleomagnetic measurements show a total opening of 80–100 kilometers for the last 4–5 million years, amounting to an average spreading rate of 2 centimeters/year.

With the activation of the spreading process, the active tectonics shifted to the central trough. On the edges and flanks of the old graben, the seismic and tectonic activity is subsiding considerably. Deep-reaching gravity block movement on sliding planes, as apparently already seems to be the case in the southern part of the Red Sea, may increasingly contribute to a gradual stabilization of the steep edges of the graben flanks.

The northeast drift of the Arabian plate caused by the spreading process is overlaid by a pronounced sinistral shear movement along the Aqaba – Dead Sea system. This plate shift has already been recognized and correctly interpreted by L. PICARD (1937) and others. More recent work confirms the previously calculated total movement sum of more than 107 kilometers. R. FREUND et al. (1970) postulate a two-phase movement, which they compare chronologically with the two rifting periods in the Red Sea.

Y. BARTOV et al. (1980), however, used the absolute age of dikes parallel to the rift to show that the total movement is younger than 22 million years. Our own work in the Midyan region also indicates a postevaporitic beginning of the shear movement (H.-J. BAYER et al., 1983). The connection between drift and lateral movement, and spreading and shearing, is the subject of new work in progress.

With the reactivation of tectonic processes at the beginning of the Pliocene, volcanism also picked up. We find young alkali basalts in the graben, on the present-day coastal plain, and on the flanks, as well as on the graben shoulders, and there as far as 200 kilometers away from the graben margin. The latest volcanic events have been recorded from history. They also indicate tectonic activity continuing up to the present, particularly on the Arabian side of the Red Sea rift. This shows off the large influence of the rift mechanism for the Quaternary geology of this area.

2. Regions of Investigation

2.1. Midyan Region

2.1.1. Topographical and Geological Introduction

(H.-J. Bayer, H. Hötzl, A. R. Jado, F. Quiel)

2.1.1.1. Geographical position

That part of northwestern Arabia named Midyan is directly adjacent to the NNE-trending Gulf of Aqaba (Plate I, see insertion at back cover). The southern end of the Gulf from Ra's Ash Sheikh Humayd and eastward, the coastline runs almost east–west marking the location where the Red Sea opens up. With the mountain range limiting it in the east, this area has the shape of an elongated triangle. The south side is 70 kilometers long whereas the western side, along the Gulf of Aqaba, is 165 kilometers.

Because the region extends far into the Red Sea, it is referred to in the literature as the Midyan Peninsula, especially in comparison to the neighboring Sinai Peninsula. The distinction from the hinterland, however, is very unclear, especially as this area also includes the northwestern spurs of the Hejaz Mountains running directly through it. We thus prefer to avoid the term peninsula, and name it here the Midyan Region. According to geographical coordinates, this region extends from 27°30′ to 29°30′ N latitude and 34°30′ to 35°30′ E longitude (Plate I).

The landscape of this region is predominantly mountainous; the highest peaks are in excess of 2,500 meters and are found on the eastern edge in the spurs of the Hejaz Mountain range. But even within the actual triangular region near the Gulf of Aqaba, occasional elevations attain heights of 2,000 meters. In places, the mountains come right down to the Gulf. A wider strip of coast has developed south of Aqaba; it is taken up by vast talus fans falling to the coast.

Only in the south is there a larger planation. This is the Ifal Depression, a broad aggradation surface opening fanlike from Al Bad' to the south to the Red-Sea coast; together with this coastline it forms a nearly equilateral triangle 50 kilometers on a side. Besides Al Bad', the oasis settlements of Magna on the Gulf of Aqaba and Aynunah on the wadi of the same name in the southeast should be mentioned.

In antiquity the Midyan Region was famous for its mineral resources, especially gold and silver, which were mined by the Nabateans. Occasional ruined settlements and burial sites bear witness to this period.

2.1.1.2. General geological situation

The Midyan Region belongs to the crystalline Shield area of the Arabian plate. In its eastern part it represents the direct continuation of the Hejaz Mountains range paralleling the coast. The Precambrian rock series reach NNW to the Gulf of Aqaba, where they sometimes face younger sediment series on the Sinai Peninsula across the way. The direct continuation of the crystalline rocks is present but, shifted more than 100 kilometers to the SSW, in the southwestern part of Sinai Peninsula. The reason for this is to be found in the Aqaba – Dead Sea – Jordan shear system.

The Precambrian crystalline rocks consist of a thick sequence of metasediments and metavolcanic rocks, penetrated by thick plutons mainly of granitic but also of basic composition, and by dyke swarms. The metasedimentary and metavolcanic enveloping series are primarily classified as belonging to the Hejaz Formation (M. SAHL and N. FOTAWI, 1982). In the course of the subsequent orogenetic event they were even transformed into amphibolite facies. Synorogenically intruded granites, along with diorites and gabbros that cover extensive surfaces in the Midyan area, have been more or less intensively affected by pressure and movement. The commonly veined monzogranites are younger; they are especially frequent along the Aqaba coast. There are also discordantly penetrating calc-alkalic granites that are numerous in the eastern mountain range (Jabal Al Lawz).

The crystalline sequence of the Midyan and Sinai regions is already present on the northern edge of the Nubian-Arabian Shield. Here at the northern end of the Red Sea, the appearance of the Precambrian series is above all a result of the young axial culmination of the earth's crust in the preliminary phase of the graben formation and the resultant erosion. After the opening of the Red-Sea graben, this area was elevated as a graben shoulder to a steep escarpment and the erosion surface tipped like a desk top to the east. The preserved Cambrian-to-Permian sandstones of the original sediment deposit are only exposed some 85 kilometers east of the present-day escarpment, which is already outside of the Midyan region in the area of the Hisma Plateau.

The previously uniform and continuous outcrop of Precambrian parallel to the coast in the area of the graben flank today has not only been shifted more than 100 kilometers by the left lateral Aqaba system, but has also been broken up into a block mosaic by a large number of individual faults. Late Tertiary, mainly clastic but also carbonate and evaporitic sediments were deposited in these subsidence zones and partial grabens created by young tectonics on both sides of the Gulf of Aqaba. Today they are mainly distributed in the southwestern part of the Midyan region in the surroundings of Jabal Tayran and Jabal Al Musayr (Plate I, and Figs. 7 and 8).

R. A. BRAMKAMP et al. (1963) introduced the term Raghama formation, after the Jabal Ar Raghamah south of Al Bad' in the Midyan region, for the Tertiary sediments overlying the crystalline in the northern part of the Saudi Arabian Red-Sea coast. This term was used very commonly in the literature, although the respective stratigraphic extent was not known. M. M. A. BOKHARI (1981) provided a somewhat more detailed description of the Tertiary sediment sequence in

Fig. 7. The central Midyan area with Jabal Hamdza covered by Oligocene limestone (background right) and the Jabal Ar Raghamah (left), with its evaporitic Miocene sequence. Detailed geologic section see Fig. 8. In the foreground Wadi Telah a tributary of the Wadi Al Hamd. (Photo: H. HÖTZL, 1982).

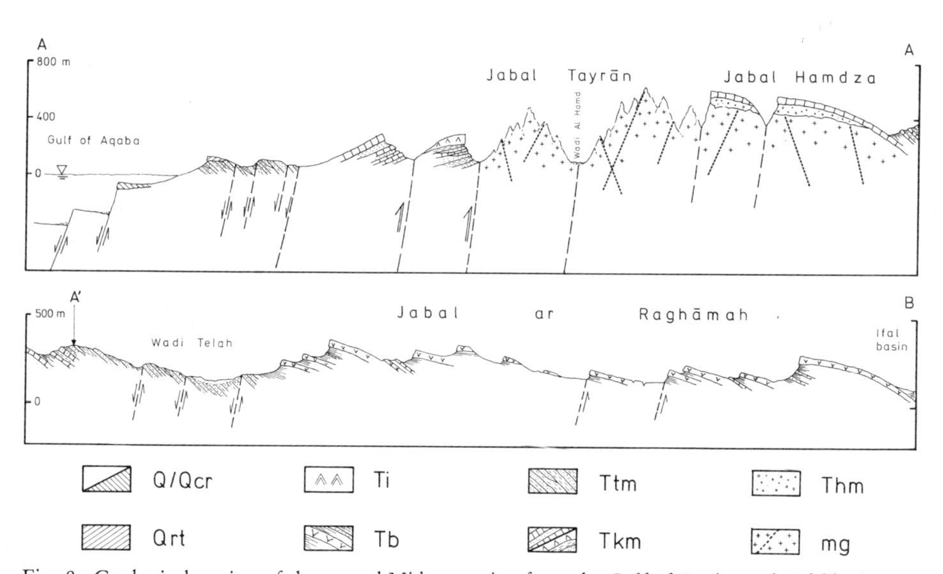

Fig. 8. Geological section of the central Midyan region from the Gulf of Aqaba to the Ifal basin (ex: R. STEPHAN, 1984). W–E profile: Upper profile (A–A): western part, lower profile (A′–B): eastern part (*Q/Qcr; Q* Quaternary alluvial and beach sediments, *Qcr* recent coral reef; *Qrt* raised Pleistocene reef terrace; *Ti* Pliocene clastic to evaporitic Ifal Formation; *Tb* Middle to Upper Miocene evaporitic Al Bad' Formation; *Ttm* Lower Miocene clastic Telah member; *Tkm* Upper Oligocene limestone and gypsum of Al Kils member; *Thm* Oligocene conglomerate and sandstone of Al Hamd member; *mg* Precambrian monzogranite with dykes).

the Midyan region and suggested raising the term to Raghamah Group. This he divided into the Jabal-Tayran and Al-Bad' Formations. Although an exact classification is not possible, he assumes for this group in general a Miocene age and at the same time contrasts it to the Ifal Formation, which is generally assumed to be Pliocene.

Using foraminifers, C. DULLO et al. (1983) developed a stratigraphic classification of individual strata series. For the first time, Upper Oligocene could be demonstrated for the middle carbonate series (Wadi Al Kils Member) of the Tayran Formation, proving an earlier onset of sedimentation than had been previously assumed.

The uppermost parts of the evaporitic Al-Bad' Formation belong to the Upper Miocene (Torton). Only for the presumably Pliocene Ifal Formation are there as yet no stratigraphic indications as to how far it extends upward.

As geophysical studies have shown (M. M. A. BOKHARI, 1981), these Pliocene, mainly clastic sediments can reach thicknesses of more than 2,000 meters within Ifal Depression. In contrast, the Quaternary deposits are relatively thin, and generally described as individual gravel terraces, extensive alluvial cones and, in the southwest, uplifted marine terraces.

2.1.1.3. Young tectonic evolution

The Midyan region is located in the immediate area of overlapping of the two decisive tectonic structures for northwestern Arabia: the Red-Sea graben and the Aqaba – Dead Sea – Jordan shear system (cf. section 1.2.). There is as yet no clear concept of the causal relationship of these two mechanisms of movement, or of their reciprocal effects. For the tectonic shaping of the Midyan region, the horizontal shear of the Aqaba system is the more important (H.-J. BAYER et al., 1983). The entire sinistral strike slip movement along this structure, which is demonstrably younger than the Neogenic dikes in Sinai and in Saudi Arabia, is estimated at 110 kilometers (A. M. QUENNELL, 1956, R. FREUND et al., 1970). Stratigraphic marks found on the Sinai show that there were two phases of strike slip activity, with recent and persistent tectonic activity for the younger phase. According to R. W. GIRDLER and P. STYLES (1974), S. A. HALL (1977), and D. J. NOY (1978), this younger phase is within the period between 5 million years ago and the present. The annual strike slip rate calculated from this amounts to at least 0.54 centimeters. M. EYAL et al. (1981) and G. STEINITZ (1978) have a different concept of the time involved for the entire movement which would require higher shift rates of 2.2 to 3.5 centimeters per year.

The lateral shear caused a further extension of the rearranged graben structure, with the opening of rhombically defined, en-echelon parts of the graben (Fig. 9). On the Gulf of Aqaba, the lateral strike slip itself includes a 60-kilometers shear in what today is the sea, and 50 kilometers in highly fault-structured shear belts on both sides of the Gulf. This extensive shearing along a number of paths of movement results in a whole system of block faults that in turn go through their own wedging or detaching movements. The shearing in this entire region is characterized by tremendous horizontal paths of movement as well as large interlacing faults, in places with wide mylonite zones. Microtectonic structures

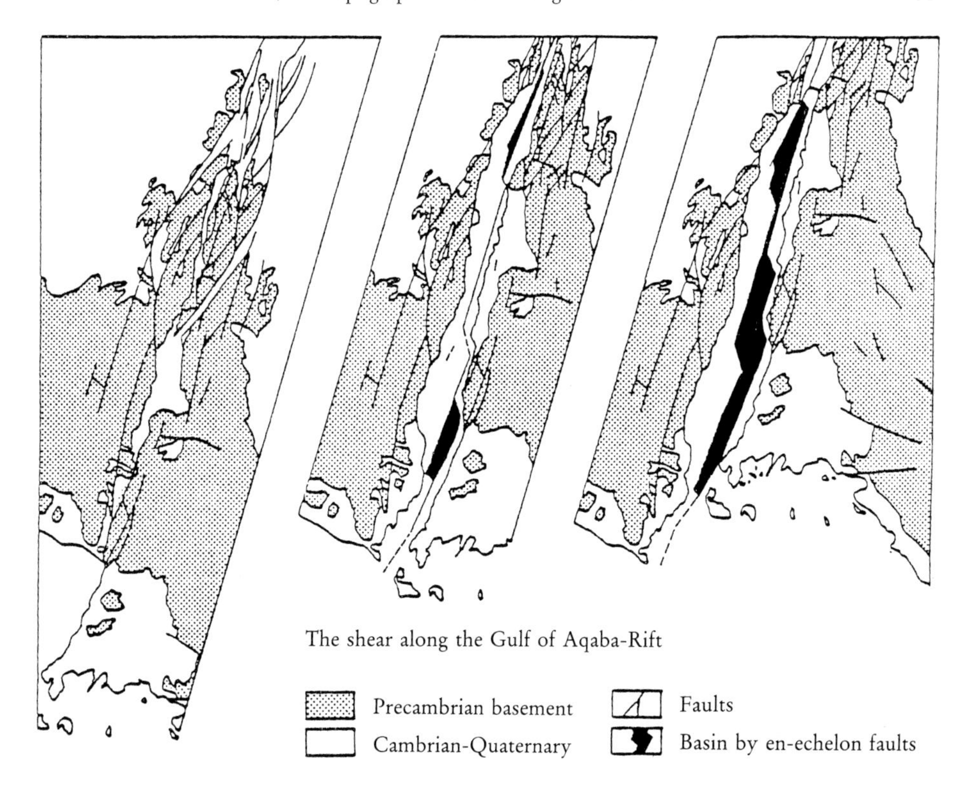

Fig. 9. The shear along the Gulf of Aqaba with the development of the en-echelon basin structure (after R. FREUND et al., 1970 and M. EYAL et al., 1981).

include a variety of tensional and compressive deformation phenomena. Thus, for example, step downfault forms and folded upfaultings may be found in close proximity.

On the basis of block tectonics, the Midyan region could be divided into its southern third and the main northern part. The later represents the more uplifted basement portion, which disaggregates into equivalent, opposed blocks.

The southern third shows a division into three parts from east to west. In the east there is again the more uplifted basement as the direct continuation of the Al-Hejaz Mountains. Both have its orientation and its western boundary parallel the Red-Sea system. The middle section contains the triangular Ifal Depression with Young Tertiary sediments with thicknesses exceeding 3,000 meters. This means a vertical shift as opposed to the eastern block amounting to 6,000 meters and more. The aerial geomagnetic measurements (M. M. A. BOKHARI, 1981) show a highly variable relief under the sediments, with actual intercalated small basins between single anticlinal structures. The Ifal basin seems to have been subject to a very rapid, irregular subsidence. The fault structures under the Ifal basin responsible for the depression can no longer be determined with certainty. One may, however, assume that the overlapping of the Aqaba and Red-Sea directions caused a wedge fault block.

In the west, toward the Gulf of Aqaba, there is the Tayran-Musayr block. It is less uplifted than the northern basement block and consists of two large basement uplifts (Jabal Tayran and Jabal Al Musayr), surrounded by the Oligocene-to-Miocene Raghama Group. In spite of the internal shearing, this block may still be viewed as a uniform anticlinal upwarp with a shallow axial pitch to the SSE.

2.1.2. The Quaternary Along the Coast of the Gulf of Aqaba

(S. S. Al-Sayari, C. Dullo, H. Hötzl, A. R. Jado, J. G. Zötl)

2.1.2.1. The Jordanian coast

In the Aqaba area, the coast today has been changed by a variety of human activities such as harbor-building. In the hinterland, however, a variety of old coastlines may still be seen. G. M. Friedman (1966) found a subrecent reef west of Aqaba within the tidal zone. Carbon-14 values of 4,770 ± 140 years B. P. place it in the Later Holocene. This reef within the tidal zone is today subject to increasing reworking and does not appear as a morphological step.

There are, however, a number of distinct marine terraces on the shallow slopes around Aqaba. From Aqaba down to the new container port, the coast shows only a few relics of marine sediments; these belong to the lower, 8-m terrace. The outcrops are much better exposed south of the container port. Approximately one kilometer south of the marine station, the coastal plain is covered by four morphological terraces down to the Saudi-Arabian border. Wadis and basement hills create only a few interruptions.

The top of the lower terrace is 8–10 meters above the mean present sea level (m. s. l.). This position fits well with other observations at different locations on the Gulf of Aqaba (J. Walther, 1888, U. K. Cimiotti, 1980) as well as on the Red-Sea coast (C. J. R. Braithwaite, 1982). Terraces at lower elevations appear mainly as relics. It is difficult to determine whether these terraces represent a separate sea-level cycle, or are only erosional relics due to unfavorable outcrop situations. In comparison with the terrace sequences at Ash Sheikh Humayd in Saudi Arabia, these lower terraces seem more likely to be relics on the basis of complete marine sequences from m. s. l. up to elevations of 10 meters and more.

The second morphological terrace is between 12 and 13 meters above m. s. l. Generally, the fossil shoreline is only indicated by beachrocks and such larger bioclasts as corals and shells. The steep sides of the wadis show that this second terrace is younger than the 10-meters terrace, as the layers of the former overlay those of the lower terrace. For this reason, these first and second morphological terraces may belong to the same cycle of change in relative sea level.

There is an unconformity between these first two terraces and the third one. Beachrocks or gravels from the shore of the younger terrace are in onlap contact with the reef carbonates of the third terrace, which is now 16 meters above m. s. l. The fourth and uppermost terrace is at 30 meters above m. s. l. (Fig. 10).

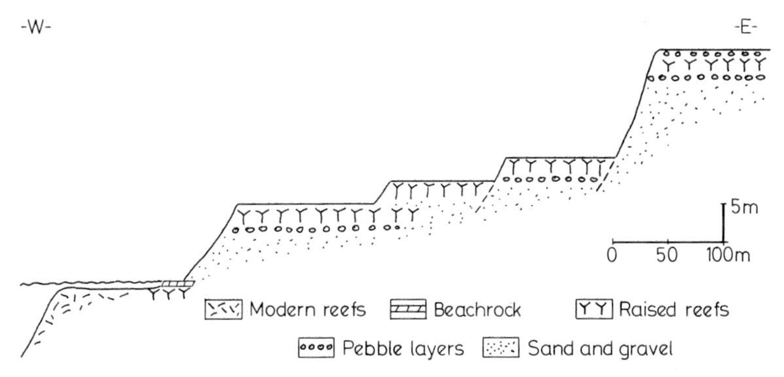

Fig. 10. Schematic cross section of the marine terraces south of Aqaba, Jordan.

2.1.2.2. *The alluvial fan around Haql*

The basement retreats some 10 kilometers south of Aqaba, and the widening seaboard between Jazirat Fir'awn and Ra's Suwayhil As Saghir is covered over a length of nearly 70 kilometers by extensive alluvial formations. They begin at the adjacent, steeply uplifted basement blocks and fall sharply down to the coast. The coastline covered by the talus fan is generally 6–10 kilometers wide. In the hinterland of Haql and Al Humaydah, the region including the wadis Al Mawrak, Al Humaydah and Umm Jurfayn, extends up to 20 kilometers inland, attaining a height of 450 meters a. s. l.

The entire hinterland area is laced with pronounced N-S to NNE trending faults. In addition to the left-handed horizontal components following the Aqaba system, there are also considerable vertical shifts. These are mainly responsible for the straight boundaries between the basement and the young talus fans. Satellite and aerial photographs show clearly that even the young clastic sequences have been affected by tectonics. A uniform aggradation surface within dissected alluvial fans in the hinterland of Haql is thus to be explained as a graben-like collapse of a block, which apparently only sank after dissection of the main fan and then was evened off by subsequent sedimentation.

The field work done so far has not permitted closer examination of this section of the coast, although it would provide information on youngest tectonics. It was only possible to cover the Wadi Umm Jurfayn along the road from the coast to Al Bad' or Tabuk, respectively. The terrace-like dissection of this talus fan seen there is described below.

Wadi Umm Jurfayn runs at a northwest slant, unlike the other wadis, which are arranged perpendicular to the coast. In the basement region, persistent sedimentation of the wadi has virtually prevented terrace formation, but the wadi is deeply incised into the old aggradation surface (Fig. 11).

Although the basement is exposed on the left side, the right orographic side borders on a talus cone reaching far back. From the general situation it is a formation that has developed in an excavation zone in the main terrace body. The fan shows a division into an older and a younger part. The main wadi cuts into the older section with a terrace edge five to six meters high; the lower part of this

Fig. 11. Sediment sequence of the main terrace of Wadi Umm Jurfayn, south of Haql. Silty and sandy material of dryer period overlain by coarse gravels and sand. (Photo: H. HÖTZL, 1978.)

older section is dissected by a dense net of narrow channels. They are adapted to the base level of the main wadi. The recent sedimentation caused by the numerous small channels in the upper part of the fan is so pronounced that the older sections of the fan have been almost completely covered by gravel.

Outside of the basement, in the area of the actual alluvial coastal zone, the Wadi Umm Jurfayn is accompanied on both sides by thick terrace bodies. In the upper section the terrace step is approximately 8 meters, in the middle section 12–14 meters, and thins out toward the coast to 6–8 meters. Compared with the uniform inclination of the main body this is to be explained by repeated changes in the base level on the coast. When the sea level is low, there is first an increase in regressive deep erosion; when the sea level then rises, this causes increased sedimentation in the lower course.

Today, the wadi channel over the entire area down to the coast is being filled. There are no deep outcrops here to provide information on the extent and age of the wadi fillings, but it may be said that with the last glacial low-water mark, deep erosion in the channels increased. The persistent Holocene sedimentation began with the eustatic rise in sea level.

Besides the thick accompanying terrace body of the main fan, there are no further intermingled terraces in the lower section of the Wadi Umm Jurfayn, with the exception of isolated, more pronounced highwater marks, especially at the mouths of tributary wadis. This is all the more remarkable as in the upper section of the wadi, the above-mentioned older talus cone was aligned with a distinctly higher sedimentation level in the wadi. The lack of further terraces here can be due to the relative narrowness of the wadi, which generally leads to truncation over its entire breadth in erosion phases.

There are still remnants of one such intermediate terrace in the broad Wadi Al Humaydah, which reaches the coast at Haql (Fig. 12). Erosion relics are found for example on the right side only three kilometers above the coast. More impressive is, however, above all the estuary fan on the coast belonging to this terrace level. It corresponds to the 12-meters marine terrace, whose reefs developed on the delta fan, and in places were then slightly covered with gravel. During a later eustatic decrease in sea level – whereby here the last Ice Age must be assumed – the main wadi channel cut itself into the northern side of the fan, and thus protected it largely from erosion.

This fan of some 3.5 kilometers in length is divided into sections by the old braided channels. With the decrease in sea level, they were activated by regressive erosion and then they dissected above all the originally continuous reef. On the fan itself, the channels flatten out backwards relatively quickly in accordance with the small catchment area.

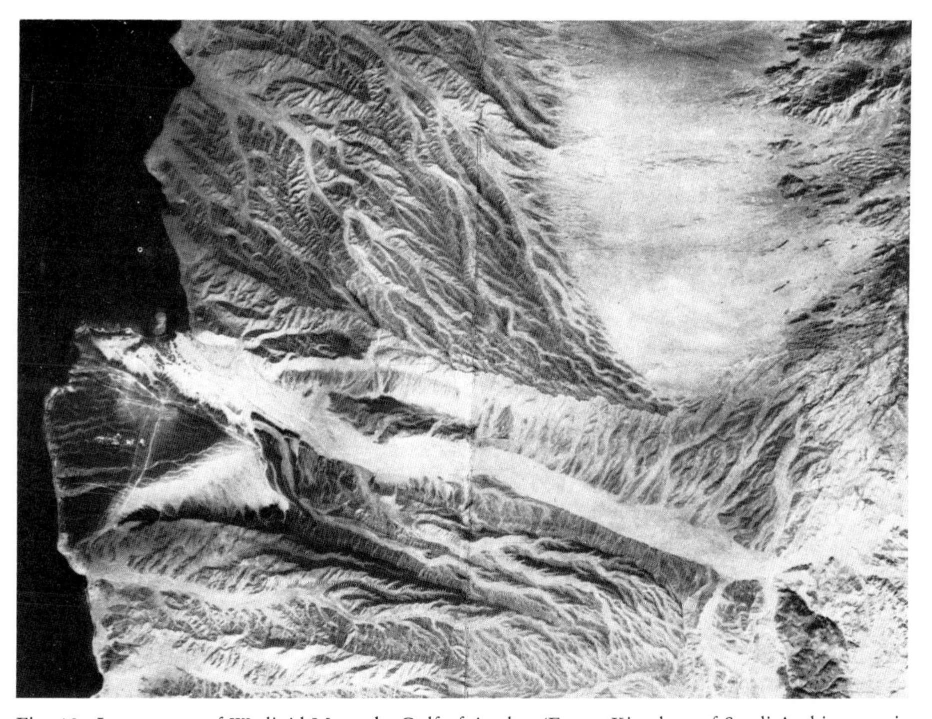

Fig. 12. Lower part of Wadi Al Mawrak, Gulf of Aqaba. (From: Kingdom of Saudi Arabia, mosaic map 1 : 50,000 sheet 20, 21). In the upperpart of the photo the old alluvial fan is dissected by a fossil dense local drainage system. Recent debris from the basement causes new covering and planation. Recent sedimentation also occur in the deep cut main wadi channel. Parts of an interim wadi accumulation is preserved in a less dissected small terrace body on the right bank (dark color) and in the fan at the wadi mouth. This fan is bordered by a small raised barriere reef. More information is given in the text.

As mentioned above, in Wadi Umm Jurfayn there is only the terrace body created by the dissection of the main surface. This alluvial surface falls relatively steeply to the coast after beginning at the edge of the basement some 12 kilometers from the coast at an altitude of 420 meters a. s. l. It is hard to say to what extent it was originally adjusted to a level different from that of the current base level as its lower section was later reworked a number of times by changes in sea level. The aerial photographs show flattened areas in zones parallel to the coast upon which the channel system of the higher surface empties, while farther down a new channel system develops whose depth increases as it approaches the coast. More detailed field studies to support the information from the aerial photographs are unfortunately not available.

Generally, this previously uniform aggradation surface has been considerably overprinted by both sedimentation and erosion. In the upper areas, which are directly adjacent to fault scarps, previously formed surfaces are even today being covered continously by gravel. Toward the coast, one sees increasing dissection by the channel system developed on the surface. These channels may either continue directly on to the coast, or turn slightly to empty into the deeply incised wadis from the basement.

Thus, all that remains of the original surface are a few generally narrow strips. These may be recognized by their covering of pebbles with especially dark desert varnish and may thus be easily distinguished on the aerial photographs from the otherwise slight eroded or young gravel-covered surfaces.

The terrace structure exposed in Wadi Umm Jurfayn also shows the formation of the thick, wide alluvial fan as a result of a longer period during which gravel was laid down, or of intermediate periods of erosion. The upper part of this terrace body consists of a sequence up to 12 meters thick of coarse gravels to well assorted boulder gravel layers (Fig. 12), whereby particularly coarse materials appear both at the base and the top.

In the upper section of the wadi near the basement, only these higher sediments are exposed in the terrace body. Their stratification is somewhat flatter than the recent wadi gradient. For this reason, the deeper sequence is also exposed in the middle and lower sections. In contrast to the upper portion, it consists mainly of fine clastic yellowish sediments. These include silts, fine sand and occassional gravel layers. In addition to sabkhah structures, dune remnants may also be seen in these sediments. The superimposition by the coarse, mainly fluvial series was followed on a clearly developed relief with occasional deep channels. They were filled with the coarse sediments. Here as well, there are unfortunately no indications of the total thickness of the deeper series exposed in the terrace profile.

2.1.2.3. The southern coast between Maqna and Ra's Ash Sheikh Humayd

Basement and Tertiary sediments appear over almost this entire section of the coast down to the coastline. Quaternary sediments usually take up only a narrow strip often only a few meters wide, or are limited to the filling of the wadis dissecting the mountain ridges. Only in the south are the younger deposits distributed over somewhat larger areas, although these coverings are usually very thin.

These are mainly marine deposits belonging to the coastal area. There are fringing reefs accretioned on older cliffs and sometimes covering them shallowly, beach sediments and in the southern part, transitions to lagoon deposits and sabkhah sediments. While the latter generally are Holocene sedimentation, the uplifted reef terraces belong to different Quaternary stages.

Clastic continental deposits include primarily more or less recent talus formations and impressive talus fans that most often empty out of the numerous channels and ravines of the very fractured and dissected basement into larger wadis, or onto the beach. There are also the aforementioned channel fillings in the wadis, occasional wadi terraces, flood-plain sediments and relatively coarse windblown sands. In the shallow bays in the southern part, salt-encrusted fine sands and silts form the transition to the actual sabkhah sediments. Disregarding the few gravels and pebbles on the wadi terraces, the loose continental sediments are usually Holocene formations, or at least material redeposited in the Holocene. With the steep relief of the coastal area, especially the direct deep drop to the basin of the Gulf of Aqaba, the eustatic changes in sea level in the Pleistocene have led to enormous retrograde erosion as a result of the increased gradients. With these changes in the base level, the previously deposited loose material on the narrow strip of coast and in the small wadis was apparently mostly eroded and

Fig. 13. Sharm Dabbah, southern coast of the Gulf of Aqaba. The terrace ridge bordering to the sea is protected by a reef cap, while the zone behind with its loose terrestrial sediments was eroded during periods of deeper sea level. (Photo: H. HÖTZL, 1982.)

deposited as sediment in the deeper marine basin. There, geophysical methods have shown the Quaternary material to be very thick.

In the coastal area, of the Pleistocene sediments only remnants of earlier reef bodies and reef crests are retained. As structures that had formed simultaneously they were less susceptible to erosion, especially when they grew up from consolidated Tertiary sediments, or the basement itself. There is thus today a fringe of terrace-like erosion remnants running along the coastline (see Fig. 13). This confirms the spatial constancy during the Quaternary, even in such places as the southern part, where there is a relatively low relief.

Today the reef terraces often form striking isolated erosion remnants, or more or less connected, small, table-like crests. The characteristic cut (Fig. 14) shows the tectonically tipped underground with the horizontal or slightly slanted reef body superimposed as cape rock. The formation of the narrow erosion remnants is due to the original thinning-out of the reef flat in the area of the old coastline, and the interfingering there with nonconsolidated clastic sediments.

As mentioned above, these were then eroded by succeeding sea-level variations, so that depending on geology and morphology, more or less wide depressions and channels developed behind the reefs, which then were sometimes taken up by bays when the sea level later rose (Fig. 13). This will be mentioned several times in the following sections, as these back-coast depressions recur repeatedly in the entire northern section of the Red Sea, down to the Jeddah area.

Fig. 14. Outlier of marine terrace, 3.5 km north of Maqna. The lower part is built up by tilted Young Tertiary clastic sediments. The unconformable cap is formed by a Quaternary reef platform. Elevation of the outlier 22 m above m.s.l. (Photo: H. HÖTZL, 1983.)

The present-day distribution of these bays is the result of the last drop in sea level in the latest Pleistocene (Würm glacial stage), and the later increase in the Holocene.

In the coastal section under discussion here, there is great variation in terrace formation and distribution. Around Maqna, the originally continuous reef terraces have been very chopped up and eroded. Thus, from about 15 kilometers north to about 12 kilometers south of Maqna there are only occasional isolated erosion remnants. They are topped by spurs of narrow crests that limit the wide bays and wadis. North of Maqna, tipped Miocene sandstones and conglomerates form the basis of these reef sediments. To the south, these are sometimes granites, sandstones and gypsum rocks.

In general there is no terrace sequence left, as the reef covering the relatively small erosion remnants only belongs to one respective terrace level. The remnants of the 2-meters step are an exception, as they extend in front of these erosion remnants.

The remnants of the older reef terraces show individual heights of 12, 16 or some 22 meters. With all the tectonic breakage in the underground, the differences in height could be explained on the basis of corresponding shifts in an originally uniform level, but owing to the equivalency of step heights in the neighboring areas, they would seem to belong to different terrace levels.

The mouth of Wadi Al Hamdh is at Maqna. It is the main wadi in the central Midyan area and its catchment originally extended beyond Al Bad'. Today it is beheaded 5 kilometers west of Al Bad' by the catchment area of Wadi Ifal, which has a more favorable location as regards the base level. In its upper course Wadi Al Hamdh is a broad aggradation plain. There is also sedimentation in the narrow middle and lower course.

In the wadis, especially directly east of Maqna, there are still remnants of gravel terraces. The oldest terrace system might well be represented by two individual isolated occurrences of the 'boulder gravel terrace'. The one occurrence is a small hill rising 20 meters above the wadi floor in the middle of the planation of the upper course near the present-day watershed. The second occurrence is west of the stretch of basement on the northern side, about 30 meters above the floor of the wadi that there turns to the south. This terrace remnant is divided in two, and the southern part has been displaced by a fault and tipped to the west. A connection between these two occurrences can only be found in the similarity of the debris they contain (various rock components from the basement), and the general situation.

A somewhat lower terrace is rather extensive. In and above Maqna it shows continuous development over a longer stretch. The terrace step is as high as 8 meters, but quickly flattens out upward, to then be covered by younger sediments. The surface of these gravels and pebbles shows a dark patina. At least a part of the channel originally was located slightly farther to the north (F. A. R. ZAKIR, 1982), so that the present wadi was formed by a new epigenetic channel above Maqna. It seems that this terrace is related to the level of the 12-meters reef terrace, but there is no clear connection.

In the section of the wadi near Maqna there is also a younger terrace element, but it is only subordinate in appearance. It cannot always be clearly distinguished from recent to subrecent sections of alluvial cones, or highwater marks.

The wide alluvial cone on the western edge of the Jabal Tayran also shows a differentiated pattern. The older cone – now dissected – empties or melts into the main terrace mentioned above. At the fan/wadi transition there is a small local occurrence of sinter. There are also somewhat larger sinter deposits in the Maqna Oasis below the present spring outlet.

The southern section from Sharm Dabbah to Ra's Ash Sheikh Humayd shows a partially interrupted terrace crest. The protective reef plate kept this crest from being eroded, while the loose material behind it was excavated to a greater or lesser depth. These areas are, as mentioned above, sometimes taken up by bays, such as Sharm Dabbah or Sharm Mujawwan. The uplifted reefs also cover more extensive areas from Ra's Ash Sheikh Humayd to Ra's Al Qasbah, the actual southern tip of Midyan (Fig. 13, 15).

The structure of the terrace crest is again comparable to the situation around Maqna. The tipped Tertiary forms the base, but here it is usually the Pliocene Ifal series that the reef follows inland with decreasing thickness. On the eastern edge of the crest, an interfingering of beach formations with clastic continental sediments can often be seen.

The planation levels are also very different here. The steps are best developed in the south, where areas of reef terraces have been preserved. Roughly 3 kilometers southeast of Ash Sheikh Humayd Station, the following levels could be distinguished:

- −1 to 0 m Recent reef some 120 meters wide.
- 2.5 m Erosion platform about 150 meters wide, made of an older reef community; shell coverage as part of young beach formation.
- 6.0 m Intermediate step some 80 meters wide, planation made up of corals covered over with shells.
- 9.0 m Terrace about 100 meters wide with weathered coral colonies; gradual step transition or sharp edge.
- 12.5 m Terrace step about 100 meters wide, made up of reeftop communities about 1 meter thick, a shell layer 0.6–0.8 meters thick with *Venus, Cardium, Strombus, Pinna* and others, also displaced Tertiary clays.
- 17.0 m Planation after a shallow rise (reef platform), about 500 meters wide.

Remants of this sequence are also found west of Ra's As Sheikh Humayd Station (at the end of the asphalt road). There, above shifted sabkhah sediments as well as Pliocene silt- and sandstones, there is a 1-meter thick beachrock layer subjacent to another shell layer (bivalvs and gastropods) as well as a horizon rich of echinoids. This whole sequence is crowned by the coral-reef plate. Its upper edge is at 12 meters above m. s. l. Behind it there is in places a somewhat setback gravel terrace with a planation at 21 meters above m. s. l.

Farther to the north, about 3 kilometer from Ra's Al Qasba, the terrace crest immediately begins with a 20-meters cliff step. The tilted Tertiary is there covered by a reef plate up to 5 meters thick. This shows a shallow rise to the rear of 2°.

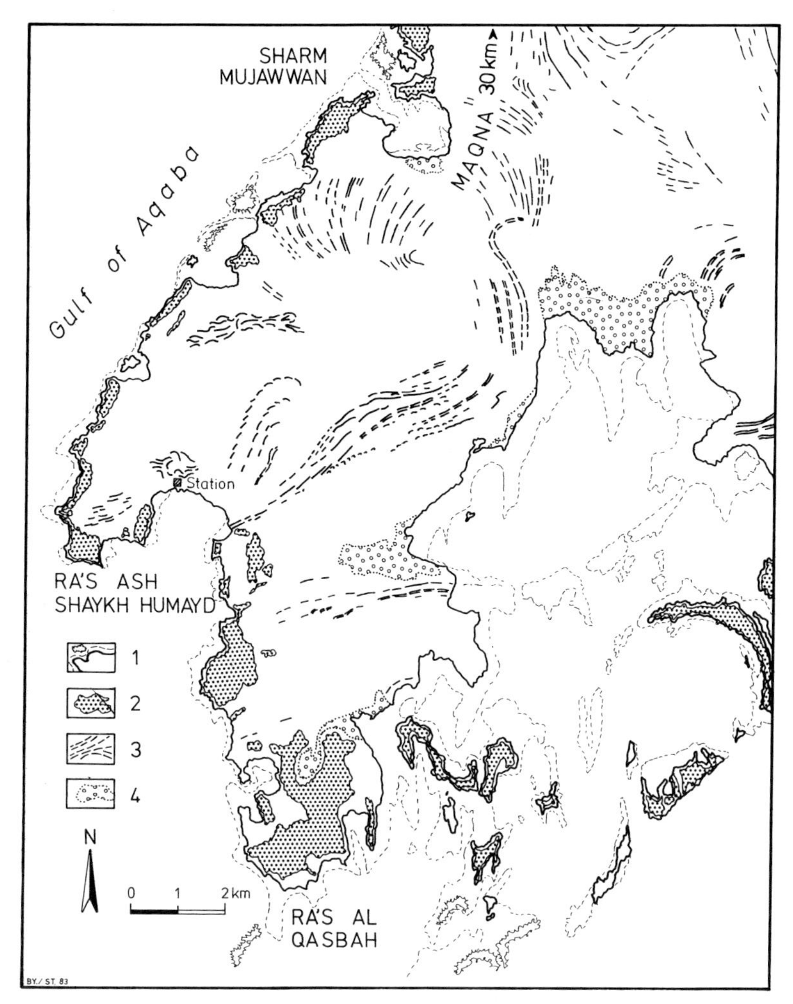

Fig. 15. Occurrence of marine terraces and recent sabkhahs along the southwestern margin of Midyan (section of the aerial photo map of R. STEPHAN, 1984). *1* coastline with modern reefs, *2* raised Pleistocene reefs; *3* ridges of outcropping Pliocene; *4* coastal sabkhahs.

Partly there is also a terrain step that leads to the 26-meters niveau; it shows littoral sediments interfingered with gravels.

North of Sharm Mujawwan, the following steps are to be seen: 2.5, 8.0, 12.0, 16.0 and 20 meters. The total breadth of the crest is somewhat more than 1 km, and on the inland edge we again find the transition from beach rock to gravels. The steps mentioned here are mainly erosion surfaces in the older reef body. Younger reef deposits are often difficult to separate, but apparently are pronouncedly in retreat. The 20- and 16-meters steps have been displaced by a tectonic fault which is then covered by the nearly horizontal 12-meters terrace. The 8-meters step shows typical beach debris and in places a beach dune that slants shallowly down to the 2-meters step.

This situation remains about the same going north until the gypsum crest (Jabal Al Kibrit) approaches the coast, about 2.5 kilometers north of Sharm Dabbah. In places on the steep cliffs, the reef flat is exposed with varying thickness and facies in the 12-, 16- and 20-meters steps. On the back edge of the crest, the interfingerings with the terrestrial gravels keep turning up.

2.1.2.4. Composition and zonational patterns of the modern and raised reefs

Living coral reefs are developed along the coast of the Gulf of Aqaba and the Red Sea. According to H. MERGNER and H. SCHUHMACHER (1974) two main types of fringing reefs can be distinguished: fringing reefs in a very near-shore position, and fringing reefs with a lagoon. There are many transitional types, but in general they all can be referred to these two main types.

The present morphology of the living reefs is predominantly a result of erosion from the last glacial marine low watermark (G. GVIRTZMAN et al., 1977). Before steep cliffs, there are only fringing reefs with very small lagoons, or none at all. We find the same situation on the flanks of old flooded canyons. In those areas where the sea floor is more flat or even horizontal over longer distances, fringing reefs with lagoons have developed. Only such relic structures in the reef crest as pools and radially arranged furrows indicate the old erosional patterns, which become increasingly covered by active reef growth (Fig. 16 e). Beside these fringing reefs, there are also modern reefs in some places, which already have the character of barrier reefs. They are more common south of the Tiran Strait.

Fringing reefs with larger lagoons show the most complete sequence of different sedimentary environments within shallow water. The beach region is characterized by gravels and sands of different composition, depending on the rocks of the hinterland. Bioclasts of corals and mollusks occur as well. Together with gravels and sands, these skeletal grains form beachrocks, which show a gentle inclination towards the sea (Fig. 16 e). Besides recent beachrock formation, H. MERGNER and H. SCHUHMACHER (1974) reported cemented iron nails within the beachrock; erosion also occurs, leaving small tidal pools. There is a narrow beach channel zone, in which mollusc shells accumulate. The next zone, the sea grass zone, is characterized by carbonate sands and by the sea grass *Halodule*, on which large foraminifera live, belonging to the genus *Marginopora*. The coral rock zone consists of dead corals surrounded by cemented carbonate sands. Only few octocorals occur, along with occasional colonies of *Stylophora*

Fig. 16. a) Recent surf notch, predominantly caused by microbiological erosion. On top there is a small erosional plane, which was formed by the Flandrian transgression. 100 m south of Sharm Mujawwan. b) Section of the second morphological terrace on the Jordanian coast. *1* microatoll zone, *2* seagrass zone, *3* coral rock zone, *4* bioturbated layer with *Callianassa* burrows, *5* reef branched corals of the front reef community at the top. c) Front reef community of a modern fringing reef with *Porites* colonies and *Millepora dichotoma* (arrow). North of Maqna 2 m below sea level. d) Fossil sea grass zone with molds of aragonitic molluscs. Third terrace of the Jordanian coast. e) Recent fringing reef north of Ra's Ash Sheikh Humayd. Pools and furrows (white) within the reef crest (dark) are relics of the erosional patterns of the last glacial low stand sea level. Note the partly eroded beach rock in the foreground. f) Front reef community of a raised reef with branched corals, predominantly *Porites*. Lower terrace south of Jabal Al Kibrit, near Sharm Mujawwan. (Photos: C. DULLO, 1982.)

→

a
5
4
3
2
1
b
c
d
e
f

pistillata. Among the echinoids, *Diadema* and *Tripneustes* are common. Towards the microatoll zone, more scleractinian corals occur. Among the branching forms, *Acropora* and *Stylophora* are predominant, but such massive forms as *Favia, Favites,* and *Goniastrea* are more important. The backreef edge is characterized by large colonies of *Platygyra lamellina.* The reef crest is barren. Dense coral crops are developed only at channel or furrow margins, as well as at the margins of pools. The reef front also exhibits dense colonies of scleractinian corals and hydrozoans, among which *Millepora dichotoma* is typical for the reef front community (Fig. 16 c). For detailed information about the recent carbonates of the Red Sea see G. FRIEDMAN (1968), H. MERGNER and SCHUHMACHER (1974, 1981), and H. BRAITHWAITE (1982).

The raised coral reefs exhibit the same zonational patterns as the modern reefs. In addition to the lateral facies patterns, vertical facies development can be studied as well. The lower terrace along the Jordanian coast of the Gulf of Aqaba exhibits at its edges typical front-reef communities (Fig. 16 c). This is indicated by numerous colonies of *Millepora dichotoma* and branches of *Porites.* Spines of the echinoids *Phylloacanthus* and *Heterocentrotus* also characterize this zone. Both echinoids are common at the reef crest edge towards the open sea. The morphological step of the lower terrace still seems to show the old relief. Only in wadi canyons formed during the last marine low water mark (G. GVIRTZMAN et al., 1977) other facies zones are outcropping. Especially the second and third terraces in Jordan exhibit thicker vertical sequences.

In the second larger wadi canyon south of the Jordanian marine station the following sequence can be seen. The marine sediments of the second terrace start with the microatoll zone, which is characterized by colonies of *Favia, Favites,* and *Goniastrea.* This zone is overlain by mollusc shells, which predominantly occur as molds due to aragonite leaching. Those accumulations of shells are known from the sea grass zone (H. MERGNER and H. SCHUHMACHER, 1974, C. DULLO, 1983 a). The next layer belongs to the coral rock zone, indicated by a few branches of *Stylophora pistillata* and spines of *Diadema.* There then follows a bioturbated sand layer with numerous *Callianassa* burrows, which in the 'Recent' occur in the sands of wadi deltas. The top is characterized by the reef crest facies, which is overlain by branched corals and hydrozoans, indicating the seaward edge of the reef crest (Fig. 16 b). In general, the development of the second terrace reflects a more regressive character in its lower and middle sections. The top, however, is more transgressive. On the other hand, this development demonstrates the reef growth model of H. MERGNER and H. SCHUHMACHER (1974). The third terrace exhibits different vertical facies sequences as well, but this sequence is generally more transgressive. The fourth terrace is more eroded and not as thick as the lower ones; vertical successions therefore are scarcely developed.

Similar cycles have also developed in the lower terrace between Jabal Al Kibrit and Ra's Ash Sheikh Humayd. The top of the lower terrace there is between 9 and 12 meters above m. s. l. The raised reefs do not show the various types of lagoonal facies as seen south of Aqaba. They consist, like the modern reefs of this area, of a beachrock zone and a short zone with skeletal sand, which interfingers with the well-cemented reef crest. Only few colonies of *Stylophora pistillata* and *Favia, Favites* or *Goniastrea* are present. The reef crest itself varies in width between 10 meters and 50 meters. In contrast to the Jordanian terraces, these terraces exhibit more reef-front communities at their seaward step. The vertical sequences of the northern terraces exhibit a more regressive character, except for their uppermost part (Fig. 17 a), whereas the sequences of the southern terraces have a purely transgressive character (Fig. 17 b). It is difficult to refer these facts

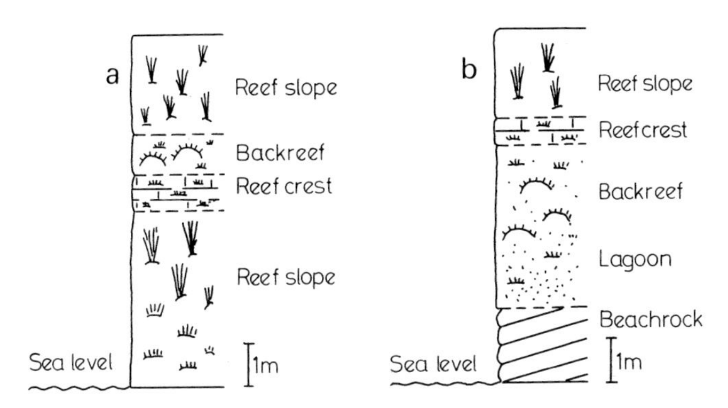

Fig. 17. a) Section of the lower terrace north of Sharm Mujawwan exhibiting a pseudoregressive sequence in the lower and middle part. b) Section of the lower terrace north of Ra's Ash Sheikh Humayd, exhibiting a transgressive sequence.

only to a relative change in the sea level, as they are contradictory. It seems more likely that these phenomena reflect erosion and construction cycles as shown in the model of H. MERGNER and H. SCHUHMACHER (1974). A regressive sequence then would indicate prograding reef-front communities. This progradation can continue up to the point where the relief of the sea floor steepens rapidly. The reef growth became increasingly restricted to the upper part of the reef front, leading to overhanging cliffs. Those cliffs are more easily destroyed by erosion. New front-reef communities can settle above those eroded reef crests. If the prograding reef growth is quicker than the relative sea level change, a pseudoregressive sequence of this sort can develop (Fig. 17 a), although the whole reef was built up during a transgressional phase (Fig. 17 b).

The higher terrace between Jabal Al Kibrit and Ra's Ash Sheikh Humayd is between 16 and 20 meters above m. s. l. Due to huge amounts of debris covering the flanks of this terrace, its marine carbonates are not well exposed. The facial patterns therefore cannot be given. Raised reefs 30 meters above m. s. l. are only recorded as tiny relics.

2.1.2.5. Diagenesis in the raised reefs

A comparison of the quantitative biotic composition of the Recent carbonates with their uplifted counterparts shows a great loss of aragonite fossils in the marine terraces. The aragonite fossils are not only replaced by calcite or occur as molds, but they are also destroyed by compaction within the carbonate sand. They are therefore excluded from the fossil record (Fig. 16 b). Only thick-shelled molluscs are preserved in aragonite, exhibiting relics of ancient colors. In contrast, calcitic molluscs are not leached and still show their primary microstructure.

The Jordanian terraces show more or less the same degree of diagenetic alteration in the different layers, except that the lower one is characterized by more thick-shelled aragonitic molluscs. In contrast, the terraces between Jabal Al Kibrit and Ra's Ash Sheikh Humayd are characterized by a sequence of progressive diagenesis of aragonitic structures (C. DULLO, 1983 b). In the lower terrace, which is not a real terrace but a fossil shoreline with uncemented bioclasts (1 meter above m. s. l.), there has already been a replacement of aragonite cements by blocky calcite in the interseptal spaces of corals. The primary microstructure of corals and molluscs is still preserved. Only a few branching corals exhibit leaching of the skeleton. The next horizon (9–12 meters above m. s. l.) is characterized by a replacement of the coral skeleton by calcite, whereas the molluscs are unaltered. The aragonite cements are completely replaced by calcite in this layer. The diagenesis of molluscs starts at first in the higher terrace (16–20 meters above m. s. l.), in which corals are already transferred into calcite. This diagenetic sequence can be referred to the different influence of organic substances during aragonite precipitation, to the density of the crystal fabric, as well as to the degree of saturation of the meteoric water with respect to calcium carbonate within the diagenetic environment.

Comparing both localities, the lower terrace of the Jordanian coast and the second terrace (16–20 meters above m. s. l.) near Ra's Ash Sheikh Humayd show the same diagenetic facies. The geological setting of the terraces is the most important factor influencing the degree of diagenesis. The Jordanian localities are very close to the Precambrian basement, which in a short distance reaches altitudes exceeding 1 000 meters. The hinterland of the terraces between Jabal Al Kibrit and Ra's Ash Sheikh Humayd consists of evaporites and clastic sediments of the Tertiary Al Bad' Formation and the mountains in this area are not as high. Due to the higher mountain range on the eastern margin of the Gulf of Aqaba in the north, there are higher amounts of meteoric water, leading to an aggressive diagenetic environment. The localities at the southern end of the Gulf are affected by lesser amounts of meteoric water, as they are much farther from the mountains. Additionally, superficial meteoric waters, running to the Gulf, are reduced in quantity by percolation in the clastic sequence of the Al Bad' Formation. On the other hand, these waters became saturated by leaching the evaporites, leading to a reduced aggressivity. The degree of diagenesis is thus lesser in these terraces than in their Jordanian counterparts.

2.1.2.6. Age and stratigraphic position

Age determinations of the different terraces are still lacking. Preliminary hints can only be given by comparisons with other localities. The lowermost fossil reef, which is recorded on land by a fossil shoreline can be referred to the Flandrian transgression. Radiocarbon dates of this horizon from the Sinai Peninsula indicate an age of 4,770 ± 140 years (G. FRIEDMAN, 1965). Similar results were gained by A. DABBAGH et al. (chapter 2.4.2. in this volume) for the lower most terrace along the southern Red-Sea coast. In Maqna this subfossil and already well cemented reef is situated in the intertidal zone with corals typical for the reef front. Landward, these reef limestones are partly covered by beachrocks. A similar situation can be seen on the Jordanien coast (Fig. 10).

Radiocarbon dates of the higher terraces are spurious due to diagenetic alteration of the carbonates (conf. H. HÖTZL et al., chapter 5.4. in this volume). Except from the lower most reef terraces all others seem to be too old for radiocarbon age determinations. For control one unaltered *Tridacna*-shell from the 12 meters-platform northwest from Ra's Ash Sheikh Humayd was prepared for age determination. The ^{14}C-analysis gave a value of 0,57 ± 0,75% – modern which only means that the shell is older than 35,000 years.

Therefore ^{230}Th/^{234}U dates are more trustworthy. The lower terrace (9–12 meters above m. s. l.) on the Sudanese coast was dated by L. BERRY et al. (1966) to be 91,000 ± 5,000 years old, using this method. G. GVIRTZMAN, G. M. FRIEDMAN (1977) reported three terraces from the southern part of the Sinai Peninsula, which have been dated at about 110,000 years (lower), 200–250,000 years (middle) and more than 250,000 years (upper). For this reason, the lower marine terrace in the Gulf of Aqaba can be referred to the younger transgression, in comparison to the Mediterranean stages of Pleistocene stratigraphy.

2.1.3. Contributions to the Quaternary Geomorphology of the Ifal Depression

(E. BRIEM, W. D. BLÜMEL)

2.1.3.1. Geological structure

Geologically, the Ifal Depression named after the main Wadi Ifal is the northern, terrestrial extension of the Red Sea graben system (Plate I, see insertion at back cover). The eastern side is limited by the nearly straight NNW–SSE striking main fault line (in the direction of the Red Sea), and the uplifted Precambrian basement at altitudes exceeding 2,000 meters. The NNE–SSW striking western limit is formed by upward arching of probably Miocene evaporites which cover individual fragments of the basement. The spacious structures show an acute-angled latticing of the Red Sea and Gulf of Aqaba directions; this lattice is mainly responsible for the morphological formations.

Since the Oligocene the Ifal Depression has been a sedimentation trough, catching littoral and terrestrial deposits the thickness of which is estimated to exceed 3,000 meters. There are three units involved, which we divided, according to the evaporites (Al Bad' Formation, Miocene) so conspicuous in the terrain, into pre-evaporitic (Jabal Tayran Formation, Oligocene – Miocene), evaporitic and post-evaporitic series (Ifal Formation, Upper Miocene – Pliocene). The pre-evaporitic complex is composed mainly of littoral, sandy facies, while the post-evaporitic series are terrestrial, mainly reddish-colored complexes of lightly indurated, clastic sediments (M. M. A. BOKHARI, 1981; C. DULLO et al., 1983).

In contrast to the Geological Map 1 : 500,000 (R. A. BRAMKAMP et al., 1963), these probably Pliocene sediments cover almost the entire surface of the Ifal Depression; they are only obscured by eolian sands or thin fluvial deposits. The Pliocene deposits are superficially missing only in the area of the recent drainage courses of the wadi systems.

The clastic series of the Ifal Formation have been subjected to great tectonic stress. Presumably, in the Late Pliocene or Pleistocene, there were block shattering and limited tectonic block tilting along with a general tendency to uplifting in the entire Ifal triangle. Morphologically, these lifting processes in the Quaternary caused a general erosive overprint of the young Tertiary series, which have persisted up to the present, interrupted only by brief accumulation periods. Some of the sediment groups are sharply tilted, but for this reason appear as shallow hogback complexes piercing through the eolian sand cover.

Depending upon its intensity, the tectonic overprint along the long axis of the Wadi Ifal is clearly divided into eastern and western halves. The uplift was definitely more intensive in the western than in the eastern part. There are no extensive remnants of older terrace systems in the western part; the entire region is distinctly dominated by erosive forms, creating a mainly structural relief. In the entire eastern part we find terrace remnants in characteristic forms. The eastern part has thus been relatively inactive in more recent geological history, showing a rather undisturbed development of a terrace sequence.

The morphological studies were particularly concerned with the survey of this terrace system. The objective was to recognize and date, if possible, tectonic movements in fault or flexure zones and to point out geomorphic consequences.

Fig. 18. Oldest terrace generation ('boulder gravel terrace') north of El Aynunah in the SE part of the Ifal triangle. (Photo: W. D. BLÜMEL, 1982.)

2.1.3.2. Terrace systems

We found three generations of terraces in the area under study. They have remained especially well preserved in the piedmont of the eastern half of the Ifal Depression, the foreland of which is particularly protected from erosion (Figs. 18, 20, 21, 22).

There are remnants of the oldest terrace generation at a level of about 250 meters at the edge of the mountainous area. There is a gravel layer as thick as 20 meters in places with very coarse pebbles which at the edge of the mountains north of El Aynunah may exceed a diameter of 2 meters (Fig. 18). The gravels show a colorful spectrum of mainly crystalline and metamorphic rocks from the Precambrian basement upon which the hinterland is built.

The virtual absence of gravels from the overlaying sediments indicates that, during the formation of this gravel terrace most of the Cambro-Ordovician sandstones in the direct overlaying had been eroded away.

The terrace surface consists mainly of the typical desert stone pavement, from which the pebbles protrude. Rubble and gravel have a dark patina and are often coated with desert varnish. The eolian shaping increases toward the depression and the wind has carved the massive pebbles. Therefore, this is a degraded terrace profile with residual gravel, showing that the original gravel group was thicker. In the few places where the entire profile is freshly exposed, most of the gravels are weathered at their cores and disintegrate when touched. Thus, the oldest level was exposed to severe weathering and erosion (Fig. 19).

Fig. 19. Core weathering, desintegration and residual detrital pavement of an old terrace (eastern part of the Ifal Depression). (Photo: E. BRIEM, 1982.)

The gravel group covers a capture surface cutting the mainly gravelly-sandy sediments of the Tertiary tilted blocks. We can reconstruct a probably Early Quaternary phase of the denudation which formed a level at about 230 meters at the edge of the mountainous region; fluvial action then covered it with flat alluvial cones.

The next deepest terrace level extends at the edge of the mountains at about

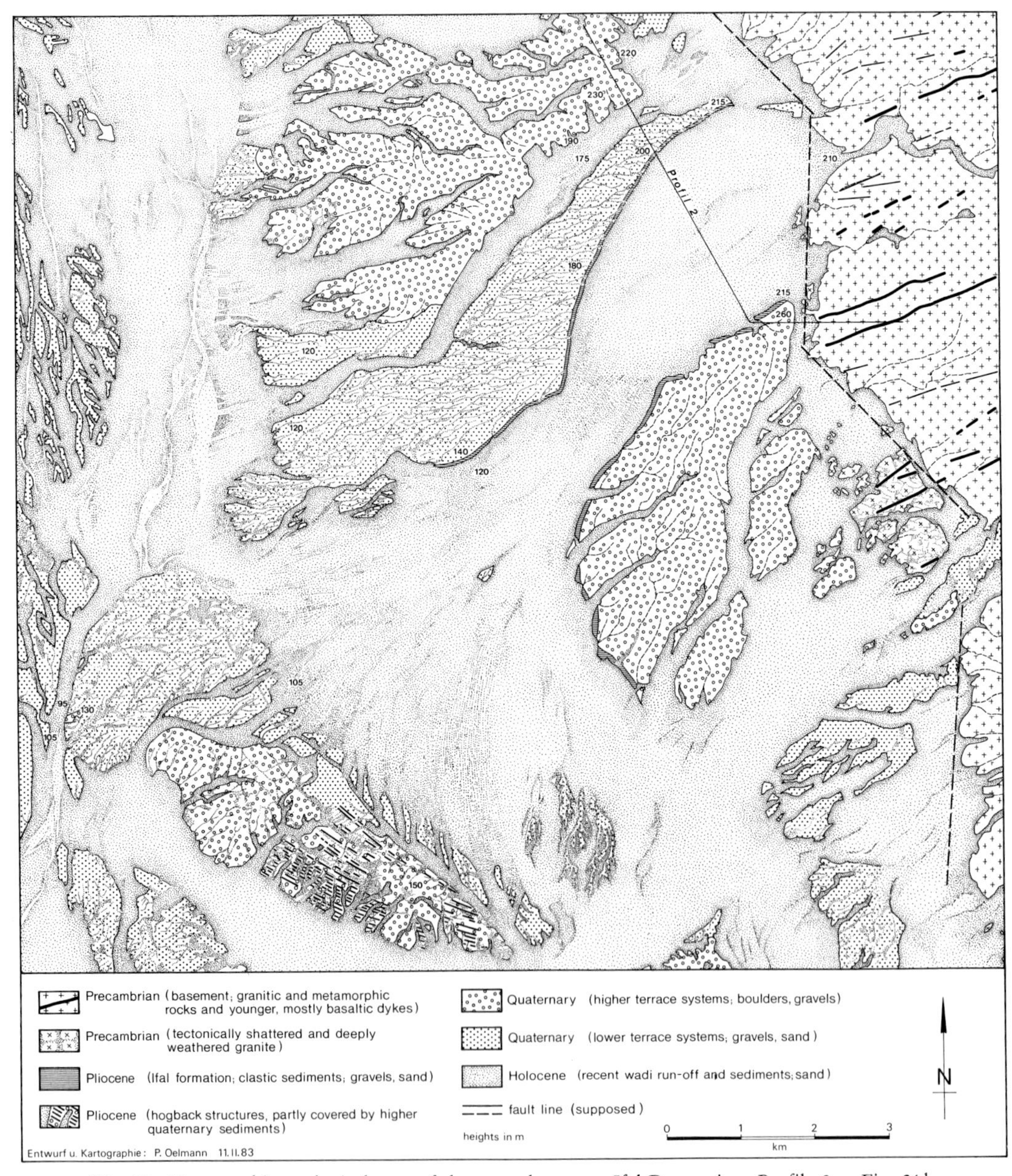

Fig. 20. Geomorphic-geological map of the central, eastern Ifal Depression. Profile 2 = Fig. 21b.

215 meters. The structure of the terrace is identical to that just described; above a capture surface at 195 meters there is a gravel deposit maximally 20 meters thick in interfingering flat alluvial cones. The gravels are far less coarse, but still reach a diameter of 20 centimeters and more. Crumbly or core-weathered gravels are not to be found; the entire gravel group shows only slight reddish weathering. The residual rubble on the surface and the patina are also developed here; therefore older and younger generations of terraces cannot be distinguished in aerial photographs (Figs. 20, 21).

The next youngest event has apparently been a powerful phase of erosion which – with the exception of a few remnants – lowered the surface of the ground by 50 to 60 meters. This was followed by an aggradation of at least 20 meters.

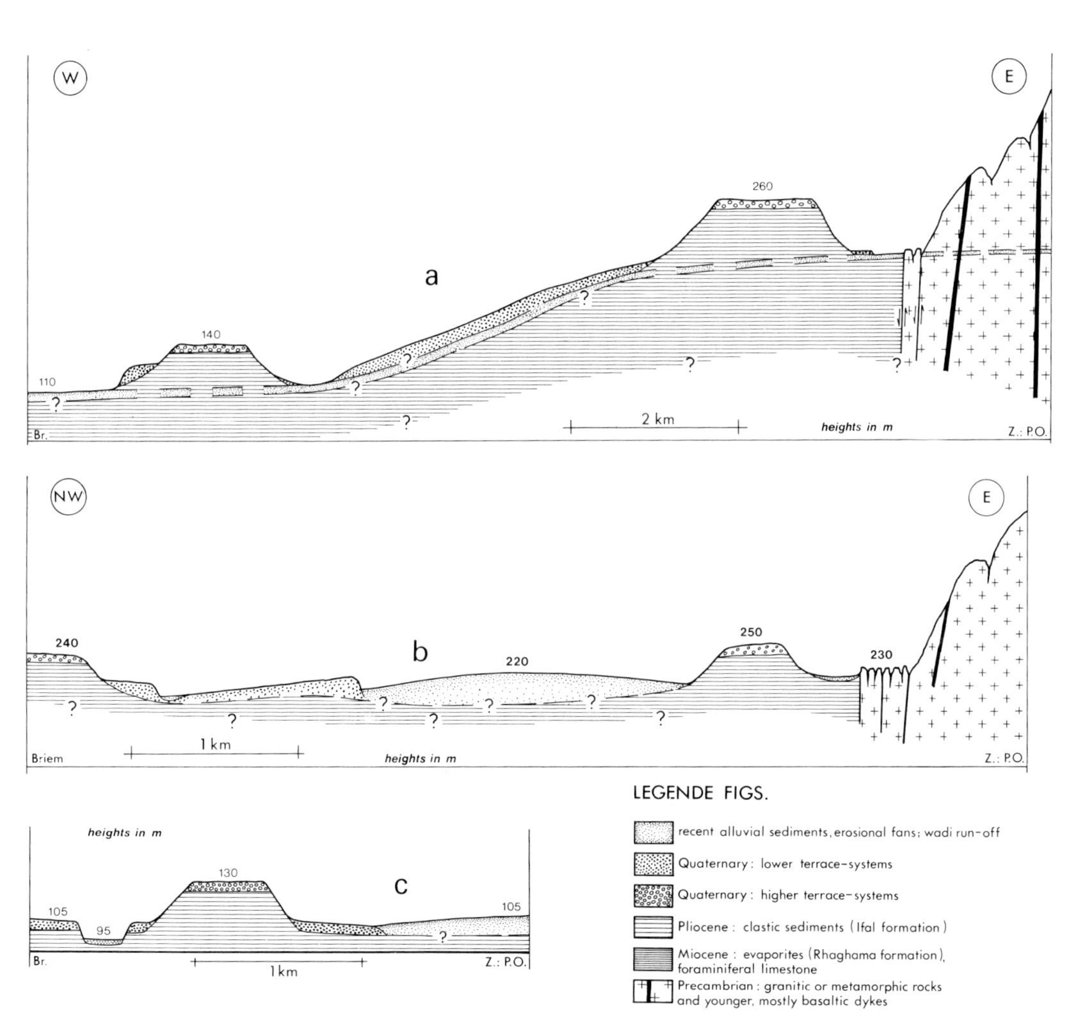

Fig. 21. a) Terraces: length-profile of the east Ifal Depression; see Fig. 20, approximately continuation of the E–W part of profile 2; b) Terraces: cross-profile of the east central Ifal Depression, profile 2 of Fig. 20; c) Terraces: cross-profile, Wadi Ifal, central Ifal Depression (lower left margin of Fig. 20).

The following cycle of erosion and accumulation provides the transition to the recent situation. Hogback-forming erosion along the main drainage channels dissected the older levels down to far below the present-day ground level. Well borings showed that the following accumulation reaches average thicknesses of at least 15–20 meters (Fig. 21). Subrecent to recent sedimentation is much finer than all the older sediments. The material is sandy to gravelly and light in color.

The recent accumulation and the two older terrace systems have a characteristic appearance in the entire area under study. Interesting are two thick terrace complexes of limited extent which appear at Al Bad' and El Aynunah in the breakthrough sections of the wadis. Here, up to 20 meters of fine grained, loess-like but well layered sediments are exposed, which J. BÜDEL (1954) and other authors described as a marly sand terrace in Northern Africa and Sinai.

The presence of thick, fine grained deposits in narrow valleys or ravines (for example near El Aynunah) is a problem in view of the understanding of the flow process, which cannot be resolved without more precise investigation. We shall limit ourselves to mentioning the problem, and refer the interested reader to the literature.

The other terrace systems are beheaded by the coastline (see Plate I, insertion at back cover). Cut off straight, the gravels of the second terrace generation and the tilted Tertiary underground form a cliff some 30 meters high which is particularly evident in the southeastern part of this area. Major tectonic activity, with vertical shifts of more than 30 meters, is to be found in the graben probably up to Middle and Young Quaternary age (see Fig. 22).

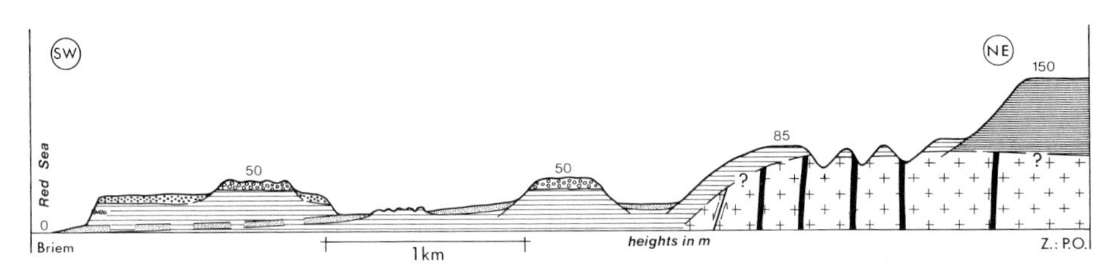

Fig. 22. Geological cross-section of El Aynunah region (south-eastern Ifal Depression).

In the same area, there is an older coastline some 20 meters above the present sea level, which is easily recognizable by the fossil-rich "beach rock", dissected out as monadnock. Variations in the sea level play a decisive role in the formation of terraces in this area. Besides climatic eustatic changes in sea level morphological developments and interplay of tectonics are important.

Starting with the minimal variations in sea level, it must be assumed that erosion took place in the coastal region during the low-water periods in the glacial stages, at least if one presumes intermittent moister phases. Viewed cyclically, accumulation occurs with the change to a warmer climate or with decrease in water discharge.

The three demonstrable cycles could be interpreted as belonging to the glacial stages, and the transitions to the interglacial stages, with the erosion phase during the decrease in sea level and increase in run-off and with the accumulation phase at the end of the last glacial stage in the transition to an interglacial period. In addition it has to be considered that detailed investigations in other arid areas showed that changes in precipitation are not necessarily parallel to cooler and warmer periods. On the other hand, the genesis of terrace sequences can be determined by tectonical movements independent of climatic-geomorphic conditions. These hypothetical considerations require closer examination through sample studies and intensive observation of the terrain.

2.1.3.3. Features of the Pre-Quaternary underground

The Geological Map 1 : 500,000 of Saudi Arabia, Map I-200A (R. A. BRAMKAMP et al., 1963) shows, by the symbol "Qt", that the terrace alluvium (gravel, sand and silt deposits) in the south-opening triangular structure of the Ifal Depression (see Plate I, insertion at back cover) is of Quaternary formations. This generalized classification – most likely based on aerial photographs – deserves correction. Terrace or alluvial-fan deposits of this kind are to be found on top of the eastern part of the Ifal Depression. They begin at the western edge of the crystalline basement and belong to wadi systems that reach far to the east and northeast in the basement. There are different generations of alluvial fans or terraces (see above). The youngest episodic run-offs flow to the southwest as tributaries of the Wadi Ifal from the north (Plate I).

Considerably less space is covered by Quaternary accumulation forms on the eastern margin of the upward-arching evaporite series (vicinity of the Wadi Sha'ib An Nakhlah). In the south and southwestern part of the Ifal Depression particularly, outcrops of tilted Late Tertiary layers are found. They have been discordantly beheaded by erosive processes and are presently dissected out by selective denudation and erosion as weak hogback structures (Fig. 23). Small surface areas also consist of these Tertiary clastic sediments in the central and eastern part of the area under study; the relatively thin cover of terrace sediments was cleared off.

Both on the ground and in the aerial photographs, the surface of the Tertiary outcroppings often cannot be differentiated from Quaternary gravel coverings.

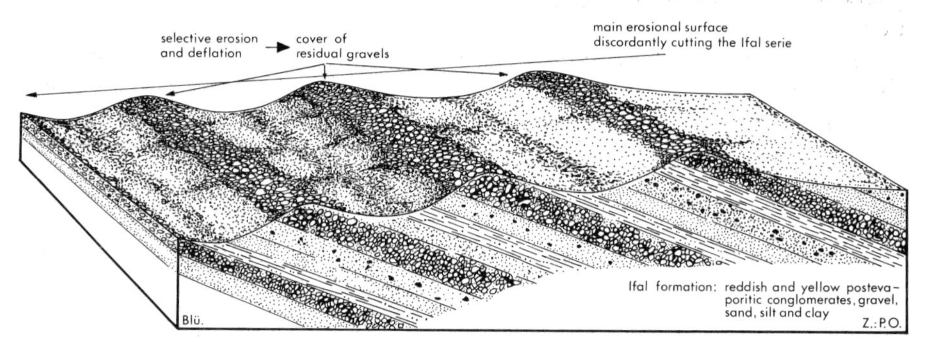

Fig. 23. Structural forms ('hogbacks') produced by selective erosion and residual accumulation of gravels on outcropping tilted layers of the Ifal Formation (especially in the southern and western parts of the Ifal Depression): Strata with a larger content of gravels are morphologically hard compared with beds of finer clastic material.

The reason for this is to be found in the partly conglomerate facies of the post-evaporitic young Tertiary Ifal series which, with selective removal of the finer gravels and sands, leaves a superficial scattering of residual pebbles that closely resemble the younger deposits of the alluvial fans. Both substrata also have very similar sediment spectra because the Tertiary and Quaternary riverlet channels drain the same area. The post-evaporitic Late Tertiary clastic rocks of the Ifal series as well as the discordantly overlying Quaternary fan- or terrace-generations come from the wadi drainage area north and east of the Ifal Depression (dominantly crystalline basement).

The mainly reddish but partly multi-colored clastic Ifal layers show different degrees of chemical weathering. The formation is made up of alternating layers of gravels, sands and pebbles. From the paleogeographical point of view, the layers of debris, fine sand, silt and clay should, in particular, be mentioned. Apparently, they represent climatic-morphological phases of the Upper Tertiary of more humid or seasonally-humid character. The finely granular layers are presumably tropoid soil formations of fersiallitic or ferallitic type[1]. In the basement area and on the deposited layers in the Ifal Basin, soil or weathered material were formed in periods of relatively slight morphological activity. They have been eroded from the basement during periods with pronounced semiarid morphodynamic and were deposited over the already existing basin fillings. Coarser pebble layers suggest that chemically less affected rocks (near the respective former base of weathering) were carried off and contributed to the Pliocene Ifal Formation as gravel and pebbles. These alternate layers of basement material in various stages of weathering have a composite thickness exceeding some hundreds of meters. M. M. A. Bokhari (1981) mentions 3,000 meters. It is quite possible, however, that this figure is too large, because with the tectonic fractuation and lack of stratigraphic leads, an additive determination of thickness is very problematic.

The oldest of the Quaternary alluvial-fan generations, deposited discordantly on an underground which was already tectonically shifted and partly re-eroded, has a direct connection with the earlier genesis of the post-evaporatic Ifal Formation. The coarse blocks, with diameters of some meters (Fig. 18), appear to be "woolsack" forms such as would develop with deep chemical weathering and granular disintegration along systems of joints and highly permeable rock sections.

The convincing congruence of the course of the valley and the tectonic lines in the basement, which are seen in aerial photos, render this assumption probable. The "boulder gravel terrace" (Fig. 18), the oldest alluvial-cone generation, represents the root of downward weathering dating from the Tertiary. The wadis followed these easily cleared lines. Accentuations or concentrations in Quaternary precipitation in this arid to semiarid area led to excavation of these crystalline boulders or "woolsack" forms, which were already subcutaneously rounded. The wadi system had meanwhile expanded considerably and was able to collect very rapidly the large water masses, which occur with this type of climate, and could also transport these remarkably large blocks into the Ifal Depression.

---

[1] Details of the intensity of disintegration and the sediment-petrographic habitus will be published after completion of analysis.

2.1.3.4. Indications of Quaternary tectonics

Late-Pliocene and Quaternary tectonics have impressively faulted, tilted and bent the Ifal strata. The total geomorphic situation shows that the youngest and most intensive movements in the Ifal Depression are related to the upward arching of the evaporite series in the west ("Raghama Group", M. M. A. BOKHARI, 1981).

Currently, a "marginal trough" seems to develop farther on in the immediate vicinity of the emerging gypsum ridge. This is apparent in the change in course of the Wadi Sha'ib An Nakhlah (Fig. 24). The wadi and its partly very irregular course shifted virtually in parallel direction with the west. This, thus, shows the areas with the most pronounced downward movement. Degeneration of some sections of the course indicates rubble overload and very young accumulation activity as a result of especially pronounced local depressions. Straight edges in

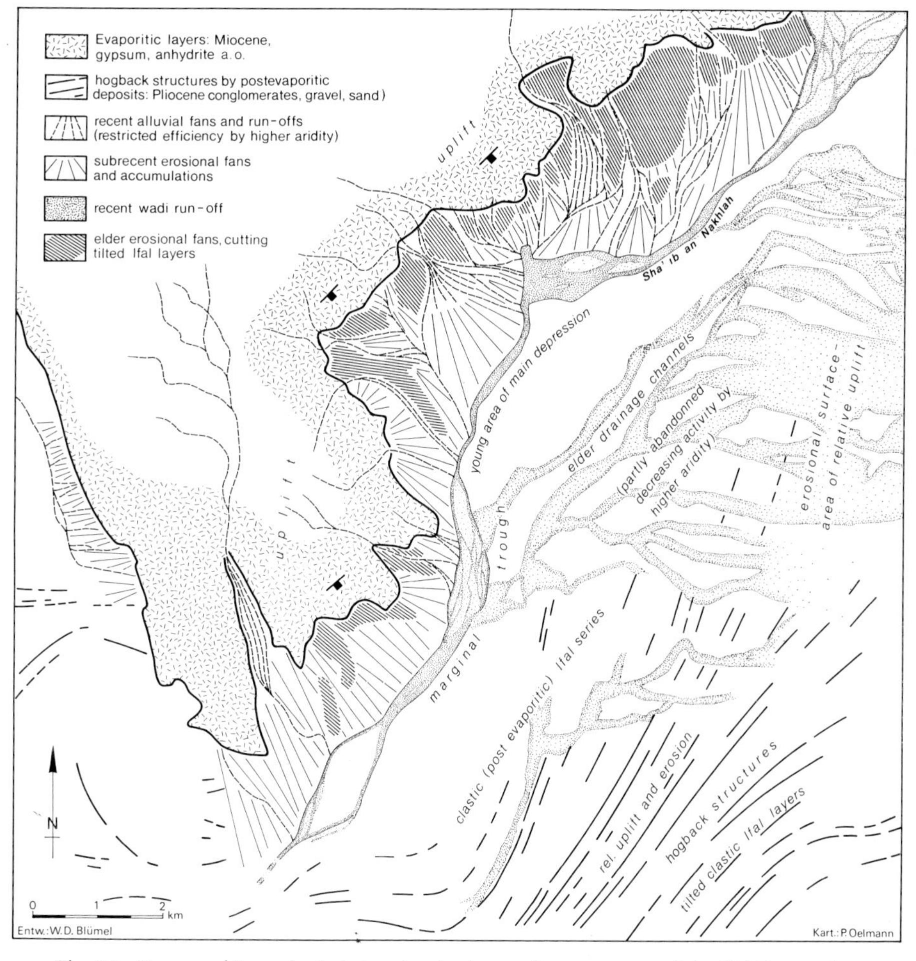

Fig. 24. Geomorphic-geological situation in the southwestern part of the Ifal Depression.

sedimentation sections and bends in the course of the wadi bed are also, very likely, expressions of young tectonic movements (faults, flexures).

The erosion area with the post-evaporitic Ifal layers described above is located east of this marginal trough. Its west flanks drained into the tributaries of the Wadi Sha'ib An Nakhlah (Fig. 24). There is scarcely any connection left between these courses and the wadi mentioned. Observations in other parts of the Ifal Depression also permit the conclusion that precipitation intensity in the recent past to the present has generally decreased. There are very few references for chronological fixations of the older wadi courses, for shifts in the recent wadi, and thus for the subsidence velocity in the western part of the Ifal Depression.

Further indications and geomorphological results of recent lifting or relative sinking processes are found in the immediate contact area between the emergence of the evaporitic Raghama series and the clastic Ifal series. Figs. 25 and 26 clarify the situation with a number of relief generations. Numerous small wadis extend from the arched evaporite layers into the outcropping Pliocene clastics deposits. Lateral erosion has beheaded the partial nearly vertical outcropping Ifal strata and subsequently covered it with an accretion mainly of gypsum detritus and a scattering of basement pebbles. The origin of the basement pebbles (mainly reddish granite, granite porphyry and basalt) cannot be explained unambiguously. In this area two wadi chasms are connected with the outcropping basement complex farther to the west (Plate I, see insertion at back cover). The pebbles from this area may have been spread far out laterally as an alluvial fan on the edge of the depression. But because the pebbles are well rounded it would seem more likely that they are residual gravel from the beheaded Ifal series, which often show conglomeratic consistence.

This oldest erosion and alluvial-fan generation (terrace 4 in Fig. 26) shows a dark patina and is in contrast to the younger layers. Progressive and/or intensified lifting forced the wadis to intersect owing to the resulting gradient distribution, particularly in the contact area of the two series.

Fig. 25. The geomorphic situation on the SW-margin of the Ifal triangle: In the background (left) the up-warping layers of the evaporitic part of the Bad' Formation in connection with different terraces dissecting the tilted and faulted clastic layers of the Ifal Formation. (Photo: W. D. BLÜMEL, 1982.)

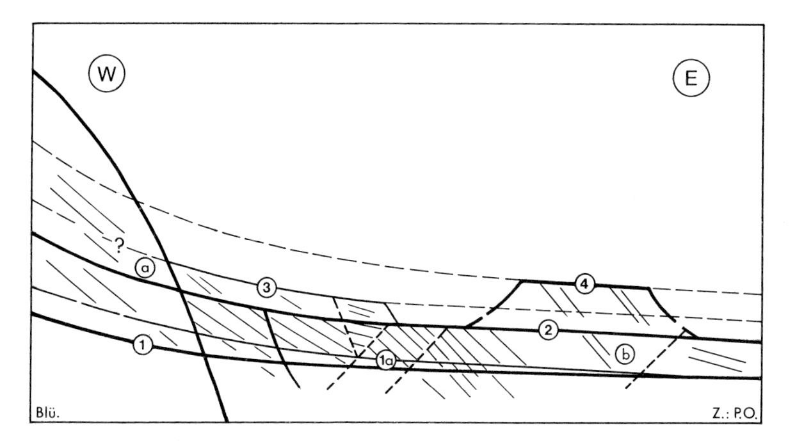

Fig. 26. Schematical cross-section through geomorphic generations east of Jabal Al Kibrit. *a)* Upwarping evaporitic layers (Ifal Formation – gypsum and anhydrite). *b)* clastic "red beds" of the Ifal Formation (post-evaporitic conglomerates, sands, loamy strata). *1* recent wadi run-off / flood-plain; *1a* elder Holocene terrace (only few meters above actual run-off); *2* Young Pleistocene terrace; accumulation of gypsum debris (surface 7–10 m above recent wadi). *3* remnants of terrace only few meters above (2) maybe no individual generation in common sense but only remains of the eroded terrace (4); *4* "butte", remnants of the oldest (preserved) terrace.

A second generation was cut sectorally into the older alluvial fan and also covered with a terrace layer which is thin only near the elevated area (terrace 3 in Fig. 26). Both generations exist only in table-mountain remnants and related forms (Figs. 25, 26). The vertical distance of niveau (3) from generation (2), of presumably Late Pleistocene age, is only a few meters.

Terrace (2) can be observed in near mountains as rather well preserved and widespread. Its surface lies 6–15 meters above the recent wadi course. The terrace accumulation consists of gypsum debris and limestone fragments and gains thickness with increasing proximity to the delivery hinterland.

Farther away, younger alluvial-fan sediments overlie previous accumulations. Basement gravels are not included. The recent main catchment area of the numerous small wadis involved is within the evaporite outcroppings.

The youngest relief generation is formed by the recent wadi course with some only slightly distinct subrecent terraces (1 a in Fig. 26) and obsolete high-water areas. Here as well, the increasing aridity of this region is documented by more concentrated regional run-off. The sediment load consists of gypsum, sand, sandstone pebbles and calcareous gravel, with a few assorted pebbles from higher terrace remnants. In some places the older deposits are covered by recent wadi sediments and disappear under the surface. In other places, the wadis flow in relatively narrow beds only a few meters deep to the main Wadi Sha'ib An Nakhlah.

In contrast to the morphological situation, particularly on the east side of the Ifal Depression, the three main relief generations on the west side are probably younger and of much smaller size. An absolute time mark could not be established. Their genetic circumstances also differ. They are clearly produced tectonically by the pronounced lifting of the evaporites and the relative sinking in their eastern margin. Eustatic variations of the sea level, such as those that affected the sequence and development of the large eastern alluvial fans/terraces (see first part of this paper) do not occur here.

The northern part of the Ifal Depression (where the triangular structure south of Al Bad' flanked by the evaporite series and basement narrows) includes, as a further example of Quaternary, lifting processes of alluvial-cone sediments caused by the upward movement. An outcrop was found with dragged-up structures in the form of rather coarse and well-rounded pebble deposits. These non-hardened sediments belong to an alluvial-fan generation the remnants of which begin on the eastern slope of the Ifal Depression and extend over large areas. In the middle part of the cross-section the alluvial-fan surface is buried by younger sediments from the Wadi Ifal. The pebble groups which incline from the east side reappear on the west side oriented in a different direction in harmony with the vertical movement of the evaporites.

Other young tectonic movements (vertical and horizontal faults) could be found in numerous outcrops in the outer wadi banks in the tilted Ifal layers. They extend into the terrace body, and particularly into the middle level (terrace 2 in Figs. 26, 27). These movements did not, however, produce distinctly measurable

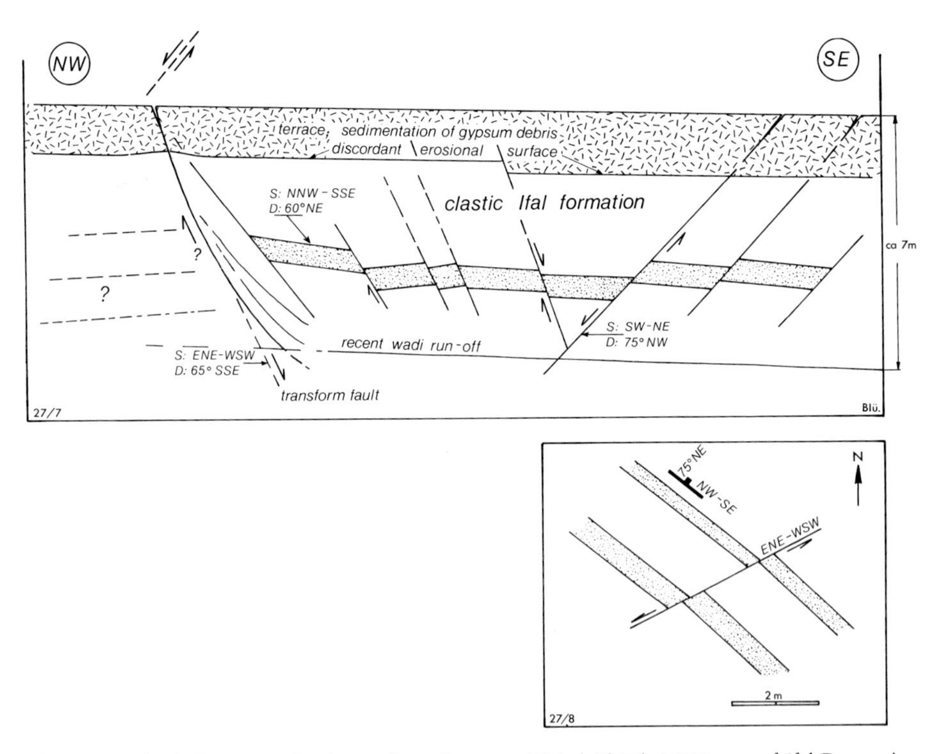

Fig. 27. Geological cross-section by wadi erosion east of Jabal Al Kibrit/SW part of Ifal Depression (see Plate I). Vertical and horizontal faulting in the Pliocene Ifal Formation (gravel, sand, silt; debris). Faults seem to continue as fractures within the terrace accumulation (dominantly gypsum debris) but do not show any distinct vertical movement on the surface. The layers are dislocated along ENE – WSW – striking small transform-faults.

Table 3. *Morphological and tectonical development of the Ifal Depression*

Time in 1000's years	Alpine – Europa	Mediterranean	Eastern Ifal Depression Midyan: Morphological Features – Niveaus of the Terraces (Coastline / Center / Mountain Border)	Eastern Ifal Depression Midyan: Tectonical Activities
		modern conditions →	0 / 95 / 195 Accumulation (Sand, Gravel)	
0	Holocene		0 / 80 / 180 Erosion (channel)	
	Würm	Tyrrhenian		
	R/W (Eemian)		5 / 100 / 200 Flat Lower Terrace system	
	Riss	?		
	M/R (Holsteinian)	Milazzian	Beach-Rock, 20 m Beach Terrace ?	Vertical Displacement
500	Mindel	?		
	G/M (Cromerian)	Sicilian	30 / 105 / 215 (20) / (100) / (195) Main Lower Terrace system	Fault Throw (Coast) > 30 m
	Günz	Emilian	?	Warping of the Terraces; Uplifting of the Ifal-triangle, compression
1000	D/G (Waalian)	Calabrian	55 / 140 / 250 (50) / (135) / (230) Higher Terrace system dominantly Erosion	
	?		?	
	Donau			
1500	?		Erosion (?)	Faulting, Block-shattering, Flexures
2000		Astian		
2500	Biber (?)		Ifal-Formation mostly reddish clastic Sediments	Rifting-Subsidence (Central Graben); Spreading of the Ifal-triangle; Uplifting (Graben shoulders)
3000		?		

morphological steps or other forms but can be seen in the various thicknesses of the terrace accumulation (Fig. 27). The outcropping of relatively slight fault throws (of the faults we observed) were reflattened by erosion processes.

At the same time, horizontal movements are as notably expressed as vertical movements in a highly sheared strata complex. The amounts of transposement of the steep dipping Ifal layers are found in a small area in the decimeter/meter range (Table 3).

The order of magnitude of the sum of the movements could not be determined. The presently dominating direction of the joint system strikes NNE–SSW to ENE–WSW, with the evaporite series moving in the southwestern direction, and the Ifal Depression relatively or absolutely northeast of it. In other places movements go only in the opposite direction. Detailed observations are necessary to answer these questions and to gain more definitive information than could be given in this preliminary report.

2.2. Coastal Region from Dhuba to Yanbu al Bahr

2.2.1. General Topographical and Geological Considerations

(H. Hötzl)

2.2.1.1. Topography and morphology

The coast from Dhuba to Yanbu Al Bahr is more than 400 kilometers long. This segment thus covers most of the northern part of the Saudi Arabian Red Sea coast. Its topographical and geological structure lacks uniformity. While there is a relatively uniform, broad coastal plain farther to the south, here the flat coastal strip is interrupted by occasional mountain ridges striking down to the sea (Fig. 28).

More extensive young aggradation areas are found around Yanbu, between Umm Lajj and Al Wajh, as well as south of Dhuba, where they are related to the Azlam trough, which there aproaches the coast at an acute angle. The crystalline basement, however, extends to the coast south of Umm Lajj, in a segment north of Al Wajh, and directly to the north of Dhuba.

As is to be expected for the edge of a tectonic graben, the coast has a very straight course without larger indentations. In the northern section of the Red Sea rift, this straightness is particularly true in the far greater section from Umm Lajj to the north up to the end of the Red Sea at the Midyan Peninsula, and it has a mirror image in the Egyptian-Sudanese coast on the other side. The trend of 150° is identical to that of the entire graben.

Only south of Umm Lajj does the coast project about 30 kilometers to the west and then continue for 80 kilometers along the 150° course. North of Yanbu Al Bahr, an initially flatter course of some 130° indicates the beginning of the transition to the gently curving seaboard especially characteristic of the southern section of the graben down to the Yemeni border.

The mountain ridges mentioned above give this section of the coast a pronounced morphological structure. Even though the actual escarpment with its general climb to more than 500 meters is farther into the hinterland, the occasional en-echelon blocks create a pronounced relief. The tectonic displacement of the Young Tertiary series and the intensive terrace formation usually give even the young aggradation surfaces extending to the coast a distinct structure.

A striking contrast to the straight course of the coast is the deviant bathymetric line immediately before it. A mere look at the 200 meters isobath shows significant deviations. This isobath seldom runs parallel to the coast; underwater ridges and troughs repeatedly show the conjugated 70° course indicated by

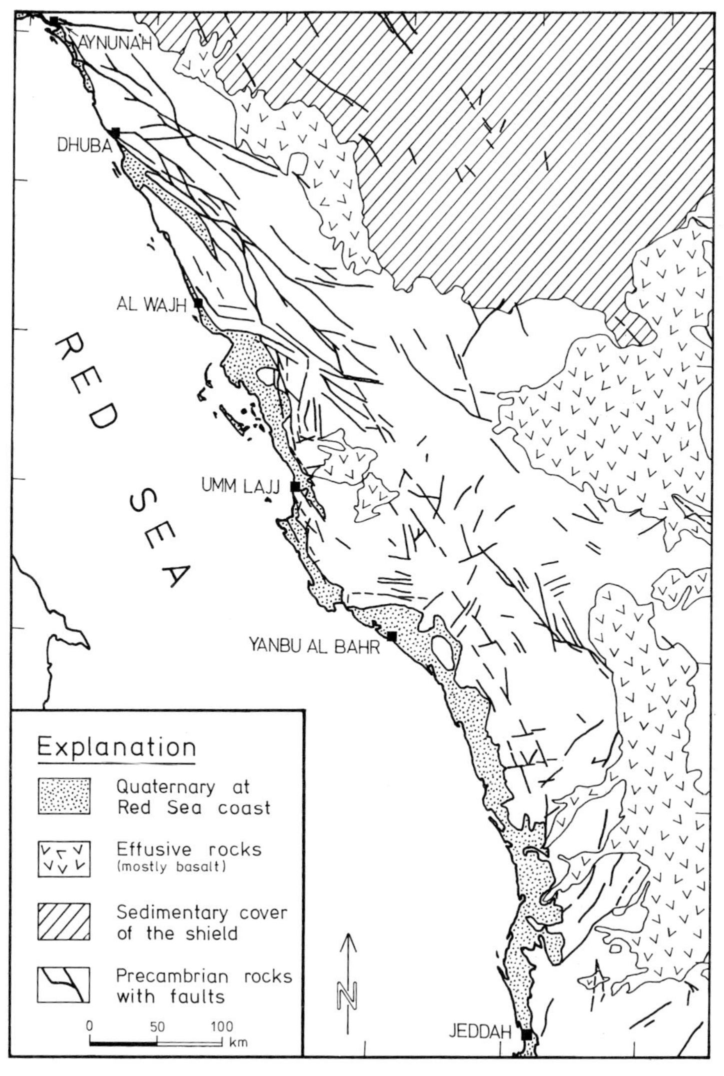

Fig. 28. Saudi Arabian coast line of the northern Red Sea with tectonic features of the hinterland.

transform faults. The deviations are not as great in the flatter areas above 100 meters. Most noticeably from Al Wajh to the north, the arrangement of the reef belts shows that this area of extremely shallow water runs nearly parallel to the graben. Only south of Al Wajh is there a very pronounced reef extending from Hanak to Mashabih Island and marking the 130° course oblique to the coast.

In spite of this differentiation, the foregoing summary of this long northern section of the coast is justified by the geological build up and rather uniform structural plan. This includes the fact that throughout almost the entire course of the coast, Young Tertiary sediment sequences are included in the block tectonics.

2.2.1.2. Geological structure of the hinterland

The hinterland involved here belongs exclusively to the old crystalline shield. Accordingly, the geological structure includes the various, mostly granitic, intrusive bodies and metamorphic encasing series.

According to J. DELFOUR (1977) and G. F. BROWN et al. (1963), the Precambrian rock sequence in this northern section of the Shield is divided into rock complexes of various ages.

Older Basement: The older basement includes gneisses, amphibolites, quartzites and mica-amphibolites. Migmatites with pegmatite and aplite veins emphasize the very great changes in the rocks, which apparently were repeatedly subjected to metamorphic stress. Granites and diorites more than 1,000 million years old are included in these rock series.

Urd Formation: This sequence is characterized by ultramafic and ophiolitic rock series with superimposed, highly metamorphic clastic to volcanogenic sequences. Granitic as well as ultramafic intrusive bodies penetrated this rock series some 700 million years ago.

Hulayfah Group: This group is also composed of an originally alternating sequence of sedimentary and volcanic rocks. Basal conglomerates, graywackes and sandstones dominate, along with marble inclusions. The middle section is made up of volcanic series. Acidic tuffs and tuff lavas dominate, along with compact rhyolites and ignimbrites. The upper part of this group is taken up by thick clastic sediment series.

The entire series has been subjected to severe tectonic stress and folding. Moreover, the rocks have been changed to green-schist facies. Then there is the fact that thick and extensive intrusive complexes have intruded into the series.

As far as the mineral components are concerned, the entire spectrum from granite to gabbro is represented. Absolute age datings show an age from 640 to 630 million years.

Hadiyah-Murdana Group: This is composed of basal volcanic-rock series, as well as superimposed sediment series in the form of metavolcanic rocks, phyllites, sandstones and conglomerates.

Jibalah Group: This youngest sequence of thick intrusive bodies consists primarily of peralkaline and alkaline granite plutons, including the striking Mandabah ring structure in the Yanbu hinterland. Radiometric datings showed a value of some 520 million years. Slightly metamorphic schists and metavolcanic rocks appear in this complex as enveloping series.

All in all, the Precambrian rock sequence of the shield reflects a variety of orogenic events. These involved folding, metamorphosis and tremendous intrusions. For information on the exact distribution of the series, the reader is directed above all to consult the 1 : 500,000 geological map by G. F. BROWN et al. (1963).

The only younger rocks in this crystalline hinterland are Tertiary to Quaternary basalts. The largest closed basalt area here is the Harrat al Uwayrid, which is more to the outside of the area under study. It extends far more than 200 kilometers nearly parallel to the graben and at the height of the escarpment.

In the actual hinterland closer to the coast, in the zone in front of the main escarpment, there is only one larger closed basalt sheet. This is the Hallat Abu Nar (Harrat Lunayyir) east of Umm Lajj and covering a surface of nearly 160 square kilometers. A few tongue-like flows reach the coastal plain just north of Umm Lajj and occasionally can be followed all the way to the sea.

Other young occurrences of volcanic rock are the small, flow-like basalt ridge in the vicinity of Jabal Salajah some 70 kilometers southeast of Umm Lajj, as well as the two wide basalt tongues of the Jabal An Nabah reaching down to the sea some 40 kilometers north of Yanbu Al Bahr.

2.2.1.3. Geological structure of the coastal plain

The course of the coast and of the coastal plain on the Arabian side of the Red Sea show a remarkable morphological contrast. In the middle and southern sections, the coasts have more of a broad sigmoidal curvature but with a rather uniformly wider coastal plain; the northern section shows a comparatively straight course. Here, however, the coastal plain shows considerable variation in width; at most it is 40 kilometers wide, e. g. south of Al Wajh, whereas elsewhere the basement approaches the coast directly. This results in a distinct spatial division formed by young sediments in the northern section of the coastal plain.

The pronounced Quaternary sedimentation surface in the area of Yanbu is the direct continuation of the more or less uniform coastal plain in the middle section around Jeddah. Only slightly farther to the north, the outcrop of the basement projects about 50 kilometers to the west and the coast plain narrows down to less than six kilometers. It is somewhat wider only south of Umm Lajj; at Umm Lajj, however, it almost completely disappears for a small stretch owing to the retreat of the coast back to the pronouncedly faulted edge of the basement. This southern coastal plain is divided into sections by a basement outcropping only some 40 kilometers south of Umm Lajj, as well as by the two basalt tongues of the Jabal An Nabah extending down to the sea. The rest of the coastal plain is taken up by young Tertiary sediments that usually form a gently rolling landscape, as well as Quaternary terrestrial aggradation sediments and young reef terraces near the present-day coast.

The interruption of the coastal plain at Umm Lajj is due to a narrow, horst-like basement block that outcrops at an acute angle to the somewhat retreating coast. North of Umm Lajj, to about 20 kilometers south of Al Wajh, the coastal planation increases in breadth. Here it has the approximate shape of a longish right triangle; the hypotenuse is the coast itself and its short sides are the faults that run north–south and west-northwest–east-southeast and bound the basement. Again, young Tertiary sediments outcrop contiguously to the faults, as well as on occasional smaller exposures within this triangle. The far greater part of the surface, however, is taken up by extensive Quaternary fluvial terrace bodies. Up to about 40 kilometers to the north of Umm Lajj there are occasional basalt flows, covered in places by sand dunes. These are the flows from the young volcanism of the Hallat Abu Nar immediately to the east.

There is again only a narrow strip of coast from Marsa Martaban bay, 25 kilometers south of Al Wajh, to Sharm Marra, 65 kilometers north of Al Wajh. Most of it is taken up by Tertiary blocks arranged parallel to the coast, and a zone only one–two kilometers wide is made up of Quaternary reef extending in front of the blocks.

The Quaternary aggradation became more extensive only in the vicinity of the Azlam graben. This trends northwest, at an acute angle to the main graben, and with its young filling forms the coast for about 40 kilometers from Dhuba to the south. South of Dhuba, Tertiary sediments are usually exposed only in a narrow zone at the edge of the main graben. North and south of the mouth of the Azlam graben the hilly basement extends to the coast.

The Quaternary in this entire coastal area will be treated in detail in the following subsections. For the sake of completeness, the Tertiary stratigraphic sequence will be discussed here. In the northern section of the Saudi Arabian Red Sea coast, the pre-Pliocene component today is usually summarized under the concept of the Raghama Group (previously known as the Raghama formation, M. M. A. Bokhari, 1981). As shown below, the entire sequence is to be classified as Upper Oligocene to Pliocene.

Owing to the mainly littoral nature of this sediment sequence in the border zone of the graben, facial differences may be seen over very small areas. In addition, there is the syntectonic deposit related to the graben event, which also intensified the rapid changes in the sediment sequence. As yet, the Tertiary has not been uniformly described; what is particularly lacking are detailed stratigraphic studies to allow parallels to be drawn for larger areas. In general, a division into three series is assumed; this is often indicated lithologically by thick gypsum and anhydrite in the middle series.

The pre-evaporitic complex (Oligocene – Middle Miocene) consists of various clastic series with more or less frequent intercalations of fossil-rich carbonate layers, sometimes of reef limestone. In the middle section (Middle to Upper Miocene), the evaporites may also be represented by marine marls. The post-evaporitic series (Upper Miocene to Pliocene) is again more clastic, but here as well as there may be inclusions of fossil-rich limestones, sometimes lumachelle layers and reefs, or generally thin evaporites.

For the southern part of the section under consideration here, the area around Yanbu, there is a general map, scale 1 : 250,000, with an exact description (C. Pellaton, 1979).

2.2.1.4. *Main tectonic structures*

A look at the course of the coast in the northern section of the Red Sea shows it to be remarkably straight. This straightness clearly demonstrates the north-northwest (150°) trend of the graben structure and underscores its taphrogenetic origin with the course parallel to both the central graben trough and the African side. The section of the coast under discussion here from Yanbu Al Bahr to Dhuba is straight only as far as Umm Lajj. The southern part from Umm Lajj to Yanbu' Al Bahr shows the first of the sigmoid curves which are then typical of the farther middle and southern sections of the Red Sea coast.

One naturally tends to expect a parallelism between the faults of the contiguous mainland and the main tectonic direction given by the coast. This is the case only to a certain extent. There are indeed parallel structures, especially in the aforelying Tertiary fault steps, e. g. at Al Wajh, but closer observation shows that even the coast, despite its rather uniform course, is made up of faults coming from different directions. These structures, which are especially distinct in the crystalline shield, are generally old and in some cases were later reactivated.

In the area under study, the southeast–northwest (135°) trending Najd fault system is particularly characteristic. This sinistral-wrench fault system (described by G. F. Brown, 1970, and 1972) dissects for a length of more than 1,300 kilometers the entire Arabian Shield and shows horizontal amounts of displacement within the Precambrian totaling more than 200 kilometers. The coast reaches this fault bundle between Al Wajh and Al Muwaylih north of Dhuba. It runs at an acute angle to the direction of the Red Sea. The young activation of this Precambrian system becomes clear in the Azlam graben, which was downwarped as a result of parallel faults. It is filled with Cretaceous to Quaternary sediments up to 1,000 meters thick (F. B. Davies, 1981). The northeast–southwest (60°) trending dextral wrench faults conjugated to this Najd system are less penetrating and have smaller amounts of displacement.

The 150°-trending normal faults of the Red Sea system developed concurrent with the actual rifting in the Tertiary. This new direction also retreats pronouncedly even in the coastal area because, as previously mentioned, existing older and nearly parallel fault systems were often activated as normal faults for the extension (example: the 135° direction). The shear movement (conjugated to rifting along transform faults, which here in the northern section of the Red Sea took place at an extremely acute angle to the trough axis) also uses, with the 10° direction (Aqaba system), a pre-existing fault system (cf. F. B. Davies, 1981).

South and north of the section from Al Wajh to Dhuba, the dominance of the fault direction varies to a greater or lesser extent from this pattern. South of Al Wajh, the basement retreats more pronouncedly at one of the young-activated faults of the Najd system. The Aqaba direction determines the farther course. The step faults of the basement, in conjunction with the 135° direction here, is not only responsible for the triangular form of the coastal plain between Al Wajh and Umm Lajj, but also for the projection of the coast south of Umm Lajj. This approximately north–south course of the edge of the basement also corresponds to the equidirectional displacement of the central graben trough in this section (cf. Tectonic Map of Arabian Peninsula 1 : 4,000,000, G. F. Brown, 1972).

South of Umm Lajj to the Yanbu Cement Plant north of the bay of Sharm Al Khawr, 150°-trending normal faults are determinative for both the basement and the Tertiary. The retreat of the coast, but most especially the pronounced retreat of the basement north of Yanbu, can be seen as a consequence of the latticing with nearly east–west trending fault systems which are dominant in the basement there (Jabal Salah) (cf. C. PELLATON, 1979).

In the section north of Dhuba, and thus also north of the actual Najd fault system the 150° direction dominates. Numerous nearly parallel faults determine the step faults of the blocks from the escarpment to the graben trough. But on the other side of the escarpment, to the east, in the superimposed Paleozoic sandstones, structures parallel to these also dominate. They proceed to the northwest up to the Aqaba – Dead-Sea line.

2.2.2. Quaternary from Dhuba to Al Wajh

(S. S. AL-SAYARI, H. HÖTZL, H. MOSER, W. RAUERT, J. G. ZÖTL)

2.2.2.1. The Dhuba area

From Dhuba (27°20′ N) to Al Muwaylih, some 40 kilometers to the north, the basement approaches the coast especially closely. The remaining seaboard is 1–3 kilometers wide and is mainly taken up by dumped Tertiary sediments. For long stretches, the Quaternary itself is limited to a series of reef terraces that in places are less than 100 meters wide; they lie in front of the Tertiary series and form the actual coast.

The wadis came from the basement and ran perpendicular to the coast; when the sea level was lower they cut deeply into the Tertiary and sometimes even the Quaternary series. In the smaller wadis, the later sediment supply did not suffice to fill them completely. Fjord-like bays, such as Sharm Jubbah, Sharm Badu and Sharm Barr show the intensive excavation during periods when base level was lower.

The coastal plain broadens south and north of this section of the coast between Dhuba and Al Muwaylih. It is taken up there with thick alluvial fans and wadi channels of varying widths. Dhuba and the adjacent area to the south will be described as an example of the geological situation.

Directly from the locality of Dhuba to the south, the Precambrian crystalline retreats in a parabolic arc. From the mouth of Wadi Dama near the Azlam graben some 25 kilometers in the south, the coastal plain is thus widened from 1 kilometer to 16 kilometers. Both the reatreat of the crystalline and the course of the coast itself are determined by faults. Here, the interlacing of three fault systems is decisive (F. B. DAVIES, 1980). In the Dhuba area there is first the outcrop of a fault bundle striking from 70–100°. Both systems are faults or movement paths laid down in the Precambrian and later reactivated. An example of this is the opening of the Late Cretaceous Azlam graben parallel to the 135° direction. In the course of development of the Red Sea rift as well there were vertical movements along those old fault lines that help determine current morphology there. The third fault bundle is then the Red Sea system trending 150°–160°;

when it intersects with the other two directions, it is above all determinative for the course of the coast itself.

In the Dhuba hinterland, the Precambrian basement limits the coastal plain, together with the juxtaposed Tertiary limestone series. Morphologically, this boundary line is an especially noticeable steep step owing to the sloping limestones. Behind this, there is first a gentle upslope, then some vertical shifts at faults, and finally, at about 400 meters a. s. l. the old planation surface that is the remainder of the pre-Oligocene peneplain. The few large wadis are deeply incised. Otherwise, however, only the superficial old weathered layer is excavated, leaving only a very slightly knobby landscape that still emphasizes the originally uniform flatness of the surface.

The age of this planation and its dissection are apparent in the deposited or superimposed Tertiary reef limestones. They cover the steep slope of the crystalline down to the coastal plain like armor; a duricrust-like hardening on the surface increases this impression. A first look at the steep slope can easily give the impression of an exclusively tectonic tilt; only closer observation shows that it is a sedimentary slanted depositional pattern, that later may have been slightly tipped in places.

South of Dhuba, these limestones are preserved only as morphologically steep ribs in the flank area, but directly east and northeast of Dhuba, the flattening transgressive superimposition on the old peneplain may be seen. According to F. B. DAVIES (1980) in the area of the coastal plain these reef limestones are already covered by a Young Tertiary and Quaternary sequence that is hundreds of meters thick.

M. BIGOT and B. ALABOUVETTE (1976) place these reef limestones in the lowest Miocene. On Midyan, C. DULLO et al. (1983) succeeded in demonstrating uppermost Oligocene for facially equivalent formations. As these apparently are the first deposits made by the sea as it broke into the graben trough, a slightly higher age in the deeper parts of the basin would be thinkable as compared to the occurrence near the marginal uplifted blocks. These limestones thus show that the decisive morphological elements of this section of the coast developed with the first beginnings of the Red Sea rift in pre-Oligocene time. The situation is exactly the same in the area of El Aynunah, some 80 kilometers farther north, as well as in Umm Lajj 300 kilometers south, as shown by the analogous position of the limestones in these areas.

North of Dhuba, the younger Tertiary series sometimes show a broad face, but south of Dhuba they are generally covered by Quaternary sediments. These extend right up to the steep step in the mountain described above. The Young Tertiary consists of different terrestrial series (sands, silts, sabkhah clays with gypsum). Calcarenites with plentiful sea urchins and reef limestones become more frequent toward the coast. In the outcrop areas north of Dhuba this series is usually tipped toward the basin, or broken up by block faulting into blocks pitched at different angles. According to the profile section given by F. B. DAVIES (1980), south of Dhuba it would be not a tipped but only a tectonically sunken block that forms the underground for the coastal plain.

South of Dhuba, the Quaternary is mainly taken up by alluvial fans of varying ages. They are highly dissected, and respectively younger fans and wadi fillings

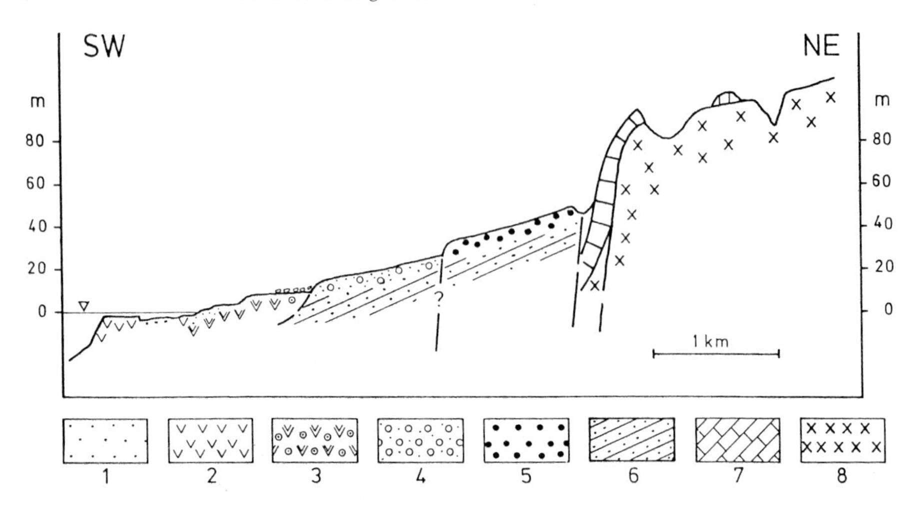

Fig. 29. Schematic cross section of the terrace sequence 4 km south of Dhuba (*1* recent lagoon and beach sediment; *2* modern reef; *3* Pleistocene reef and lagoon sediments; *4* younger gravels; *5* older terrace sediments with coarse pebbles; *6* Miocene to Pliocene sands and reefs; *7* Lower Miocene (?) reef limestone; *8* Precambrian rocks).

are found interpacked. Within the context of the Quaternary project, only general observation without detailed mapping has so far been possible, but for this area the following general classification of the terraces may be made (Fig. 29).

The oldest Quaternary talus cones which, without precise classification could very well be Upper Pliocene, are topographically the highest gravel accumulation; younger erosion has made them into the uppermost terrace step. They are mainly preserved at the immediate edge of the mountains, as well as in occasional narrow gravel ridges directed toward the coast. Remnants are also found as appositions on occasional Tertiary eminences. An example of this is the posterior terrace step near the oil storage tanks in Dhuba (Fig. 30), as well as the higher gravel ridges north of Wadi Al Amud and between Wadi Al Amud and Wadi Dama in the south. In aerial photographs they have especially dark surface colors as a result of intensive formation of desert varnish. These are mainly very coarse gravels which in places near the mountains are still tilted distinctly toward the coast (tectonic shift?).

The terrace structure shows repeated changes from sands and gravels to layers of coarse gravel. On freshly exposed surfaces under the desert varnish there is a cementing material resembling red earth within the gravels, and in places the gravels are underlaid by reddish sands. Finely clastic red sediments deposited concordant to the gravels sometimes make a considerable contribution to the terrace structure. Their chronological position has not been completely explained. In the wadi near the Dhuba storage tanks or in Wadi Al Amud, comparable sediments are at an angle and are to be placed in the youngest Tertiary.

In the wadi near the Dhuba storage tanks, the upper edge of this coarse-gravel terrace is about 3 kilometers from the coast at an altitude of some 55 meters above m. s. l.; over a distance of about 1 kilometer it falls to an altitude of 40–35 meters and ends at an erosion (fault?) step.

Fig. 30. Outcrop of coarse terrace gravels (main terrace) on slightly tilted sandy to calcareous Tertiary in the wadi east of the harbor of Dhuba. Lateral extension of the coral 2 m. (Photo: H. HÖTZL, 1978.)

The forelying, next-deepest main-terrace planation is somewhat less dissected but also preserved only in occasional gravel ridges; it begins at about 25 meters above m. s. l. and drops to about 15 meters behind the oil storage tanks. It corresponds to the wadi terrace step inserted farther up in the main wadi and rising to 45 meters near the mountains. The cut of this terrace body slightly below the narrow spot in the area of Tertiary limestones shows, going from bottom to top, the following sequence: gravels (>2 m), sands with a root horizon (1 m), gravel (0.3 m), wadi sabkhah sediments (0.2 m), and finally, gravel with intercalated sands (3 m). Farther down the wadi, coral banks are included in the terrace structure. The coastal strip going down to the sea is taken up by two marine terrace steps at 8–10 meters and 3 meters, respectively (Fig. 29). Large amounts of debris from the hinterland have, however, created narrow interfingering of the gravels with the reef or beach deposits.

Farther to the south in the section between Wadi Salma, Wadi Al Amud and Wadi Bahar to Wadi Dama, the zonal sequence of terraces described above going from the mountains to the coast is replaced by a lateral juxtaposition. As a result of the pronounced degradation in the intervening erosion phases, various isolated gravel ridges directed toward the coast were left between the wadis. This results in an intensive interlocking of alluvial-cone remnants of different ages.

In all, four main generations could be distinguished here. The two older ones are identical to the two main gravel terraces described behind the oil storage facility at Dhuba. Remnants of the oldest coarse-block aggradation are, as has been mentioned before, preserved on the northern side of Wadi Dama at the entrance to the Azlam graben near what used to be an undercut slope, the terrace step drops in places more than 20 meters to the recent wadi floor. B. F. DAVIES (1980) makes a fault responsible for the straight southern border of this terrace remnant. South of Wadi Salma, further remnants are preserved near the coast on the edge of a Tertiary deposit. There are also well developed coarse-gravel areas north of Wadi Salma between basement pediment and occasional Tertiary eminences.

The second alluvial fan generation has the nature of a main terrace. The remnants of its surface have been deeply dissected in places by later erosion and are especially extensive between Wadi Salma and Wadi Dama.

The younger alluvial fans or terraces are included in the more or less wide wadi channels; some of these are subrecent formations. The comparison of these younger sediment bodies and their classification in the accumulation and erosion cycle that produced the marine beach terraces would, however, involve further study.

The heavy sediment accretion in these large wadis, including those from the adjacent Azlam trough in the south, has generally prevented the formation of reefs here in the coastal area. Only from Wadi Samad to the north, where there are only wadis with small catchments and thus smaller amounts of sediment to deliver, are continuous reef terraces to be found.

At the mouth of Wadi Samad, the youngest element is the erosion flat at 1.80–2.0 meters. There are some slightly cemented beach dunes with shells, e. g. *Tridacna,* cemented into them. Then there is a reef body (upper edge about 8 meters above m. s. l.) that is deposited along the 10–12 meters step. This higher step shows a heterogenous and also laterally variable structure in the wadi section. Presumably Tertiary limestones containing echinoids are followed first by gravel, then reef limestones turning at times into calcarenite and finally a gravel covering of varying thickness.

The marine terraces are especially clearly developed from Sharm Dhuba to the north; the terrace sequence is repeatedly dissected by narrow inlets that developed from drowned wadis. Here again the example should be mentioned of the wadi directly on the northern edge of Dhuba some 1.5 kilometers north of the harbor bay. It does not itself have a bay at its mouth; because of its large catchment and the amount of debris it could deliver, it was able to aggrade down to the coast the erosion channel parallel to the increase in sea level.

On the northern side of this wadi, 1.2 kilometers behind the coast, the older Tertiary limestones still lie horizontally with a thickness of more than 50 meters upon the Precambrian, or lie in front of it. This Tertiary forms a narrow ridge 500 meters wide that ends in the east with an erosion step. It consists of occasional surface remnants that are as much as 150 meters in height. Most of this elevation is taken up by a lower surface that at first slants slightly to the coast and finally drops at a pronounced steep step (fault) to the marine main terrace. The latter forms a notable planation. It drops from about 45 meters above m. s. l. to 35 meters. It then passes into the 20–25-meters-terrace with a step with a difference in elevation of 15–20 meters.

Table 4. *Results of ^{13}C and ^{14}C measurements on samples from the Northern Red Sea coast. Analysis performed by GSF – Institut für Radiohydrometrie,* Munich–Neuherberg

Locality	Geographic position	Material	IRM Lab. No.	δ ^{13}C (‰ PDB)	^{14}C content (% mod.)	Assumed initial ^{14}C content (% mod.)	^{14}C age uncorr. (years B. P.)
20 km N of Sharm Ra's Ash Sheikh Humayd	lat 28°14′ long 34°39′	*Tridacna* shell 12 m-terrace	7659	+2.8	<1.3	100	>34,800
Wadi Tiryam 5 km upstream from the coastal road	lat 27°55′ long 35°24′	sinter limestone from the terrace 10 m above wadi floor	7655	−7.3	3.5±0.8	85	25,700 +2,000 −1,600
Dhuba Wadi 2 km N of Dhuba	lat 27°22′ long 35°37′	*Cardia* shells 1 m below top of 6 m-terrace	7651 b	+2.1	12.3±1.2	100	16,800 +800 −700
Dhuba Wadi behind the oil harbor	lat 27°19′ long 35°44′	coral from the reef layer in the main terrace, 1.5 m below top	7644 a	−5.6	<0.9	100	>37,600
Dhuba Wadi behind the oil harbor	lat 27°19′ long 35°44′	coral from the reef layer in the main terrace, 10 m below top	7644 b	−2.0	<0.5	100	>42,600
Al Wajh 300 m N airport	lat 26°15′ long 36°25′	massive coral, 1 m below top of 10 m-terrace	7631	−1.4	<1.0	100	>37,000
Al Wajh 300 m N airport	lat 26°15′ long 36°25′	massive coral, 1 m below top of 3 m-terrace	7634	−2.5	<1.0	100	>37,000
Harrat Qalib NW corner at the Sea	lat 25°06′ long 37°09′	*Strombus* shells from a fossil beach dune 4.2 m above m. s. l.	7638	−1.9	2.2±0.5	100	30,500 +2,100 −1,700
Umm Lajj mouth of Wadi Umm Lajj	lat 25°16′ long 37°11′	coral and shell debris from a reef limestone below a lava flow	7639	−0.8	2.4+0.6	100	30,000 +2,300 −1,800

The 10–12-meters-step varies in width and is sometimes missing entirely. Occasional sloping surfaces were measured as 9 meters. There is another intermediate step at 6 meters above m. s. l. In the wadi north of Dhuba it takes up a width of 400 meters. The 1.5–2.0-meters-step, in contrast, is only of secondary importance. It appears only as a narrow sill in the cliffs of the higher step, or in smaller indentations as an intermediate step with a width of a few meters.

In the wadi itself, a gravel terrace is above all to be mentioned; one kilometer back from the coast it shows a height of 20 meters above m. s. l. Its structure includes marine carbonate debris and an oyster-shell bank. A younger gravel aggradation seems to correspond in the coastal area to the 6-m-level.

We have no concrete age data for a chronological classification of the terrace sequence on the section of the coast described here. In the general sequence it may be compared to information given in section 2.1. for Midyan. For the chronological relationship, the basalts in the coastal area from Umm Lajj to Yanbu could also be used here (cf. section 2.2.3.). For a precise match with the Dhuba area, however, it would be necessary to determine to what extent the individual steps are true erosion or accumulation terraces or, as we have suggested, young tectonic fault steps.

Shell and coral samples were taken from the younger marine sediments for ^{14}C age determinations (cf. Table 4, samples Nos. 7644 a, 7644 b and 7651 b). As expected, the corals from the main terrace behind the Dhuba oil depot showed an age outside of the range of the ^{14}C method (>37,600 and >46,600 years). The Cardia shells taken from the above-mentioned 6-meter-terrace directly north of Dhuba from a depth of one meter under a lithothamnion bank showed a value of $16{,}800^{+800}_{-700}$ years. A tectonic shift would be theoretically possible as an explanation of this difference in level as compared to the considerably lower sea level at that time, but here contamination as a result of recrystallization seems more likely. Generally, the marine terraces are mainly erosion flats that were incised into youngest Tertiary or Old Quaternary reefs and calcarenites. It is very difficult to distinguish distinctly younger Quaternary deposits.

A tufa sample was also taken, not directly from the area around Dhuba, but from Wadi Tiryam, 70 km north of Dhuba. The sample comes from the gravel terrace at the entrance to the mountains (Fig. 31). Silty sediments are followed there by a sinter layer up to 2 m thick containing impressions of plants. The sinters are covered by sands and gravels. Coarse sands and gravels with occasional gravel banks are deposited in the wadi channel in recent time.

^{14}C dating of the sinter gave a value of $25{,}700^{+2000}_{-1600}$ years. The problems that these values present, and especially the difficulties involved in a meaningful chronological interpretation when there may have been contamination from seepage of recent precipitation, will be discussed in more detail in the final section of this book, on the chronological classification of the Quaternary.

2.2.2.2. The Al Wajh area

The geological situation in the Al Wajh area shows a number of parallels to that in the Dhuba area; this is true both for the general tectonic arrangement and for the distribution of the Quaternary.

Fig. 31. Wadi Tiryam few kilometers upstream of the coastal plain. The main terrace is built up by sands, silts, sinterlayers and gravels; on the terrace plain remnants of younger loess-like sediments. (Photo: H. HÖTZL, 1978.)

In the vicinity of Al Wajh as well as to the north of the locality, the coast is particularly straight and parallel to the steep 150° course of the Red Sea. The basement, the Tertiary blocks and the Quaternary reefs have all settled on faults parallel to this course. About 20 kilometers south of Al Wajh, the basement retreats bayonet-like at a 90°–110° fault and then after 35 kilometers returns to the steep Red Sea course (s. Fig. 28).

North of this shift, the coastal plain shows a width of 6–12 kilometers, and south of it, it is 25–40 kilometers wide. The sections are thus wider than in Dhuba but, similar to the situation there, here as well the Tertiary blocks form the narrow coastal plain preceded by impressive marine terraces.

In the broader southern part, the Tertiary is again exposed only at the edge of the basement and in a few places on the coastal plain. The coastal plain itself is mainly taken up by Quaternary gravels. The main drainage system there is Wadi Al Hamdh, and it is responsible for the accumulations and erosion. It first reaches some 160 kilometers back to the east, then turns to southeast in the area of the Najd fault system, where its catchment extends back past Al Madinah. In addition to Wadi Al Hamdh, there is also Wadi Marra, somewhat to the north but nonetheless emptying into the broad coastal zone. The debris carried by the large wadis prevents fringing reefs from being laid down directly in front of the broad coastal strip. Today, a continuous reef belt has developed some 30 kilometers away from the coast.

The narrow coastal strip to the north is actually also dissected by wadis. But their catchments are small and so they scarcely bring debris as far as the coast; as a result, very pronounced fringing reefs have been able to develop on the coast there both at present and in the past. These two sections will be discussed separately below with regard to the development and distribution of Quaternary material.

The coast around Al Wajh

Al Wajh is in the middle of the section of the coast that is 70 kilometers long and follows a uniform course from Wadi Arjah (bay of Marsa Martaban) and Wadi Antar (20 kilometers south of Wadi Thalbah). The coastal strip is 12 kilometers wide in the south and gradually narrows to 6 kilometers in the north; it is limited in the east by the plunging crystalline.

The greatest part of the breadth of the coastal plain is taken up by tilted Tertiary layers. These are usually covered by a thin Quaternary gravel deposit. Finally, a carbonate ridge forms the more narrow coastal strip; its stratigraphic position will be discussed below.

Disregarding occasional older clastic series, here as well a reef-limestone complex forms the basis of the Tertiary. In contrast to the Dhuba area or the area of Wadi Thalbah directly to the north (cf. F. B. DAVIES, 1980, 1981), however, these limestones do not appear in this morphologically exposed accretion on the basement step. In general, the entire Miocene sequence here shows a more carbonatic development, so that a precise individual classification into basal or higher parts is very difficult (cf. B. ALABOUVETTE and C. PELLATON, 1979).

In the immediate hinterland of Al Wajh the evaporites that are directly exposed farther to the south retreat. The area between the basement and the limestone ridge on the coast is mainly taken up here by Upper Miocene-Pliocene (?) clastic series. These are sands to silts with intermediate gravel layers and occasional oyster banks in the upper parts as well as inclusion of marly calcarenites and occasional reef or reef-debris banks. These clastic to carbonate series has been broken down by block faulting into single blocks tipped different ways, following faults parallel to the coast (145°–155°) or at an angle to it (70° or 120°).

Above this, there is a discordant limestone series more than 50 meters thick. In the lower part it consists of marly limestone; going upward, fine to coarser calcarenites increase, with various coquina-like shell layers and especially a variety of echinoid banks. The series usually ends with a reef flat. The thickness of the reef flat increases going toward the coast; in the lower parts of the limestone series there are reef formations containing large coral colonies in situ, as can be seen in the cuts at Sharm Al Wajh.

Apparently these limestones were deposited as coastal or near-coastal sediments in transgressive formations after the breaking into blocks and tilting of the Upper Miocene to Pliocene clastic carbonate series. Occasional, sometimes channel-shaped or lentiform gravel bodies interfingered laterally with the limestones show that, as today, there was also a local clastic aggradation through certain runoff channels in the coastal area, parallel to marine sedimentation.

The limestone sequence was affected by vertical breakage into blocks, in part synsedimentary and in part postsedimentary; this led in certain smaller areas to typical block tectonics with small grabens and horsts. Generally, the nearly horizontal stratification was retained. Slight tilting may be seen toward the coast as well as in the area of the eastern marginal blocks at the erosional depression in the east (Fig. 32). The morphological horst structure of this limestone ridge is, however, less a result of tectonic block formation than of sedimentary accretion on the western edge (reef extension) and erosive dissection in the area of interfingering with the scarcely cemented clastic deposits on the eastern edge. As a result of their rapid cementation and intensive calcret formation on the surface, the limestones proved to be especially resistant to erosion. As can be seen in very recently uplifted reefs (cf. section 2.2.3.), tremendous erosion took place rapidly on the posterior edge and the subsequent channels and depressions are quickly established. This can be seen very nicely in the bay of Al Wajh (Fig. 32). Occasional erosion remnants there document the original extent and the rapid tailing-out of the young limestone series to the east.

The age of these marine sediments is of special importance. G. F. BROWN et al. (1963) assumed a Quaternary age, based on the distinct discordance as compared to the subjacent clastic to carbonatic series that was also tipped but more pronouncedly. Recent studies also assume the same classification, first summa-

Fig. 32. Quaternary and Tertiary limestone ridge along the coast of Al Wajh. The escarpment on the land side shows block-faulting. In the depression behind remnants of tilted Tertiary and of unfaulted gravel terrace (dark ridge in the central part) are to be seen. (Photo: H. HÖTZL, 1978.)

rized by M. BIGOT and B. ALABOUVETTE (1976), then later by F. B. DAVIES (1982), B. ALABOUVETTE and C. PELLATON (1979), and C. PELLATON (1982 a, b).

There is, unfortunately, no precise data available for either the stratigraphy of the Tertiary or the chronological classification of the tectonic event. Within the context of this Quaternary project (cf. H. HÖTZL et al., section 2.2.3.) and of joint studies by the University of Karlsruhe (SFB 108) and the University of Petroleum and Minerals, Dhahran (H. HÖTZL and A. R. JADO, 1983) on young tectonics of the Red Sea graben, basalt datings were made that showed this young limestone formation to be at least oldest Pleistocene but more likely Upper Pliocene. This could also be in agreement with the reactivation of graben tectonics about 4–5 million years ago, in the course of which the subjacent Tertiary series were tipped. The limestones could be understood as a transgressive formation immediately thereafter.

The interfingering with the gravel terraces is of interest for the chronological classification. In the southern part of this coastal section between Wadi Al Miyah and the northern part of the bay of Marsa Martaban it is apparent that these limestones lie in front of the highest gravel surfaces that flattened out the tipped Tertiary blocks; their levels differ slightly. The next-youngest gravel terraces, however, already tie into a lower coastal level.

On the west side of the tablelike carbonate ridge there are a number of terrace steps; some are better developed than others (Fig. 33). These are mainly erosion

Fig. 33. Marine terrace on the west side of the limestone ridge south of Al Wajh, near the airport. The photo shows the erosional 3-m-, 6-m-, and 10-m-steps. (Photo: H. HÖTZL, 1978.)

steps. As is the case in the Umm Lajj area (cf. 2.2.3.), it is rather difficult to differentiate them clearly from young reef formations, as virtually the entire sequence is covered by duricrust that varies in thickness but still covers everything. Around the wadi cuts and the subsequent erosion depressions behind them, the terraces may be followed in a less pronounced form to the eastern edge of the limestone ridge.

In accordance with the tectonic shift, the height of the top planation of the limestone ridge varies. At Al Wajh it is at about 45 meters above m. s. l. Near the Al Wajh airport, given here as an example (Fig. 34 a), there is then the 30–35-meters-step, which has developed here as a particularly broad surface. This planation drops at a somewhat superimposed step to 20 meters above m. s. l. and the adjacent, gently sloping surface falls to 16 meters at its front edge. The step to the lower 10-meter-terrace is very sharp and steep and in places shows a reef-front development. At the airport, the 10-meters-step is preserved only as a relatively narrow planation. The coral association is again a typical reef-front formation and especially nicely developed. It was not possible to determine whether this is a younger accretion, or whether the younger erosion cut only causes it to make a fresh impression (Fig. 34).

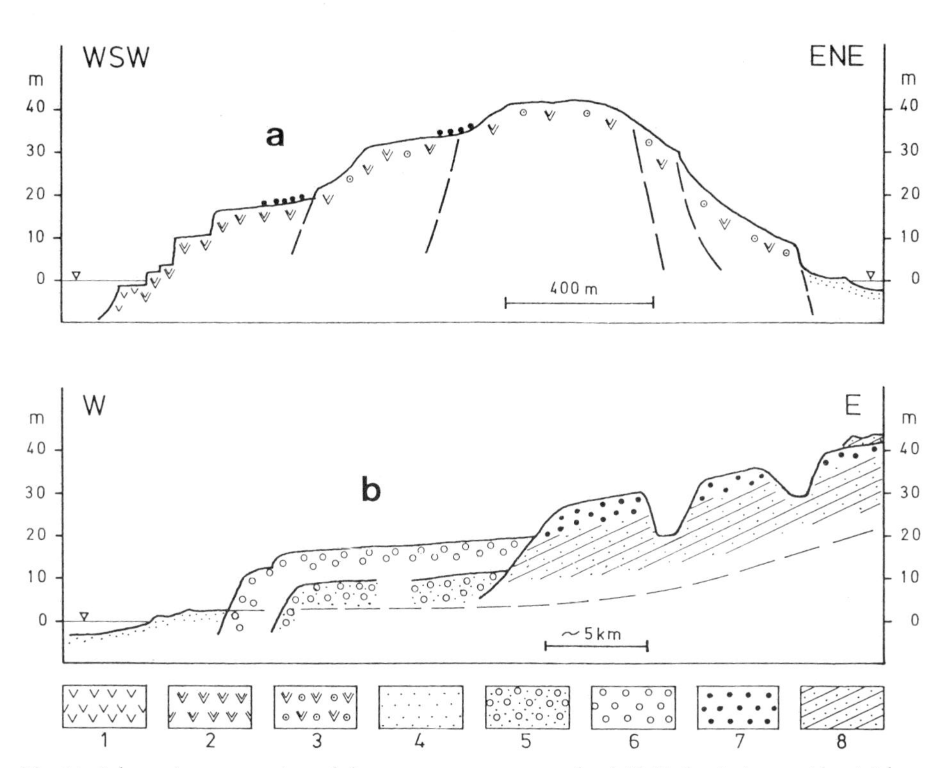

Fig. 34. Schematic cross section of the terrace sequences south of Al Wajh. a) airport ridge 1,5 km south of Al Wajh. b) mouth of Wadi Al Hamdh, 40 km south of Al Wajh (*1* modern reef; *2* Pleistocene reefs; *3* Pleistocene to Pliocene coral and calcarenite limestone; *4* beach sand and coastal sabkhah; *5* sand and gravel of lower terrace; *6* gravels of main terrace; *7* gravel and pebbles of high terrace; *8* tilted Pliocene [?] and Miocene sediments).

The lower 3-meters-terrace only forms a cornice a few meters wide. The tangential coral cuts clearly show that this is a wave-cut platform. Parts of this terrace have been excavated by a younger wave-cut platform at 1.8 meters. This youngest step is the only one that does not show a duricrust covering. It in turn is undercut by the recent wave-cut notch.

Massive corals were taken for radio-carbon measurements from the 3- and 10-m-steps, in each case about one meter under the upper edge and from a protected position. The nearly identical ^{13}C and ^{14}C values (age >37,000 years, see Table 4) indicate material from a uniform reef growth phase. This thus emphasizes the erosional nature of the 3-m-step.

This series realized at the airport can only be an example for the altitude of the terrace steps. All in all, the extent and structure of the individual terraces on this section of the coast vary considerably. If we put the 1.8-meters-step in the Flandrian regression, as its form and altitude indicate, the remaining steps would have to belong to pre-Holocene high-water periods.

The terrace series in the lower course of Wadi Al Hamdh

Wadi Al Hamdh empties 45 kilometers south of Al Wajh into the Red Sea; with a catchment extending almost to Al Madinah it is one of the most important drainage systems in the Hejaz Mountains. With its load of sediment it has made a very considerable contribution to the creation of this broad aggradation plain south of Al Wajh. Even under the present arid conditions, its waters, enriched with mud and rubble, regularly reach the coastal plain, or even the sea.

Wadi Al Hamdh was apparently developed in its present form very soon after the onset of graben formation. With the obsequent cutting back of the wadi, the Old Tertiary runoff system along the Najd fault system that drained to the southeast, was beheaded. The wadi thus contributed during the entire Young Tertiary to the dominance of clastic sediments in the forelying coastal strip.

According to B. ALABOUVETTE and C. PELLATON (1979), the oldest sediments in the graben filling – sandstones, siltstones and conglomerate sandstones – belong to the Oligocene. They are exposed in places in the hill chain in front of the basement on both sides of Wadi Al Hamdh. According to these authors, the Lower Miocene is also mainly clastic. It is made up of arkoses, sandstones and conglomerates that are also exposed in the hills and wadi cuts near the mountains.

Only in the Middle Miocene does a short marine influence with carbonates and thick evaporites dominate. But as early as the Upper Miocene (?) to Pliocene, there are again more clastic red series. In the Jabal Karkuma area directly south of the mouth of Wadi Al Hamdh, C. PELLATON (1980) reports the thickness of the Young Tertiary series as more than 2,500 meters.

In spite of the tectontic shifts in the Young Tertiary, which apparently also caused the horstlike eminence of the Jabal Karkuma, the Tertiary tends to vanish in the morphology of this broad coastal plain. This is due to younger planation and aggradation surfaces that have led to an expansive and rather uniform coastal plain. This plain is given a vague pattern by a varying network of shallow runoff channels and depressions, as well as terrace formation in the area of the wadis.

The younger aggradations are thicker only in the 10-kilometers-wide channel and in the delta fan that flows around the Jabal Karkuma and reaches a breadth of nearly 30 kilometers. The other surfaces running along Wadi Al Hamdh were generally formed in the Miocene to Pliocene and show only a thin younger gravel and sand covering. In places it is only a scattering of pebbles. C. PELLATON (1982) thus classifies these surfaces as Tertiary on the newer comprehensive 1 : 250,000 map. With regard to the morphogenetic event, the presentation by B. ALABOU-

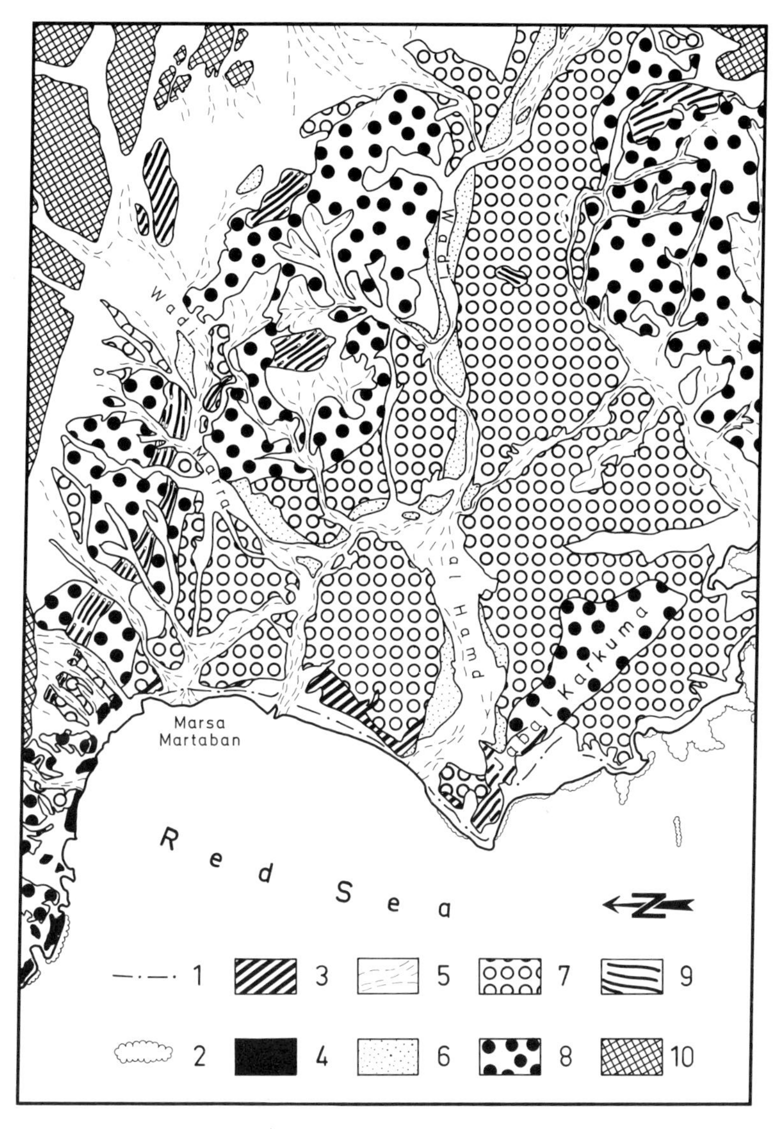

Fig. 35. Map of the mouth area of Wadi Al Hamdh with its main terraces (*1* coastline of Flandrian transgression; *2* recent reefs; *3* marine erosion platform on gravels; *4* marine accumulation terraces; *5* discharge channels mainly with sand and silt accumulations; *6* lower gravel terrace; *7* main gravel terrace; *8* old pebble plain dissected with lower plain parts; *9* tilted Tertiary sediments; *10* Precambrian rocks).

VETTE and C. PELLATON (1979) is more convenient. They speak of the older terrace surface.

With regard to Quaternary events, the chronological classification of this oldest planation, which today has an effect as a morphological element, is important. If there was a lack of more precise time markers in the fossil-rich series around Al Wajh, then this is even more so the case for the mainly clastic series in this area.

It is also only possible to conclude from a superregional comparison that the tipping of Tertiary post-evaporitic layers is related to the reactivation of the graben event in the Pliocene. It was thus not so much aggradation – especially near the mountains with wide alluvial fans – as denudation that produced planation in the sense of a pedimentation. The thin scattering of gravels on Tertiary eminences can so be understood as processed material transported over a small distance.

This pediment then ends with a relatively coarse pebble accumulation that as a result of the extensive planation is widespread, but with increasing thickness going toward the wadis nonetheless shows its connection to the runoff channels. When finer material later blew or floated away, there came to be more pebbles on the surface. Together with their intensive dark desert-varnish, they give these "old surfaces" a typical look. The dark-red color is characteristic of the terrace cuts in which older, tipped, clastic series are often exposed.

Fig. 36. Terrace Landscape north of Marsa Martaban, Al Wajh. The terraces built up by marine calcarenites and reef layers show on the top the typical duricrust weathering. (Photo: H. HÖTZL, 1978.)

On the eastern edge of the coastal plain, this old surface rises to 95 meters (Fig. 34 b). The planation of the basement immediately behind it is clearly separate and is about 120 meters above m. s. l. The old terrace surfaces are especially extensive on the northern edge of the broad coastal plain where the basement jumps to the west. The surfaces on both sides of Wadi Miyah, some 10 kilometers south of Al Wajh, are also particularly impressive (Fig. 36). Toward the coast, they interfinger on this northern edge with the young marine carbonate series described above. South of Wadi Al Hamdh, these old surfaces are especially common in the forefield of the basement island of Jabal Alqunnah. Occasional smaller remnants are also present around Jabal Karkuma, but there they are difficult to distinguish from tectonically shifted Tertiary.

The next-lower planation level forms the actual main terrace. It is an accumulation terrace that in spite of dissection is, with its very plane nature, very different from the old surface which, owing to erosion, is already dissolving into small hills. The height of the step between the two surfaces varies from 10 to 20 meters (Figs. 36 b and 37).

Where Wadi Al Hamdh emerges from the basement, the main terrace is more than 60 meters above m. s. l. It then falls to 40 meters at the beginning of the delta-like fan; on the coast, its upper edge still shows a height of 15 meters. Structurally, the terrace body mainly consists of pebbles (diameter up to 20 cm), gravels and sands. The extent of this terrace body and its connection to an earlier wide erosion channel in the vicinity of the old surface is shown especially clearly on the 1 : 250,000 map by C. Pellaton (1982). Like a trumpet, the accumulation body is at first relatively narrow, widens gradually and then passes into the open delta, presumably at a previously broken edge of the old surface (cf. Fig. 35).

At the edge of the mountains, the main terrace body has been extensively removed by younger erosion. In the middle section it is divided by a relatively narrow, deep channel and in the coastal area by broad arms of the wadi. Parallel to the formation of the main wadi channel, which has also been in recent use, there was also a further dissection of the main terrace surface by several erosion channels reaching back into the accumulation body. The transitions from the deep channels to the top planations are, however, still relatively steep, so that the plane nature of the main terrace has generally been maintained.

Another terrace body is inserted in the main wadi channel. It is also an accumulation terrace, showing intermediate filling of the originally deeper wadi channel. Of this lower terrace, only isolated surface remnants accreted on the older terrace are preserved (Fig. 37). They are formed by redeposited pebbles and gravels from the main terrace. Usually this young terrace is 6 m deeper, but the height of the step increases farther up the wadi. This is also responsible for the absence of the lower terrace near the edge of the basement, where it is apparently completely covered by young sedimentation. On the coast, this lower terrace is at 12 meters above m. s. l. In the sections of the low and main terrace parallel to the coast, they are preceded by an erosion terrace at 8 meters above m. s. l.

Sedimentation is currently taking place in the entire wadi within the coastal plain. Near the basement, the sedimentation is in the form of sands and gravels, but at the wadi mouth silts, sabkhah formations and eolian sands are being deposited.

Fig. 37. System of gravel terraces, Wadi Al Hamdh plain. In front and middle ground remnants of lower terrace; in the background the step of the main terrace. (Photo: H. HÖTZL, 1983.)

Only indirect information can be given on the absolute age of the individual terraces. Apparently the last channel deepening took place during the last glacial decrease in sea level. Whether the previous high water mark only cut the 8-meters erosion step in the lower terrace body, or whether its aggradation took place at about the same time, cannot be answered unequivocally on the basis of the data available here. With regard to the stratigraphic classification of the main terrace and the old surface, we assume, with reference to sections 2.2.3. and 2.2.5., a Middle Pleistocene (?) or oldest Pleistocene to Upper Pliocene age.

2.2.3. The Quaternary from Umm Lajj to Yanbu Al Bahr

(H. HÖTZL, A. R. JADO, H. J. LIPPOLT, H. PUCHELT)

This section of the coast is some 150 kilometers long and is remarkable for the way the 80 kilometers of its block-like middle section protrude and are bounded by faults (Figs. 28 and 38). The mainland blocks and the direction of the coast follow a fault step trending parallel to the Red Sea course (150°).

Over almost all of this section of the coast there is, with regard to the Quaternary, juxtaposition of alluvial clastic accumulations and marine deposits; these, however, show distinct spatial separation. With the exception of occasional basement and Tertiary eminences, terrestrial sediments take up most of the generally 5 to 20 kilometer wide coastal plain. The marine sediments, however, are limited to the immediate seaboard. There they take up a zone with a width of 1 to 5 kilometers.

The Pleistocene terrestrial and marine series were originally highly interfingered owing to variations in sedimentation; today, younger erosion has distinctly separated the surfaces they cover. As was mentioned with regard to the sections farther north (see 2.2.1. and 2.2.2.), in the transitional area of these two facies erosion occurs very easily, so that the rapidly consolidated marine carbonates are dissected out as ridges. The direct interfingering is now only occasionally seen in smaller areas; the two ranges of facies will thus be discussed separately in the following.

2.2.3.1. Marine terraces

The marine sediments of a coastal area must of necessity show a variety of internal facial differentiations when there are even slight changes in sea level. The total sequence ranges from open shelf basins to barrier or patch reefs with their characteristic zonation, and from lagoon formations and fringing reefs to the various littoral deposits themselves. Slight changes in environmental conditions can suffice to produce facial differences at a given location. The pronounced Pleistocene variations in sea level have only increased these facial changes in the profile sequence. As the Plio-Pleistocene fauna on the coast cannot be more closely differentiated chronologically, it has not as yet been possible to classify individual sections of a uniform profile as belonging to definite periods. There is, then, for the time being only the possibility of a morphological distinction of marine terraces of varying heights; and here, in some cases identical or similar facial development makes it difficult to determine whether it is a marine-cut terrace or an accumulation step, for example a reef structure. This differentiation is made even more difficult by a pronounced duricrust that covers everything.

In the protruding section of the coast, the marine sediments reach thicknesses of more than 50 meters, and in the direct extension on the island of Al Hasani they even exceed 100 meters. In the lower part they are made up mainly of marly-to-chalky pelagic limestones bearing many sea urchins; the upper layers consist

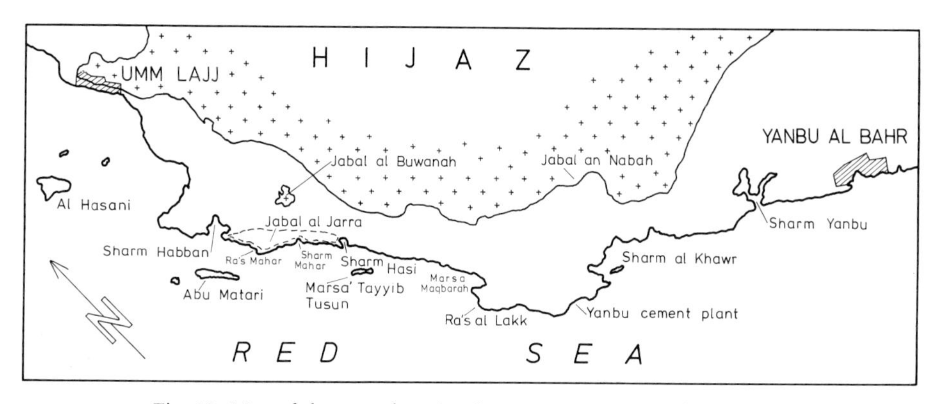

Fig. 38. Map of the coastal section from Umm Lajj to Yanbu Al Bahr.

increasingly of reef limestones. The general maps (G. F. Brown et al., 1963; C. Pellaton, 1982 a, b) and the literature (D. Laurent, 1970; B. Alabouvette, 1977; R. Vazquez-Lopez, 1981) generally assume a Quaternary age for the entire sequence. Our work showed that an Upper Pliocene or perhaps an Old Pleistocene age is to be assumed for this limestone sequence, on the basis of the elevation of the marine terrace and their connection with the gravel terraces, as well as on existing chronological data for individual superimposed basalt flows (cf. 2.2.3.3.). Only the younger terrace deposits are distinctly Quaternary.

Our work did not permit us to cover all the Quaternary sediments on the surface; only chosen locations could be visited. They will be described briefly below.

Umm Lajj

In the vicinity of Umm Lajj, the coast comes from the north and projects 30 kilometers to the southwest in three echeloned steps, each 10 kilometers wide. This is the result of the outcropping of a NW–SE-trending fault step interrupted by shallow NE-trending cross fractures. The block, which also brings the basement down to the sea, is limited by the straight coast at Umm Lajj.

Immediately south of Umm Lajj, this fault is marked by the steep upslope of the basement ridge, which is separated from the coast only by a narrow subrecent aggradation. The basement is covered in places by a Miocene limestone sequence, which dips steeply to the west with the fault (Fig. 39).

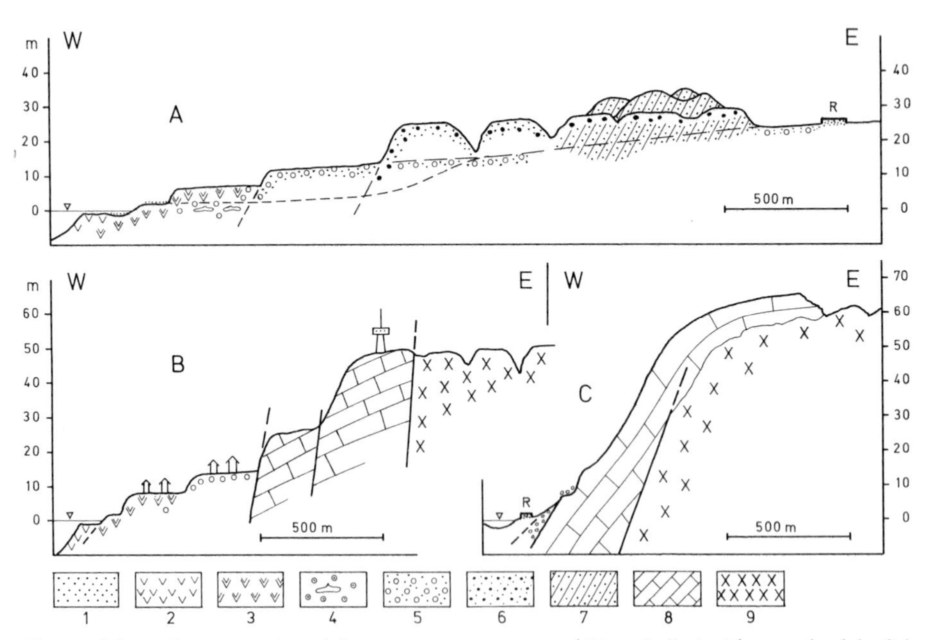

Fig. 39. Schematic cross section of the terrace sequence around Umm Lajj. A: 6 km north of the fish harbor, B: 15 km north of the fish harbor, C: 3 km south of the fish harbor (*1* calcareous sand, *2* recent coral reef, *3* Pleistocene coral reefs, *4* calcarenite with echinoids, *5* younger gravels, *6* older gravels and pepples with dark desert varnish, *7* clastic Tertiary sediments, *8* Miocene reef limestone, *9* Precambrian basement).

The limestones show not only the same steep accretion as the equivalent reefs at Dhuba and El Aynunah, but they also have the same stratigraphic position. Here in Umm Lajj, the positioning of the limestones shows the early development of the present morphology. Thus, the limestones in the entire surroundings of the mountain bulge behind the harbor are steeply positioned in the wadi section and pass into horizontal bedding in the niveau of the old peneplain, here at 80 meters above m. s. l. (Fig. 39 C). The northern subblock (altitude 50 m), today with Umm Lajj's radio tower upon it, was lowered somewhat before the erosion began (Fig. 39 B). The basement horst has also settled to the northwest and broke off directly north of Umm Lajj on a northeast-trending fault. Here as well one has the impression that the somewhat younger and mainly clastic Tertiary accreted primarly on the fault step; the morphological pattern is the result of secondary accumulation and erosion processses.

A distinct hanging notch or step 1–2 meters wide has been formed 8 meters above m. s. l. in the steep Miocene limestone slope. It was once a wave-cut notch, and in places, cemented marine pebbles are still preserved in it. Beach rocks with coral and mussel-shell remnants are still to be found, particularly in the indentations caused by steep channels. An additional, less pronounced notch can be seen at an elevation of 16 meters. Toward the north, where the basement retreats somewhat, this hanging step passes directly onto the 8-meter and 16-meter terraces. The main part of the city of Umm Lajj lies upon these two terraces. The situation is especially informative north of Umm Lajj, as a basalt flow from the Hallat Abu Nar reaches the coast there. The first determinative morphological element is the 8-meter step on the northwest corner of this spur (Fig. 39 A and 40). On the coast itself, it is made up of reef limestones and calcarenite. These materials represent deposits of reef bodies or reef debris and beach sediments, sometimes with lagoonal facies, and with many gastropod shells. The reef-limestone are located in places upon previously deposited gravels, or upon the basalt. Today there is a slight step at the transition from the old reef to the gravels outcropping behind.

The age differences in the marine sediments forming the terrace could not be determined in other sections owing to lack of information; here, however, the basalt intercalation makes such differences clear. The young reef is in places an accretion in front of the basalt ridge. But where it thins out and goes down, sedimentation takes over with a limestone series up to 5 meters thick. The basalt itself, however, has already flowed over older reef limestones.

Quaternary marine sediments could not be found in a higher morphologic position in the Umm Lajj area. Farther back, however, two distinct gravel terraces are existing (Fig. 39 A). The higher one with coarse pebbles and a remarkably dark patina has been dissected somewhat into hills owing to younger erosion channels. This terrace begins at 25 meters and shows a shallow rise going inland. Overlooking it there are occasional residual peaks that here on the northwest spur lie at about 35 meters and indicate an older 40-meter-step or a still higher planation.

The lower gravel terrace at 12 to 15 meters (Fig. 40) still shows a character of a plane surface in spite of occasional channels. It is generally parallel to the coast and lies in front of the older terrace body; it can, however, be followed farther

Fig. 40. Terrace sequence on the coast, 5 km north of Umm Lajj. In the foreground 8-m-terrace built up by calcarenite and reef layer with duricrust. The small bay marks the Flandrian erosion with surf notch at the back. The pertinent erosion plain is covered by recent beach sand. Back ground gravel terrace of the 15-m-step. (Photo: H. HÖTZL, 1983.)

back in the wider cuts of former wadis between the higher terrace body. It seems that the lava of the above mentioned basalt was flowed into an erosion channel intersected into the lower gravel terrace. The sequence of these gravel terraces and their chronological relationship to the basalts will be dealt with more closely in the next two sections (2.2.3.2. and 2.2.3.3.).

The marine 8-meter step is preceded by the 2-meter step. It shows an erosion surface of varying width. In many places it is absent, and recent to subrecent sand accretions form the transition to the 8-meter-step. In places, these subrecent sand accumulations cover the former shore platform, as can be seen in Fig. 40. Behind it, however, the former wave-cut notch, today somewhat farther inland, can still be seen.

Jabal Al Jarra

The Jabal Al Jarra is a table-like limestone ridge parallel to the coast and running some 70 kilometers from Sharm Hasi to Sharm Habban with an average width of 10 kilometers (Fig. 38). The steep step on the eastern edge follows a northwest trending fault and shows only slight erosional overprinting. Parallel faults cause slightly shifted horst and graben structures, making it somewhat more difficult to classify the erosion surfaces. In all, the ridge dips somewhat to the north; on slanted cross fractures, however, it often shows desk-like lifting.

Structurally, this ridge is made up of the carbonate sequence more than 50 meters thick mentioned above. At the bottom, marly and sometimes reddish limestones outcrop, proceeding from a sabkhah sequence. Notable here are repeated intercalations of echinoid shell layers, usually a form resembling clypeaster. In the middle part there are pure limestones in dense to calcarenitic form, often with isolated fossil layers, sometimes coral banks. In the upper layers, reef formations increase in dominance toward the coast, while on the landward side there are occasional intercalations of cross-bedded carbonate sands. The basis of this limestone sequence is only lightly consolidated fine-clastic to marly sediments. With their species-poor shell layers (sea urchins or mussels) they suggest a somewhat isolated but nonetheless purely marine coastal facies, as is also indicated by occasional sabkhah-like intercalations. A layer with large oysters (up to 25 cm) in the upper part of this series seems particularly characteristic. These sediments are exposed in the vicinity of the Jabal Al Buwanah. They themselves are underlaid by a gypsum-anhydrite sequence more than 20 meters thick. B. ALABOUVETTE (1977) assumed for this fine-clastic marly series an Upper Miocene to Lower Pliocene age on the basis of individual fossil classifications. As a result of contact due to faulting and erosion in the transitional zone, the stratigraphic situation in the adjacent limestone sequence to the west is not entirely clear.

Fig. 41. Shore terraces on the sea side of Jabal Al Jarra. The picture shows the 6-m-, 10-m-, 15-m-, and 22-m-steps. (Photo: H. HÖTZL, 1983.)

On the side of the Jabal Al Jarra facing west toward the sea, various terraces have been formed and the ridge falls down toward the sea in steps (Fig. 41). Along the course of the coast, some steps are more clearly developed than others. In some sections they take up relatively broad surfaces, while they disappear in the immediate continuation. The schematic drawing (Fig. 42 A) shows the approximate situation in the section from Ra's Mahar to Sharm Habban.

The deepest and also youngest step is the early Holocene terrace. At Ra's Mahar part of it is a narrow erosion platform, and in the bays it is a somewhat elevated beach rock.

North of Ra's Mahar, the generally narrow strand is limited by the higher 6–7-meters-terrace. Recent undercutting has made the face of the step almost vertical and it shows impressive reef-edge fauna; this, however, is not a accelerated fringing reef, but only an erosion platform. In some places, as at Ra's Mahar, there is an intermediate step in front of it at about 4 meters above m. s. l.

Going farther back, there is a steep and only slightly overprinted edge, and

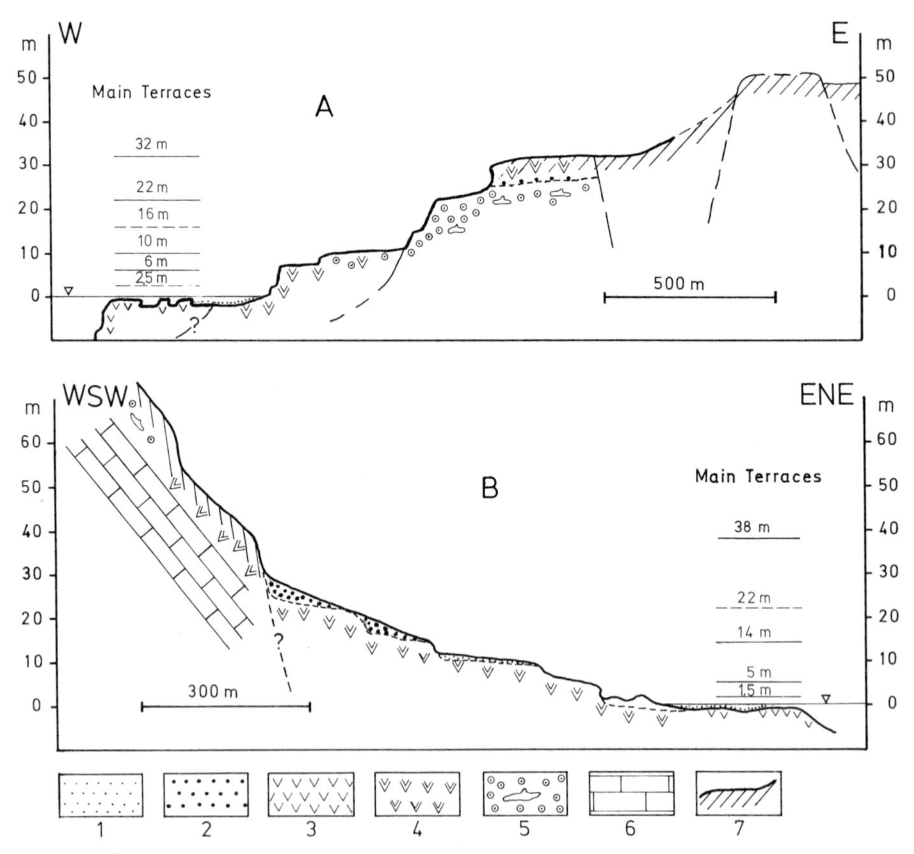

Fig. 42. Schematic cross section of terrace sequences from Jabal Al Jarra and Al Hasani Island. A) Western escarpment of northern Jabel Al Jarra, B) East coast of Al Hasani (*1* calcareous sand, *2* slope debris, *3* recent coral reef, *4* Pleistocene coral reef, *5* calcarenite with echinoids, *6* dipping of the limestone, *7* thick mature duricrust).

then the 10-meters terrace. It begins at 9–10 meters above m. s. l. and rises to the base of the next step at 11 to 12 meters. Although we cannot be completely sure here, this step does seem to have developed at least partially as a new reef structure. It is, however, cut off by younger erosion of the 6-meters-step and is being increasingly eroded by the surf.

The 16-meters terrace can still be related to the younger steps. It forms actually for the most part a narrow erosion ridge on the rise to the next-highest planation. In the aerial photograph, however, it is, together with the aforementioned terrace levels, a somewhat lighter shade of gray than the higher and darker-looking planations. This is essentially a result of the more intensive and mature duricrust of the older terraces.

The higher steps also do not show such constancy of elevation as the lower terraces. They are tipped slightly to the northwest, giving the impression that the subsequent 22-meters and 30-meters planation developed after the internal block tectonics, but was still involved in the slight general tilting. The shifts at cross faults, and particularly the resultant wide erosion gaps, have somewhat obscured the relationship of the individual levels to each other. On the northern section of the Jabal Al Jarra, there are erosion surfaces at 22, 30 and 38 meters. On the middle section these, and still higher surfaces rise to 90 meters above m. s. l. as a result of the tectonic uplift. These higher planations were generally taken up by a pronounced reef-top development with occasional transitions to backreef formations. A cut on the 30-meters-step showed that this uppermost reef flat followed a gravel layer in a beach-rock formation, which in turn lay upon an uneven erosion surface on top of the older echinoid limestones.

There is a comparable terrace sequence in the breakthrough of Wadi Rakah, which empties into the sea in the area of Sharm Mahar. It divides the Jabal Al Jarra into a northern and a southern halves. On the southern edge of the bay, the influence of fault tectonics extending up to most recent time can be seen very nicely. There, the younger reef body (7- to 16-m-step) lying on both sides of a narrow ridge of the 30-meters-step, was fractured by faults trending from 110° to 120°, along which channel-like erosion then occurred. This channel was later partially filled with tidal material. The continuity and appearance of the marine 7- and 15-meters-steps on the eastern edge of this breakthrough indicate that this wadi trench is very old.

Al Hasani

Al Hasani is an island of some 11 km^2 in the direct northwestern continuation of the Jabal Al Jarra, some 18 kilometers west of Umm Lajj (Fig. 38). It is also taken up by a ridge trending NW–SE. The highest point is at 168 meters above m. s. l. It is made up of the same carbonate sequence as the Jabal Al Jarra, here over 100 meters thick. The entire island is made up of an uplifted desk-like block tipped to the east. The flat eastern slope still includes older surface elements, but the west side is taken up by echeloned sliding blocks with back slope depressions that developed as a result of a gravitational slippage of the steep main slope.

The flatter east side is 6 kilometers long and shows marine terraces over its entire length (Fig. 42 B). The youngest step of the Flandrian transgression appears as a setback former coastline, or a flat platform reaching forward to the coast, with beginning recent undercutting. Behind it, a gently rising surface begins; it is 100 to 150 meters wide and covered with debris. Its back edge passes into the 10-meters erosion surface with a 1–2-meters step. The 10-meters planation rises over a distance of 200 meters to about 12 meters. The surface is also covered by a thin layer of alluvial limestone debris. The debris has been removed from the recent shallow channels, exposing the marine-cut platform incised in a reef-roof development.

The step starting at 14 meters rises with 3° to nearly 20 meters, then the surface again rises within a pronounced edge to 32 meters. As in the Jabal Al Jarra area, this step forms the border to the higher surfaces covered with mature duricrust. From 32 meters, the surface next rises with a uniform slope of 7.5° to 55 meters. Then there is a further steep step to a little over 60 meters. A uniformly slanted surface finally leads with 12° slope to 133 meters and, after a small buckle in the slope (8°), the highest elevation is reached at 168 meters above m. s. l.

The increasing slope of the higher old erosion surface may be understood as a tipping of the island continuing into youngest time. This is especially true for the slopes in excess of 5°, but the slope of the 14-meters-step also seems to have already been slightly influenced by this.

Tayyib Tusun

Tayyib Tusun, a wadi approximately opposite Marsa Tayyib Tusun, lies in the middle of the section of the coast between Yanbu and Umm Lajj that is 80 kilometers long and projects somewhat to the west. In the northern and southern thirds of this section, as was the case described for the Jabal Al Jarra, Quaternary marine sediments are found in a zone few kilometers wide; in the middle section, however, at Tayyib Tusun, the coast retreats somewhat, at the expense of this zone. Here, most of the marine sediments are therefore limited to a narrow strip only a few hundreds of meters wide directly on the coast.

The marine sediments are involved mainly in the structure of the 6–8-meters- and 10–12-meters-terraces. They interfinger with gravels. The upper reef layer, which thins out rapidly inland, covers the terrace in the form of a cape rock. In the terrace profile on the coast itself, some impressive coral colonies from the anterior reef zone are exposed *in situ*. These carbonate layers also show the typical reef-top and back-reef facies, as well as beach-rock formations.

The 6–8-meters-terrace is an erosional platform cut into the higher 10–12-meters-step. This older terrace is repeatedly dissected by wadi channels. Occasionally, the direct transition to the corresponding gravel terrace between the wadis may be seen. As individual cuts show, it is a matter of an intensive interfingering of the debris flowing from the land with the littoral carbonate sediments. The covering reef flat can show a slight scattering with gravel from behind, or passes into the gravel terrace with a relatively clear boundary at 0.5 meter to 1 meter.

A higher marine terrace has not been preserved on the coast itself. The study of individual wadis and the gravel ridges accompanying them, however, repeatedly revealed more or less rounded bits of coral. A more precise survey then showed that a gravel-terrace step at 40 meters above m. s. l. some 3.5 kilometers inland and already eroded, still contained occasional accretions of an old reef body. The coral colonies are located some two–four meters below the planation in the form of a sill running around the individual ridges and directly upon the gravels. In one of these "bays" between the gravel ridges, the corals pass over into an oyster bank, which in places then again passes into carbonatically bounded beach rock.

Northeast of Ra's Al Lakk, where the marine terraces again begin to take up larger areas, smaller remnants of a basalt occurrence are still present on the marine terraces. They appear to belong to a tongue-like flow on the old planation, all of which, however, has been eroded with the exception of these remnants. A K/Ar dating is being made, but the values are not yet available. In one of the two occurrences, a small hill nearly 20 meters high rising from the 15-meters-planation, the basalt forms a cap rock 4 meters thick. Below that, there is a quartz-gravel layer 2 meters thick, as well as a tipped oyster layer, coral banks and echinoid limestones.

The entire situation is not completely clear, as faults have shifted the basalts. Generally, it is compared with the conditions farther south at Sharm Al Khawr, and these gravels under the basalts are assigned to the 35- to 40-meter planation.

Yanbu Cement Plant

The Yanbu Cement Plant takes young carbonate rock from 15 kilometers north of Sharm Al Khawr for the manufacture of cement. Here this occurrence of marine sediments is limited to a strip of coast with a width of two–four kilometers. Marginal erosion has again dissected out the occurrence as a table-like ridge. With its orientation parallel to the graben, the ridge is of tectonic origin similar to the position of Jabal Al Jarra in the north. It forms the border of a block-like section of the coast which is 25 kilometers long and projects especially far out; it is limited by the bay of Sharm Al Khawr and Marsa Maqbarah.

Near the cement plant, the ridge reaches a height of 40 meters; both the top planation and the various terrace steps were formed by previous sea levels.

Here as well, in the vicinity of the cement plant, the two-meter level forms the bottom terrace step. It is essentially a shore platform incised into the somewhat older reef limestone; its maximal breadth here is 50 meters, but in many places it is completely missing. On the steep front edge this step has a height of 1.7–2 meters, and 2.5 meters at the rear. Young corals related to the formative phase of this erosional depth could only be identified beyond doubt in occasional pockets or niches.

The next-highest terrace step is six meters a. s. l. Here it has a width of up to one kilometer and is generally still a uniform, even surface. Only on the edges toward the lower two-meters step, or right next to the sea, are there shallow erosional depressions or channels, so that the upper edge in places is barely four meters a. s. l.

Farther back, this six meter planation may be limited by a steep rise, or a distinct flat upslope. This is, however, always characterized by pronounced weathering formations such as excavations and most especially a more intensive duricrust formation.

Proceeding upward, there are further terrace levels at 8 to 10 meters, and about 15 meters above m. s. l. Its surface already shows a small relief and is covered with a duricrust. There are additional terrace steps at about 20 and 25 meters above m. s. l. They vary considerably, however, in distinctness. As a rule, the edges are highly overformed and flattened; only for short stretches do they also appear as steeper edges. One generally has the impression of a considerably more pronounced and mature duricrust formation as compared to the 15-meter level, which is especially noticeable upon the top planation of this ridge at about 40 meters above m. s. l.

On the inland side, the ridge breaks down into vertical walls on an erosional margin. On the upper edge there is the overhanging hard crust, and thereunder are the slightly backweathered limestones, including marly limes in places. The erosive nature of this margin is underscored by isolated erosion remnants, as well as occasional pillars that are still jointed. The steps visible in the wall are irregular and in some places limited by bed boundaries. Their surface also bears a mature duricrust.

Planation levels as remnants of older sea levels are visible in the adjacent erosion zone to the east. Most especially, remnants of the 10-meters- and 20-meters-terraces are to be seen in the subsequent depression directed toward Sharm Al Khawr in the southeast. These terrace remnants are laced with occasional faults parallel to the graben, which are responsible for the downwarp of individual blocks.

The marine carbonates here also show distinct lateral and vertical facial differences. An understanding of the structure of the entire sequence is best obtained from the erosion remnants on the eastern edge of the ridge. Here, unlike the west side, where entire coral colonies are to be seen on the individual steps, they usually form only the very thin upper crust. Underneath, there are relatively thick arenitic limestones with occasional large fossils. Snail and sea-urchin impressions as well as large *pina* and *venus* shells are also worthy of note here.

Some three meters under the upper crust there is an oyster layer 0.6–0.8 meter thick. Below that, there is again the lagoonal development, and the fossil shells mentioned previously become less evident. The extent to which this sequence contains deposits of varying ages, or the entire stratigraphic extent of this limestone sequence with a thickness of at least 40 meters belonging to the Quaternary, are matters that cannot be resolved with reference only to the macroscopic fossil content.

While higher gravel fans north of the cement plant also reach as far as the marine-sediment ridge, the southern area, including the part with the road to the cement plant, still belongs to the erosion zone extending from Sharm Al Khawr. The younger clastic sediments have again been removed there, so that various sequences of the Tertiary Raghama Formation are exposed. Immediately south of the road, for example, Miocene gypsum is being excavated.

Sharm Al Khawr

There is no marine terrace platform in the area behind Sharm Al Khawr. Here there appears to be an old tectonic pattern for an indented coastline (cf. section 2.2.1.). The runoff from a number of wadis from the hinterland (Wadi Shahib with Wadi Al Waydan, and Wadi Al Hinu with Wadi Al Ajal, the precursors of which carried the basalt flows from the Jabal An Nabah) has repeatedly caused extensive erosion, in association with the concurrent deeper base level.

At the time of the formation of the adjacent wide six-meter reef flat to the south, the sea clearly extended farther inland. The offshoots of the two basalt ridges formed tongue-like peninsulas, on the edges of which fringing reefs were deposited, with widths of a few tens of meters. There is a very good view here of the reef assembly and their accretion on the basalts of the southern ridge where it divides into two parts (Fig. 43), as well as on the southern side of the northern ridge (Jabal Ra's Jabul). With regard to their height, they are in agreement with the 6 meters level. In the farther environs of this bay of Sharm Al Khawr, the terrace gravels corresponding to this level extend farther in the forefield of the above-mentioned wadis into the former bay and have thus prevented the deposition of reefs. The present edge of the bay is mainly taken up by recent to subrecent evaporitic, clayey-to-sandy sabkhah sediments.

The reefs deposited at the edge of the basalts are laced with isolated small faults. They cause breaking along straight fissures. These young faults usually have a 100-degree trend.

Fig. 43. Young reef accretion (6 m-step) on the basalt ridge near Sharm Al Khawr, 45 km northwest of Yanbu Al Bahr. (Photo: H. HÖTZL, 1981.)

Sharm Yanbu

From Yanbu to Sharm Al Khawr some 35 kilometers to the north, the marine sediment strip is almost completely taken up by the six-meter terrace. Today, it is still an almost completely flat, undivided plane dissected only by the channel-shaped Sharm Yanbu with its branches reaching inland.

This broad planation is mainly a uniform reef flat. It shows a facial differentiation, which can be seen best in the cuts on the western and eastern edges. Toward the sea, there was the main reef itself; and, behind it, increasingly lagoonal facies with certain shell layers and echinoderms developed. The superimposed reef flat, which on the west edge takes up the entire sediment body above the water, thins out increasingly to the east. There, most of it is taken up by relatively thick shell and echinoderm limestones, interfingered with the continental clastic series, of sands to gravels (Fig. 44).

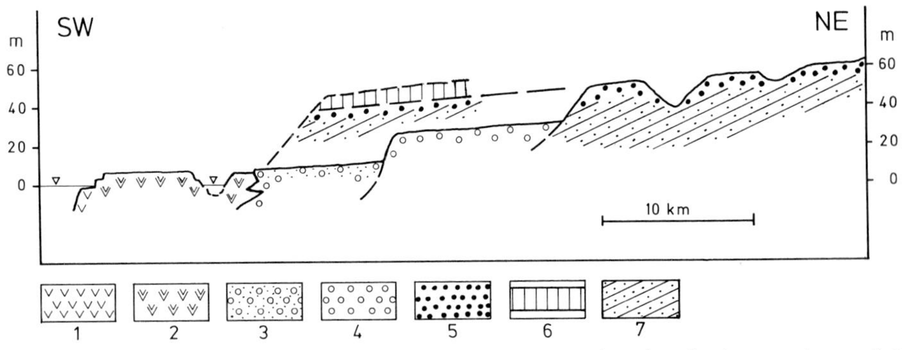

Fig. 44. Schematic cross section of the terrace sequence northeast of Yanbu Al Bahr (*1* modern reef, *2* Pleistocene reefs and lagoon sediments, *3* Young Pleistocene sand gravels, *4* gravels of the middle terrace, *5* gravels and pebbles of the high terrace [boulder gravel terrace], *6* basalt flow, *7* Tertiary sediments).

Between Sharm Yanbu and Sharm Al Khawr, the interfingering with the lower gravel terrace is especially distinct (cf. section 2.2.3.2.). It is adjusted to the back edge of the reef level with its slightly slanted surface, so that gravel can easily pass onto the former.

In all, the entire terrace body is still very well preserved and, with the exception of Sharm Yanbu, not yet dissected. This narrow, channel-shaped cut with its two branches represents the continuation of two wadi channels which today are submerged in this area. The deep, reef-free channel shows its orientation toward an intermediate, considerably deeper receiving stream. With the good morphological preservation of the 6-meters terrace, only the last glacial decrease in sea level of over 120 meters could be associated with this late channel formation.

Also in agreement with these findings is the fact that a clearly developed erosion step, usually only a few meters wide, is carved into the edges of this reef terrace. This step is 1.8–2.5 meters above m. s. l. and happens to be well developed at Sharm Yanbu. The situation there is shown in Fig. 44. Under the surface of the six-meter platform, which has hardened to a light duricrust, the attendant bleached horizon and disaggregation have led to cavetto formation very high up, only half a meter under the terrace surface; the cavettos terminate, however, at the surface of the two-meter step.

The two-meter level is essentially an erosion flat. Only occasionally can very young superimposed or deposited coral colonies be distinguished. Today, the two-meter planation also shows the first signs of a duricrust-like consolidation. It is undercut by the recent wave-cut notch, whereby it in places has been almost completely cleared out down to the present level; sometimes, all that is left of it are occasional erosion remnants.

In places, the two-meter step is still present on the landward side of the marine terrace ridge. As can often be seen in other places, the area where the firm reef flat interfingers with loose sand and gravels invites erosion. Sometimes the subsequent pattern of degraded troughs together with marine erosion have caused horst-like dissection of the marine terrace body on the seaward side parallel to the coast. When the main faults and blocks follow the same direction, sometimes only a careful comparison of levels makes it possible to differentiate these erosion forms from true tectonic horsts.

Today, the eroded depression to the rear is becoming increasingly silted up by sabkhah sedimentation.

2.2.3.2. *Gravel terraces*

The pronounced differences in relief between the more uplifted basement blocks of the graben flank and the marginal blocks lying farther down the graben cause extensive transport of rock material from the mountains to the forelying depression. The extent of the transport load and the grain size of the clastic sediments depends on the intensity of the vertical tectonic movements, on eustatic changes of the sea level, and on the respective climatological conditions. The changes in these interdependent parameters make it difficult to create distinct relationships; this difficulty is increased by the fact that precise dating of this continental series is not possible.

The section of the coast from Yanbu to Umm Lajj, especially the broad coastal plain north of Yanbu, is divided into distinct morphological units. This area will be used to show the terrace sequence from the oldest to the youngest formations. It is also included on the geological map of the Yanbu Al Bahr Quadrangle (1 : 250 000); C. PELLATON (1979) has already used this map for a preliminary classification of the terrace sequence. In contrast to our observation of a division into three steps (Fig. 44), he only distinguishes between a higher (T 1) and a lower (T 2) terrace.

Due to the faults, the basement retreats east of Sharm Al Khawr and is surrounded by the curved outcrop of the Tertiary Raghama Formation as far as Wadi Al Far'ah southeast of Yanbu. Disregarding the youngest fillings in the small wadis, the respective younger terrace sequences follow upon the roughly 50-kilometers-long curve coastwards in nearly concentric zones.

Boulder gravel terrace

Even the Tertiary chain of hills to the rear is partially covered by younger pebbles from the highest terrace level. With reference to the situation in the north, and south to the Red-Sea coast, this will be called a 'boulder gravel terrace'. The components consist of basement gravels with boulders up to 70 centimeters in size, covered with a pale-pink hematite patina. Particularly in the higher Tertiary hills farther back, which also countain coarse Miocene conglomerates, it is not always possible to distinguish them from superimposed, younger coarse clastic sediments. Some of these younger pebbles are reworked products from Tertiary sediments.

Morphologically, this oldest terrace is only partly recognizable as a uniform aggradation plain (Fig. 44). It includes occasional hills and ridges, dissected to varying degrees and eroded in places; these generally extend in front of the Raghama Formation. They are only slightly offset from this older Tertiary, or blend in with it to an indeterminate hill country. The individual hills are up to 120 meters high on the back edge of the coastal plain, and their top layers have a distinctly steeper gradient than the recent wadi floor, perhaps due to a tectonic shift. Five to ten kilometers before the coast, however, the accompanying gravel bodies disappear. Both tectonics and an earlier high-water level could be responsible for this.

The basalt tongues of the Jabal An Nabah (flow of Ḥallat Abu Nar; Fig. 45) provide information for the chronological classification of this aggradation level. The two main tongues running to the coast apparently flowed into shallow valleys sunken into this accumulation surface. These valleys were somewhat deeper in the hinterland and gradually became shallower as they approached the coast. The following profile was obtained from the cut of the new coastal road (Fig. 46), which crosses the basalt tongues that relief inversion has today turned into ridges:

8.50–15.50 m basalt, in general moderately thick, more porous in the first and the last meter.
8.00– 8.50 m red-earth soil with angular detrital components.
7.00– 8.00 m sand and fine gravels, yellowish, with occasional larger pebbles.
1.50– 7.00 m gravels, gray to yellowish-brown, with some sand lenses, pebbles mainly of granites and basic magmatites.
0.00– 1.50 m micaceous sand with occasional consolidated pebble layers, Raghama Formation (?).

C. PELLATON (1979) places the marine micaceous sand exposed at the base in the Tertiary Raghama Formation, although he believes its upper sections are Pliocene. There is a distinct discordance in the gravels and coarse block layer thereupon, which in places covers over a channel relief. A sample of the overlying basalt taken from this cut showed a K-Ar value of 1.4±0.6 million years (cf. Table 5 No. SA 81/40). It thus may be assumed that this terrace is Oldest Pleistocene to Late Pliocene, but most likely Upper Pliocene in age. At the transition to the actual edge of the mountains, this accumulation level seems most likely to be connected to a pediment surface that is divided into distinct sections but all in all makes a coherent impression. This pediment surface, however, seems somewhat uplifted in comparison to the accumulation terrace. This could be the

Fig. 45. Road cut east of Sharm Al Khawr with tilted and partly eroded (left) Miocene-Lower Pliocene (?) sand overlain by terrace sands and gravels. On the top weathered basalt flow (Plio- to Pleistocene), which was originally flowed in a flat wadi now forming a ridge by relief inversion. Total height of the rod 3 meters. (Photo: H. HÖTZL, 1981.)

result of a tectonic shift at the main fault. No terrace remnants that definitely belonged to this highest accumulation step were to be found in the major wadis (Wadi Al Ajal and Wadi Al Far'ah) investigated within the basement in the Yanbu hinterland. The absence of these gravels would indicate that the pediment surface was later cleared off almost completely.

Middle terrace

On the coastal plain north of Yanbu there are only relatively few remnants of the next-youngest terrace system. This may be one of the reasons why C. PELLATON (1979) did not give it a place of its own on the geological map. Its structure again shows mainly gravelly to coarsely clastic material with thicker pebble and boulder layers made up almost exclusively of basement material 30 centimeter in diameter and larger. This material has accumulated on the terrace surface and is covered with a nearly black patina. These surface remnants thus differ in color from the lightpink shade of the other terrace. Whole pebbles also predominate on the surface here, while the isolation effect has created more pebble fragments on the older terrace surface.

The top planation surface of the individual erosion remnants is still quite well preserved; it falls sharply at the erosional margin. The degradation thus occurred mainly as a result of changes in the base level; a reforming or dissection of the erosion remnants scarcely happened as a result of rainwater runoff – again in contrast to the higher boulder gravel terrace.

The aggradation of this middle terrace required a sea level some 20 meters higher than at present. The coastline in the Yanbu area at that time was probably some kilometers farther inland. The gradient of the aggradation surface is somewhat steeper than the recent aggradation, so that the terrace body in the hinterland is more uplifted than the present wadi floor.

The gradient of the older boulder gravel terrace is, however, even steeper than that of the middle terrace (tectonic shift?), so that it is gradually being covered by the middle terrace in the area where the coast is assumed to have been. In conjunction with this supposition, as well as observations farther to the north, the slighter extent of the younger terrace suggests that here it is mainly a matter of filling of deep channels incised in the older terrace during a subsequent period of elevated sea level.

The middle terrace may also be followed into the large wadis in the crystalline basement. Remnants are found mainly at indentations, wider spots and the mouths of wadis, where they were preserved from renewed channel-like erosion. In Wadi Al Far'ah, somewhat downstream Yanbu An Nakhl and about 45 kilometers from the coast, the terrace surface is 25 meters above the recent wadi bed, or 175 meters a. s. l.

Lower terrace

The youngest gravel terrace, called herein the lower terrace, corresponds, as mentioned in section 2.2.3.1., to the six-meter marine terrace. It is especially clearly developed in the area north and south of Sharm Yanbu near the coast. Farther inland, it is limited to wadis running between the older terraces and the Tertiary. Here the terrace step flattens out in comparison to the recent wadi floor, so that farther upstream it is eventually covered completely by recent sedimentation.

The sandy to coarse gravelly material again consists mainly of subangular to well-rounded basement components which, so far as the coastal plain is concerned, were mostly redeposited from the older terraces. The terrace surface shows the typical formation of desert pavement. The pebble components are generally somewhat smaller but show the dark patina of the next-highest terrace step, so that in the aerial photograph they show nearly the same shade of gray.

Dust-loam terrace

In the main wadis in the basement area, especially near the coastal plain, there is yet another type of terrace. This is accumulations of loess-like dust sediments, either directly airborne, waterborne, or reassorted. Their thickness, and thus the height of the terrace step can amount to several meters. Steps of 2.5–3 meters were seen near Yanbu An Nakhl in Wadi Al Far'ah. As earlier studies of the transition to the Arabian Shelf platform have shown, these sediments are mostly only a few thousand years old (H. Hötzl and V. Maurin, 1978). Since this material is fine grained and loose, it is eroded rather quickly. The preservation of the remaining remnants is mainly due to human activity; for thousands of years this fertile ground has been cultivated and protected by stone walls and plantings.

Recent sedimentation

The sedimentation in the recent wadi channels depends very much on local circumstances. There are mainly sands on the wide coastal plain. In places where older terrace bodies have been undercut, there may be flat gravel banks downstream. In the basement region, the material forming the wadi sediments depends considerably on the kinds of rocks occurring in the catchment area. Thus, nearly the entire spectrum of clastic sediments may be represented there.

Comparison of the terraces in the entire region

The terrace sequence described here for the area north of Yanbu can also be seen in the neighboring sections of the coast in a similar, if less distinct form. Deviations are due mainly to variations in the local erosion and sedimentation situation. Only wadis with small to medium-sized catchment areas empty into the curved, gradually expanding coastal plain north of Yanbu. In erosion phases, with the gradient found there, the resultant runoff leads to the creation of pronounced erosion channels.

East and southeast of Yanbu, however, there is an extensive flat coastal plain, onto which large wadis empty. Where these wadis empty, there are areas of vast degradation, and in accumulation phases there is aggradation in the form of further flat alluvial fans. There, usually only the youngest terrace is defined by the shallow channel of the recent wadi. As C. Pellaton (1979) has already mentioned, older terrace remnants are only to be distinguished as indistinct, reworked alluvialplain remnants.

North of Sharm Al Khawr, on the section some 120 kilometers long reaching to Umm Lajj, there is a relatively narrow strip of coastplane. It borders on the more pronouncedly uplifted basement block. There is a uniform, relatively steep talus fan in front of this steep step. It is dissected only by a few small wadis. As it is only sporadically gravelled over from behind, the terrace formation is mainly the result of changes in the base level determined by variations in sea level. The talus fans thus are generally divided into sections parallel to the coast. The close interfingering with the marine terraces was mentioned in section 2.2.3.1.

At a height of about 40 meters and at a distance of some 3 kilometers from the coast, the uppermost step falls 10–20 meters. In the indented and segmented

Table 5. *Potassium/Argon ages of young basalts from the coastal plain between Umm Lajj and Yanbu Al Bahr. Analysis performed by Laboratory for Geochronology, University Heidelberg*

Sample No.	Sample locality	Geographic position		Material	L. f. G. Run-No.	K(%)*	^{40}Ar (rad) $.10^7$ cm^3 STP	Ar (atm) %	K/Ar-age** in mill. years
SA 78/22 B	Wadi Umm Lajj second youngest flow near the entrance into the coastel plain	lat long	25°08′ 37°28′	fresh basalt from the middle layer of the second flow	3902 3960	0.94	0.19	95	0.51±0.35
SA 78/27 A	Wadi Umm Lajj 1.5 km E from the coast new road cut	lat long	25°07′ 37°13′	fresh/unfresh basalt of a thin flow far from its eruption center	3904 3963	0.88	0.19	96	0.55±0.38
SA 81/50 B	Wadi Umm Lajj former basalt cliff 0.5 km E from the coast	lat long	25°07′ 37°10′	fresh/unfresh basalt of a thin flow far from its eruption center	4708	0.97	0.14	96	0.4±0.2
SA 81/50 C	Wadi Umm Lajj 1 km south of the basalt cliff	lat long	25°05′ 37°10′	fresh/unfresh basalt of a thin flow overlaying coral limestone	4715	0.97	0.15	95	0.4±0.2
SA 81/40	Wadi Al Hinu 55 km NW of Yanbu E of Jabal Ras Jarbul, (Jabal An Nabah)	lat long	24°20′ 37°42′	fresh/unfresh basalt of a flow far from its eruption center	4705	0.36	0.35	93	1.4±0.6

* Potasssium content of the K-Ar-rock powder.
** Constants from 1977 (STEIGER, JÄGER, 1977)

erosional margin there are few deposits of oyster banks and coral reefs under the upper edge. The middle terrace step finds expression in occasional gravel projections in smaller wadis. The area near the coast is then taken up mainly by the lower terrace. It is then interfingered with the seven to nine-meter marine terrace and is dissected by several wadis reaching back from the sea.

2.2.3.3. Quaternary geological development of the seaboard, with attention to young tectonics

With K/Ar dating of the basalts from Umm Lajj and Jabal An Nabah (Table 5), as well as ^{14}C determinations on shells and corals from young marine terraces (Table 4), it is possible to establish individual chronological markers for young geological developments. The chronological classification of Tertiary sequences is comparatively imprecise.

The sedimentation began with conglomerates and arkoses of the Raghama Formation discordant above fault blocks. Above these follow marine sandstones and marls. C. PELLATION (1979) with reference to M. BIGOT, B. ALLABOUVETTE (1976) assums a Miocene age. A Pliocene age is supposed for the upper part of the Raghama (?) Formation with clastics, evaporites and carbonates that follow the deeper series with a distinct discordance. From the type locality, the Raghama Group is Oligocene to Miocene (C. DULLO et al., 1983; cf. also the report on the geology of the Midyan region in this volume, section 2.1.).

With the end of the marine sedimentation there were new tectonic movements of presumably uttermost significance for the change in sedimentation. The deposit of the oldest terrace sediments on the tipped, older Tertiary sequence is, as has been shown, probably of Upper Pliocene age. The basalts from the Jabal An Nabah dated at 1.4±0.6 million years flowed into wadis that were already cut into this oldest terrace surface. The recurring volcanic activity is also an indication of the tectonic activation of the area at the margin of the graben.

After the outflow of these basalts there was a further dissection and degradation of the uppermost terrace body. It is difficult to identify distinct causes for this, but apparently changes in the runoff and base level were significantly involved. The intensive dissection and reworking of the old terrace surface indicates increased precipitation. The deep, channellike dissection and later renewed aggradation of the middle terrace body with its orientation toward a higher base level indicate changes in sea level. The extent to which vertical block movements were involved in this can only be determined by a comparison over a large area. A slight antithetic tipping, persisting throughout the entire period, could be expressed in the steeper gradients of the respective older terrace surfaces.

The age of the middle terrace is narrowed down by datings of the two basalt occurrences (Table 5). It is definitely to be located after the Jabal An Nabah basalt flow (1.4±0.6 million years), but before the formation of the Umm Lajj basalts (0.4±0.2 million years). The basalt from Umm Lajj flowed out of the Hallat Abu Nar (Harrat Lunayyir), an area with recurring volcanic activity during the Plio- and Pleistocene, to the coast, using a wadi that was already cut into this terrace (Fig. 46).

The middle terrace, as mentioned above, indicates a sea level 20 meters higher

Fig. 46. Quaternary basalt flow of Hallat Abu Nar (Harrat Lunayyir) running down a mountain wadi to the coastal plain, east of Umm Lajj. The flow is dissected by young wadi discharge. (Photo: H. HÖTZL, 1978.)

than today. The later erosion and dissection of this gravel accumulation was mainly a result of a subsequent lowering of the base level. The well-preserved surfaces on the individual erosion remnants show that the climatic conditions have not changed considerably since then.

The youngest gravel terrace is then the lower terrace. Its age can be estimated mainly due to the direct connection to the seven-meter marine terrace, which is in part a result of vertical tectonic movements in certain sections occurring between 5 and 10 meters above present s. l. Although there are as yet no absolute age data for this marine terrace in this area (^{14}C only shows an age greater than 35,000 years), this step may nonetheless be classified as belonging to the last marine high watermark before sea level decrease of the last glacial.

Owing to this later decrease in the base level by nearly 100 meters, some of these wadis, re-eroding from the sea, have dissected and isolated this youngest terrace surface with, for example, extensive degradation in the hinterland of Sharm Yanbu as a result. Smaller faults covering this terrace surface underscore the vertical tectonic movements seen in the differing elevation.

The postglacial increase in the Early Holocene led to partial filling of the deep erosion channels and, as the new base level, is thus responsible for the sedimentation present today in the wadi channels. The marine high watermark was attained some 6,000 years ago (^{14}C age of shells and calcareous sinter: 4,400–7,000 years; cf. Table 4).

2.2.4. The Hydrochemical Composition of the Groundwaters of the Coastal Area at the Mouth of Wadi Al Hamdh

(W. KOLLMANN)

2.2.4.1. *Introduction*

In the period between 5 and 11 February 1978, five water samples were drawn in the Quaternary alluvial-cone deposits in the delta of Wadi Al Hamdh and the basement of the hinterland from Al Wajh to Hanak (Fig. 47). Samples of water from the Red Sea were taken in the bay of Sharm Munaybarah to evaluate a potential sea-water intrusion farther along the coast.

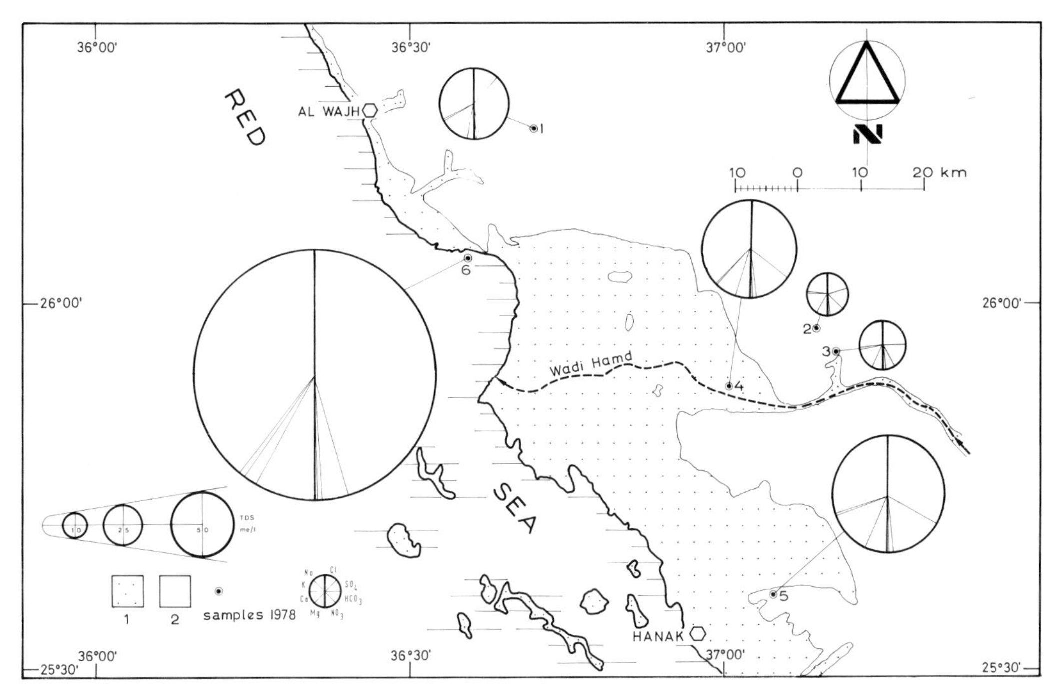

Fig. 47. Location of water samples collected in the area around Wadi Al Hamdh (*1* Quaternary and Tertiary deposits, *2* Bacement).

Samples Nos. 1–3 are joint groundwater from metamorphic and plutonic rocks. Although crystalline rocks are usually rather insoluble, the total dissolved contents from 1.6–4.7 g/kg are quite high.

Sample No. 4 comes from the silty-sandy-gravelly groundwater body in the Wadi Al Hamdh aggradation. The high salinity could basically be due to two causes. On the one hand, the ionic composition similar to the water from the Red Sea (Fig. 48) would seem to represent a model case of saltwater intrusion (Hydrogeology Department, Centre for Applied Geology, Jeddah, The Kingdom of Saudi Arabia, 1975). On the other hand, with the considerable distance to the Red Sea shore of some 35 kilometers, a leaching of sabkhah evaporites would seem even more probable. An increasing amount of salty clay in a gravel fan running out into evaporation basins (A. W. RUTTNER and A. E. RUTTNER-KOLISKO, 1972) seems more in accord with the hydrogeological situation. It should be mentioned at this point that these statements only concern the closer vicinity of sample and must by no means be true of the lower course of Wadi Al Hamdh.

In sample 5, the high salinity and water hardness are due to Miocene gypsum, sandstone and limestone in the catchment area. The existence of aquifers that are very isolated, locally independent and of very different structure can be further substantiated by location and position in the mixture diagram.

2.2.4.2. Composition of the waters

As mentioned above, the total mineralization of the crystalline waters represented by samples Nos. 1, 2, and 3 is unusually high. But while the geographically neighboring samples Nos. 2, 3 are uniformly a Na-Ca-Cl-SO_4 water type in their relative composition, No. 1 is very different, both relatively and absolutely. The composition is nonetheless also determined by Na-Ca-Cl-SO_4 ions, but in a different internal relationship (Fig. 48).

Water samples Nos. 4 and 5 may be stratigraphically differentiated, but they are hydrogeologically similar owing to the presence of evaporites and are both highly mineralized with 9.7 and 9.9 g/kg total dissolved contents, respectively. The leaching product of evaporative sediments produces a water type dominated by Na- or Ca-Cl-SO_4.

In comparison to the average sea-water concentration according to CULKIN (in G. MATTHESS, 1973), the water sample from the Red Sea (No. 6) was slightly more concentrated and was an Na- or Mg-Cl type.

Distribution of trace elements

In addition to the slightly elevated lithium contents to be seen in magmatites (G. MATTHESS, 1973), samples Nos. 5, 6 from evaporite deposits and Red-Sea water show even higher Li concentrations (Table 6). It is remarkable that leaching of evaporitic sediments can produce more lithium than is to be found dissolved in sea water. It is also to be noted that the sea water is by no means saturated with

Table 6. *Trace element concentrations and silicic acid content in coastal waters between Al Wajh and Hanak (all data in ppm)*

1978	Li^+	Sr^{2+}	Mn^{2+}	F^-	SiO_2
1	0.06	7.5	0.01	2.6	20.0
2	0.03	2.0	0.01	0.8	121
3	0.04	2.2	0.01	1.3	88
4	0.09	7.5	<0.01	0.6	27.0
5	0.32	18.5	<0.01	1.0	21.0
6	0.24	~8	0.05	0.8	–

strontium, fluoride and silica. The values are actually considerably higher in sabkhah and crystalline waters. In contrast to other areas described, no homogeneities and regularities as functions of the lithological structure of the catchment areas are to be found, which in any case could be due to pollution. In addition, the small number of samples taken in geologically very different places does not permit any general conclusions.

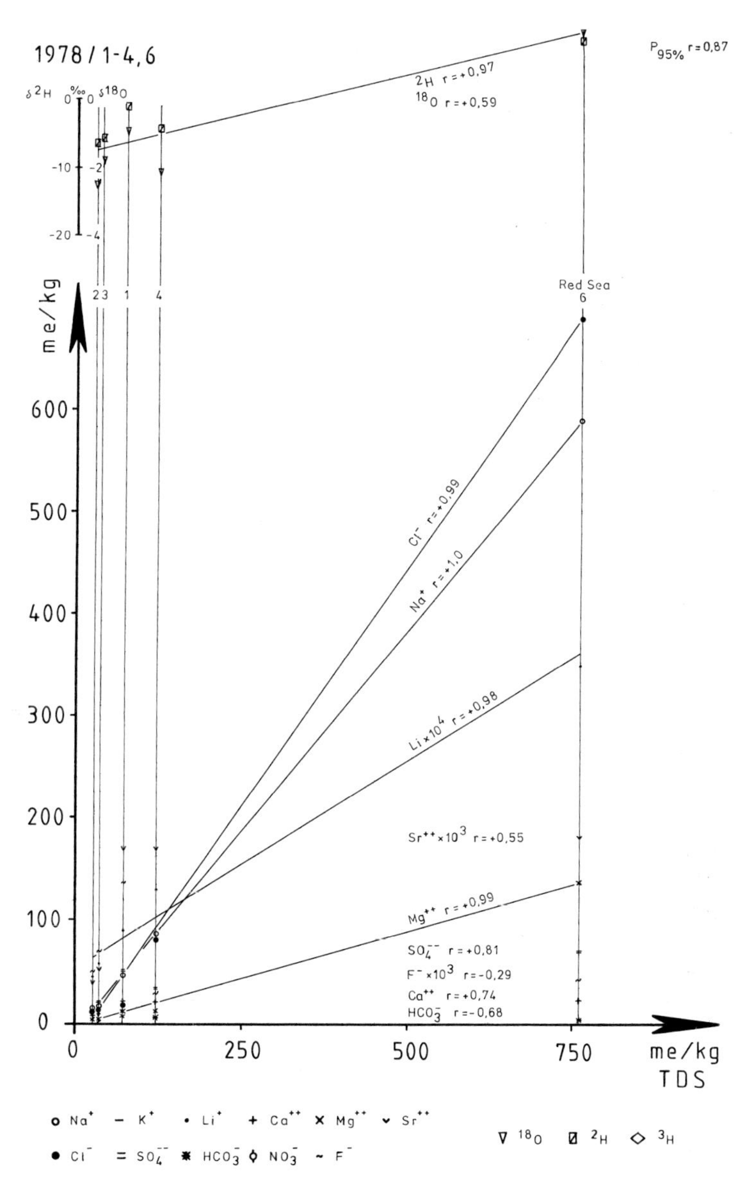

Fig. 48. Chemical composition of the groundwater from the coastal plain in Wadi Al Hamdh (Nos. *1–4*) and of water from the Red Sea *(6)*, compared to determine the possibility of saltwater intrusion.

2.2.4.3. Ion-exchange processes

With the exception of sample 1, the effect of ion-exchange processes as the reason for a change in the original primary water can be excluded a priori. Base-exchange indexes around ± 0 and salinity ratio quotients around 1 do not give any indication of secondary changes.

In sample 1 from the crystalline basement of the hinterland around Al Wajh (Wadi Arjah), the positive base-exchange index

$$\text{pos. } I_{BE} = \frac{(Cl^- - (Na^+ + K^+))}{Cl^-} = -1.6$$

as well as the negative base-exchange index

$$\text{neg. } I_{BE} = \frac{(Cl^- - (Na^+ + K^+))}{(SO_4^{2-} + HCO_3^- + NO_3^-)} = -0.5$$

show high negative values. This can be taken as an indication that alkaline earths in the primary water have been exchanged for alkali from the rock, perhaps upon contact with and passage through loam-filled joints in the crystalline. B. Hölting's test (1980) may be used to reconstruct the primary water and thus the actual catchment area. The derivation of the sodium-sulfate water from a former gypsum water may be verified for sample No. 1 by the ratios:

1 sulfate hardness : noncarbonate hardness = 2.45 : 1

2 $Na^+ : Cl^- = 2.6 : 1$

or 3 $(Na^+ - Cl^-) : SO_4^{2-} = 0.6 : 1$

This proves that sabkhah sediments are leached out from alluvial deposits in the upper wadi catchment area and explains the unusually high degree of mineralization for crystalline waters.

2.2.4.4. The question of mixture phenomena

The small number of water samples and their great distances demand much more water samples and more detailed investigations of local areas.

Study of a mixture process between intruding Red-Sea water (No. 6) and crystalline waters (Nos. 1–3) from the basement, which could take place in alluvial material on the coast (No. 4), involves a number of difficulties. The mixing model (Fig. 48) shows a significant linear correlation of the parameters Mg^{2+}, Li^+, Na^+, Cl^- and ^2H with the total mineralization.

First, sample No. 5 from Tertiary evaporites would have to be excluded on principle from consideration, as it is a matter of an independently reacting hydrogeological system. It must also be taken into consideration that the difference in concentration between the relatively slightly mineralized groundwaters and the Red-Sea water is large, and no intermediate mixture sample is available for comparison. The possibility of a simple two-component mixture is contradicted by the fact that the strontium, sulfate and fluoride contents do not show a linear relationship to the total dissolved contents. The distribution of stable isotopes is also more likely to be random. There is by no means an enrichment of oxygen-18.

In the case of increasing sea-water components, this would have to coincide with a higher mineralization of the groundwaters.

In spite of significant correlative relationships among individual ions in the total mineralization in the sense of a mixture model, the assumption of an isolated groundwater in the Quaternary coastal deposits, at least in the closer vicinity of sample No. 4, is permissible on the basis of the objections mentioned.

An enriched concentration toward the lower course by superficial crystalline waters flowing into the sand-gravel body is also out of the question. The arguments mentioned also speak against this mechanism, which A. W. RUTTNER and A. E. RUTTNER-KOLISKO (1972) found in arid parts of eastern Iran with evaporation processes effective down to the groundwater level. In this case as well, the two stable environmental isotopes should indicate a highly significant linear enrichment in comparison to sample No. 4.

2.3. Region Around Jeddah and Its Hinterland

2.3.1. Geology, Geomorphology and Climate

(P. HACKER, H. HÖTZL, H. MOSER, W. RAUERT, F. RONNER, J. G. ZÖTL)

2.3.1.1. General description

The city of Jeddah, located at 21°29′ N latitude and 39°10′ E longitude, is roughly halfway between the Gulf of Aqaba and the Bab El Mandeb Strait. There is a relatively wide break between the recent coral reefs along the east coast of the Red Sea and the 200 meter isobath extending near the cost (G. F. BROWN and R. O. JACKSON, 1963). This natural topography is favorable for the location and development of a harbor. Jeddah in fact owes its prosperity to the fact that it is one of the few places along the Red Sea coast where a large harbor could be located.

A general view to the isobaths and isohypses shows a channel with variable gradient striking E–W from Wadi Fatimah into the Red Sea. This channel is diverted somewhat to ENE by the talus cone of Wadi Fatimah. The sea floor preceding the channel has an average gradient of 6.5 percent at depths of 500 to 1,000 meters. Ten kilometer west of Jeddah there is a steep terrace with a gradient of approximately 20 percent, which is generally limited by the 500- and 200 meter isobaths. From the terrace to the coast, the channel has an average slope of only 2.5 percent, with a buckle immediately before the present-day coast. In comparison, the last seven kilometers of the lower Wadi Fatimah show a slope of 1.3 percent; from there on upward the gradient is only 0.3 percent. Although the profile B–B′ on the geological map 1 : 500,000, sheet I-210 A (G. F. BROWN and R. O. JACKSON, 1963), does not cross the isobaths at a right angle, it may be taken for granted that the calculated gradients are average values which include smaller steep terraces and flatter stretches.

The Jeddah area under investigation included parts of the coast between

21°00′ and 23°00′ N latitude and the most important wadis with their drainage basins, which in part extend back beyond the escarpment of the Shield falling toward W (Plate II, see insertion at back cover).

The principal villages and oases are situated within the main drainage channels. Although the climate of the area is generally arid, here it is modified greatly by the prevailing topography. In the summer (May – September), the lowlands are hot and rather humid (more than 60 percent humidity), but lack precipitation, whereas the highlands (relative humidity about 30 percent) may receive rain when moist air masses from the SW move northward. The prevailing winds are from the NW to W. In the winter, winds are from N to NE and temperatures are moderate. Sandstorms, rainfall and occasional thunderstorms occur when cold air masses cross the Red Sea and invade the Jeddah area. The transitional seasons are generally hot and humid, and sea breezes are rare. Hot storm winds from S and E may, however, be followed by thunderstorms with local flooding.

The lithological character of these catchments is principally Precambrian plutonic and metamorphic rocks with gabbro, diorite, quartz diorite, granodiorite, adamellite, granite, schists, amphibolites, and the intervening alluvial valley fillings. The Precambrian rocks are usually well exposed. The metavolcanic rocks outcrop only sporadically in the northern part of the area under study, but they do so more frequently in the south, especially in the Arafat Group. These rocks form the most prominent relief in that area.

Apart from the numerous vents, vast areas in the N and NE are covered by basalt flows from the Harrat Rahat. These flows lie discordantly on the Precambrian strata, or on the hard and soft Tertiary/Quaternary sediments.

Tectonic events have helped mark the courses of the main valleys and their tributaries. The main wadis are generally related to young NE-striking faults and block faults, whereas the tributaries usually run along the strike of the steeply dipping fractures in the Precambrian rocks. These fractures evolved along with the Red Sea. The tectonic pattern provides the structural control and is, of course, important in studying the hydrological problems related to the development of local and regional surface and subsurface drainage systems. Not only tectonic events, but also the Red Sea (as the deepest drainage level), and lithological changes, are responsible for the drainage directions. The geological terrain, and particularly the lack of extensive sedimentary rocks limit the potential aquifers to the alluvium of the wadis and plains. Water-bearing layers of sand, gravel and boulders occur there at different depths. Several water-bearing horizons may be observed in the course of drilling. Acquifers are generally interconnected and a single water table is the rule in most of the catchments, although data obtained from wells such as those drilled in Wadi Murwani show intercalations of relatively impermeable silt and clay layers.

The outcropping plutonic, volcanic and metamorphic rocks merely serve to replenish these aquifers after excessive precipitation. Replenishment occurs from sheet flows, or from joints after a relatively short interflow, because the storage capacity of these aquifers is very limited, and because there are no thick veneers, or any extensive and interconnected joint system.

The ITALCONSULT report (1967) mentions that the recharge to the western drainage aquifers amounts to five–seven percent of the annual rainfall. In 1967, this infiltration stood in contrast to ground-water yield from the wadis in the Jeddah quadrangle of about 105 million cubic meters/year.

2.3.1.2. Recharge areas between Jeddah and Al Kura

This hydrographically, geomorphologically (and tectonically) distinct area is bounded by Wadi Fayidah in the north, Wadi As Suqah in the northeast and by the lower course of Wadi Fatimah and the coast between 21°20′ and 21°50′ N latitude.

The short wadis proceed directly to the Red Sea and their sources are not more than 35 kilometers from the coast. The longest are Wadi Muraygh and Wadi Bani Malik (Plate II, see insertion at back cover). The watershed of Wadi As Suqah, which runs from SSE to NNW, strikes roughly parallel to the coast. On the landward side, the plain is bounded by a mountainous region underlaid mostly by folded and faulted Precambrian rocks.

The geological map (G. F. BROWN and R. O. JACKSON, 1963) shows a block consisting mainly of Precambrian granites, granitegneiss, andesites, diabases, greenstones, whose orographic isolation would be hard to explain without tectonic influences.

The common Tertiary volcanic rocks are found only sporadically on the edges of the mountainous area; most remarkable is the Al Harrah basalt flow bounding the mountainous region in the north, and extending to the vicinity of the coast. East of a cut near where the Madinah road some 50 kilometers north of Jeddah crosses the basalt flow, outcrops show that the basalt lies on gravels. These gravels represent a previous fluvial valley filling. The basalt flow appears today as a low ridge which followed a former valley whose flanks were eroded (probably also under the effect of the sea). The width of the flow varies from several hundred meters to two kilometers; the flow transverses the coastal plain over a distance of 20 kilometers, and is tens of meters higher than the plain, providing an excellent example of an inverted relief. K/Ar dating of a rock sample taken two kilometers NE of the basalt flow crossing the Madinah road at a present elevation of 40 meters showed an age of 4.0±0.8 million years[1]. There must have been a valley here in the Pliocene, into which the basalt flowed to the west (compare the chapter on the catchment area of Khulays and Usfan). The western off-shoot of the flow dips under the coastal sediments and is surrounded by coral reefs at its foot.

The small wadis have steep flanks, but the volume of the valley filling does not provide water storage sufficient for larger settlements. An old aqueduct (Kanat), which is neglected at present, once extended from Wadi Fatimah to Jeddah. Now, without the help of the desalination plants, even the large wells in the extensive wadis in the northern and southern mountains around Jeddah would not suffice to meet the demands of the city.

At present, none of the small wadis in the Jeddah mountains reaches the coast. Wadi Muraygh, the northernmost one is, however, so deep that sea water can be

---

[1] Potasssium-Argon age determination by H. J. LIPPOLT, Laboratory of Geochronology, University of Heidelberg, Lab.-Nos. 3918 and 3949.

found as far as 10 kilometers into the strip of coast below the Jeddah mountains, depending on the wind direction and intensity, and extent of tides. This is the first conspicuous indication of a post-glacial maximum in sea level in this area.

The coastal plain in the Jeddah area can be divided into two parts: a lower part to the west, with a surface two to five meters a. s. l. (Fig. 49), and a higher part to

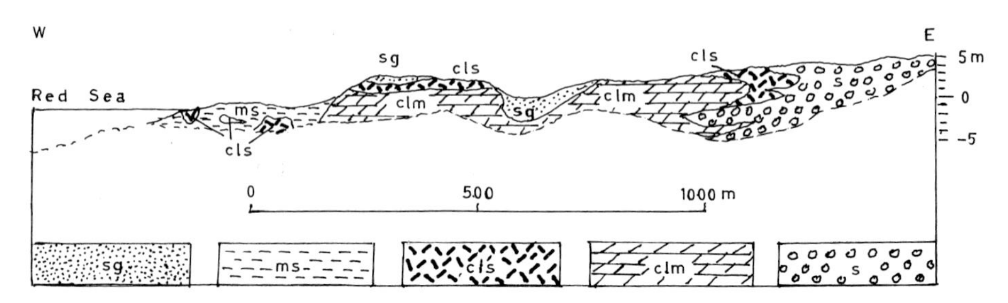

Fig. 49. Geological W-E profile of the city of Jeddah. (After D. LAURENT, et al. 1973.) (*sg* Loose sand and gravel; *ms* muddy sand mixed with shell debris, *cls* slightly indurated detrital limestone or fragments and blocks of coral limestone in a sandy matrix, *clm* massive coral limestone, *s* coarse sand with brown silt.)

the east nearly 20 meters a. s. l., composed of Tertiary and Precambrian rocks (Fig. 50).

G. F. BROWN (1960) categorizes coastal surfaces with heights of 6, 10 and 30 meters and ascribes the elevation more to the vertical uplift of tectonic blocks than to variations in sea level.

P. G. MORRIS (1975) reports that from the Pliocene on into most recent history, clays, siltstones, thinbedded limestones and gypsum were deposited. In some places, this sequence is covered by lava.

The lower coastal plain shows very distinct zones parallel to the coast which are made up of muddy sand mixed with shell debris, slightly indurated detrital limestone, or fragments and blocks of coral limestone in a sandy matrix, clean loose sand and gravel, massive coral limestone, and coarse sand with brown silt. As Fig. 49 shows, this zonal structure is very distinct vertically and horizontally; the layers overlap and interfinger in places.

The "Engineering Geology and Hydrology of Jiddah" map (D. LAURENT et al., 1973) is based on the evaluation of some 400 excavations, natural cuts and drill holes. Their depths, unfortunately, rarely exceed five meters, and average only three–four meters. With carbon 14 dating, however, useful information may be obtained.

The youngest layers (Fig. 49) are eolian and have been dissolved over short stretches by sheet floods resulting from heavy rain. All the materials came from basement rocks and are of igneous origin. Locally, they may contain small amounts of calcareous debris. Occasionally, the sand has been subjected to diagenetic processes and has been cemented with a calcareous substance. The average thickness is around two–three meters, and may reach up to five meters in places (Legend, D. LAURENT et al., 1973).

Large areas along the present shore consist of muddy sand mixed with shell debris (Fig. 49, "ms"). D. LAURENT et al. (1973) described this sand as composed of igneous particles (quartz, feldspar, etc.) mixed with a lot of shell debris and a little dark-gray mud. Pure clay was observed in only one drill hole at a depth of about 10 meters. As shown in Fig. 49, this material is restricted to the area along the shore and extends below sea level.

The strata of slightly indurated detrital limestone fragments and blocks of coral limestone in a sandy matrix (Fig. 49, "cls") are partly enclosed in the "ms" but also cover remarkable areas at a certain distance from the coast. D. LAURENT et al. (1973) distinguish two types of material here, one a detrital limestone (lithic calcarenite or lithic calcirudite) and the other, fragments or meter-sized blocks of limestone occurring in a sandy matrix. There is, however, no doubt that both of these materials resulted from the destruction of an old coral reef.

These actual reefs are now massive coral limestone (Fig. 49, "clm"), several meters in thickness, and have sinuous subvertical joints spaced several meters apart. A few cavities or large potholes several meters in size filled with sandy material have been found. The coral limestone does not always seem to be thick. It is underlain by silty sand and gravel in several places at depths greater than three–four meters.

Farther inland, at least one kilometer from the present shore, on the average, coarse sand with brown silt dominates. Angular-to-subangular sand and gravel of basement origin and igneous nature are mixed with brown silt in various quantities. Sheet floods probably created the deposits. This material is generally moderately compact, and is very compact from the surface to a depth of one meter. It is commonly more than six meters thick, but to the west, the thickness decreases to less than two meters (D. LAURENT et al., 1973).

The higher shore plane probably developed after the Pliocene movements in the fault zone that marks the eastern edge of the Red Sea Rift (G. F. BROWN, 1960).

A Quaternary age is assumed for all the layers shown in Fig. 49.

The most important evidence of the development of the coastal zone in the vicinity of Jeddah in the Quaternary is provided by the areas of coral limestones of varying ages.

First, there are the coral limestones shown in Fig. 49, which are already consolidated (and karstified in places).

At the time of the survey in 1976, there was an undeveloped outcrop in the form of a terrace remnant some six meters high in the city of Jeddah, located about 1.2 kilometers south of the Austrian Embassy and 60 meters from the coast. There was a clear distinction between an exposed marine reef in the lower part (classified as "clm" in Fig. 49) and, on the top, cross-layered sands and slightly consolidated conglomerates which in their lower parts were mixed with coral fragments.

Carbon-14 dating of the coral remnants gave an age of more than 33,500 years B. P. (Table 7, No. 5904). This means an origin in the Pre-Würm when there had been a longer high sea-level period, combined with the development of an abrasion plain. The interfingering of the coral limestones with coarse sand and layers of gravel and cobbles (Fig. 49, "s") indicates a period of heavier rainfall.

Table 7. *Results of ^{13}C and ^{14}C measurements on samples from the Jeddah area. Analysis performed by GSF – Institut für Radiohydrometrie, Munich–Neuherberg, Federal Republic of Germany*

Locality	Geographic position	Material	IRM Lab. No.	^{13}C (‰ PDB)	^{14}C content (% mod.)	Assumed initial ^{14}C content C_0 (% mod.)	^{14}C age uncorr. (years B. P.)
Sharm Abhur 10 km N new airport of Jeddah	lat 21°45′ long 39°09′	coral *(Porites)* from the 5 m-terrace	5901	—	<0.5	100	>43 200
Sharm Abhur 10 km N new airport of Jeddah	lat 21°45′ long 39°09′	shells from the 5 m-terrace	5902	+0.8	<0.6	100	>41 000
Wadi Arran 10 km E Usfan	lat 21°56′ long 39°25′	calcified woodstem from gravel terraces	5903	−9.5	48.1±2.5	100	5 900 ± 400
Jeddah old harbor E of Ash Sharafiyah	lat 21°30′ long 39°10′	coral *(Porites)* from 6 m-terrace below gravel layer	5904	—	<1.6	100	>33 500
Wadi Fatimah 3 km NE Haddah	lat 21°29′ long 39°34′	snail shells from the silty accumulations on the terrace plain	5906	−7.7	90.5±1.8	100	800 ± 200
Wadi Abu Saww 10 km SE Masturah	lat 23°03′ long 38°55′	calcite cement from gravels 3 m below wadi floor, old gravel pit	5913	−0.9	17.3±0.8	100	14 100 ± 400

Today, the appertaining niveau in Jeddah lie 5–10 meters a. s. l. When the sea receded, "s" and "clm" were abraded and assorted to form "cls".

The cavities, potholes and sinuous subvertical spaced joints mentioned by D. LAURENT et al. (1973) are, in part, karst phenomena. The karstification must, however, have taken place after the sea had receded, as sea water does not dissolve limestone, and karstification depends on the effect of fresh water (A. BÖGLI, 1978).

As mentioned above, the lower part of the slope of the basalt ridge running from the northern mountains of Jeddah to the sea is also accompanied by reef corals. A carbon-14 age of 17,000±300 years B. P.[1] was established for these corals, which appeared too young to be caused by contamination.

2.3.1.3. The large wadi systems in the hinterland of Jeddah

The remaining hinterland of Jeddah from the coastal plain to the watershed between the Red Sea and the Arabian Gulf is formed by three large wadi systems and their catchment areas. These are the Khulays catchment area, the Usfan catchment area, as well as Wadi Fatimah and Wadi Na'man with their tributaries. In this entire area, the main watershed is not so pronounced as in the adjacent southeastern area, where the escarpment of the Shield west and south of At Taif bounds the catchment area of the Red Sea like a wall (compare Plate II, see insertion at back cover, and Fig. 50).

The four main wadis, i. e. Wadi Khulays, Wadi Usfan, Wadi Fatimah and Wadi Na'man constitute a chain of parallel drainage networks of different sizes within the southern Hijaz quadrangle, and have a common drainage trend from east to west toward the Red Sea.

The drainage basins show typical arid and semiarid hydrophysiographic features, characterized by numerous narrow and shallow wadi segments of ephemeral hydraulic flow systems with limited and rather irregular replenishment. In the vicinity of the Harrat Rahat in the northeast and the escarpment west and northwest of At Taif, the relief energy and, therefore, the erosion rate are considerable, whereas downstream, toward the sea, energy decreases, leading to areas of accumulation.

The factors controlling the morphology are the differences in lithology, weathering, selective erosion, climate, and folding and block faulting. Downstream, broad zones of Quaternary flood plains alternate with the gentle slopes of isolated low hills or ridges, but upstream, the slopes become steeper and dissected.

The blurring and alteration of the main watershed result from the tremendous lava plains of Tertiary and Quaternary basalts from the Harrat Rahat. The flows reaching toward the west also have considerable influence on the geomorphological development. In the uppermost Wadi Fatimah, the large difference in elevation between the erosion base of the Red Sea and the highlands east of the

---

[1] NRI – 572 Institute for Radium Research and Nuclear Physics, Austrian Academy of Sciences, Vienna, Dr. H. FELBER.

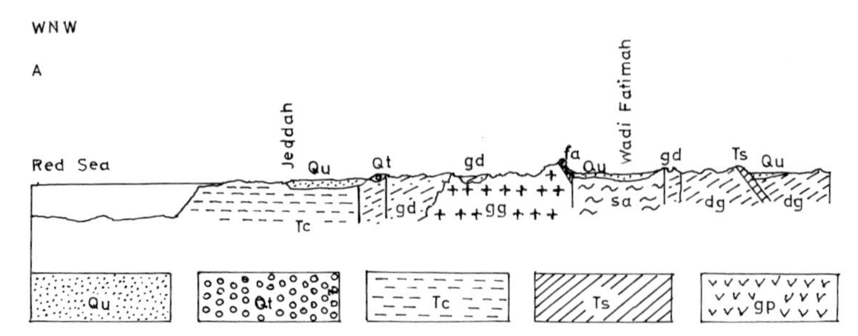

Fig. 50. Cross section Red Sea Coast-Taif (section A-A' in Plate II). The steep escarpment is bordering sharply the deeper hinterland of Jeddah. The escarpment is here more than 100 km away from the coast line (*Qu* gravel, sand silt and clay, *Qt* Quaternary terrace, sand and gravel, *Tc*

escarpment takes effect, and the regressive erosion of the uppermost erosion gullies reaches back behind the face of the escarpment (Plate II, NW of At Taif).

Khulays basin and its catchment area

The Khulays catchment has a total area of 5 330 square kilometers (H. A. AL-NUJAIDI, 1978). It is the largest wadi system in the extended vicinity of Jeddah (Jeddah quadrangle). The largest drainage systems which come together in the Khulays basin are the Wadi Murwani and its numerous tributaries, which dominate the northern half of the area, and in the south of the catchment area, the system of the Umm Sidrah Oasis and the Wadi Arran. The Wadi Murwani catchment is mainly in Precambrian rocks. Only its uppermost roots reach back to the closed-off flows of the Harrat Rahat, the erosion remnants of which, however, also cover territory in the west. It finally flows into the tectonically depressed basin (Plate II and Fig. 51).

Despite the widely different ages of the volcanic rocks, it is unlikely that the basalt flows of the Harrat Rahat once covered the basement, all the way to the Red Sea, and that the Precambrian rock was only later uncovered by erosion. In the catchment area of Wadi Murwani, the basalts may be divided morphologically into the vast basalt sheets in the plateau region, and the flows running toward the west following pre-existent valleys. Tectonics and erosion have today changed the topographical positions of these flows, which also contain morphogenetic information.

At the northern edge of Khulays basin, the offshoots of a basalt flow today form a ridge. The greater resistance to erosion of the basalt in comparison to the Precambrian basement rock has led here, as in the Al Harrah flow north of Jeddah, to a relief inversion. In both places, outcrops attest to an earlier fluvial valley filling under the basalt.

Upvalley from the embouchure of Wadi Murwani in the Khulays basin, these basalt flows lie on the remnants of erosion terraces on either side of the valley (Fig. 51). As happened with the erosion of the piedmont, the wadi was eroded selectively on the border between the more resistant basalt flow and the Precambrian rocks. About 10 kilometers upstream, the gravels are located under the

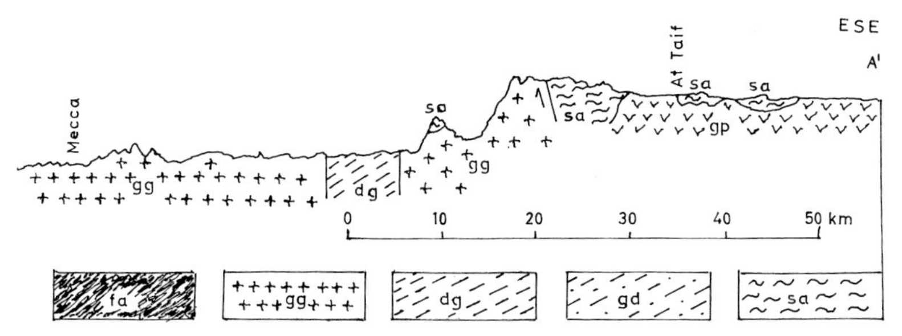

consolidated clastic rocks (Miocene), *Ts* Shumaysi Formation (Eocene), *gp* granite (Precambrian), *fa* Fatimah Formation, *gg* granite and granite gneiss, *dg* diorite and granodiorite, *gd* andesite, *sa* amphibolite schist). After G. F. BROWN et al., Map I-210A 1 : 500,000, 1963.

Fig. 51. Aerial view of Wadi Murwani. Basalt flows running in former wadis covering now terraces and ridges. (Photo: J. G. ZÖTL, 1976.)

topmost flow, 60–70 meters above the present valley floor, i. e. nearly 250 meters a. s. l. In general, it may be assumed that in this section even the first flows followed a preexisting valley.

Beyond 22°22′ N latitude, i. e., in its upper third, the wadi flows in Precambrian basement which is characterized by highly dissected flanks, numerous steep steps, and a valley floor that is only developed in certain places.

In no place do the wadi and its tributaries carry water year-round. Only after heavy rainfall (e. g., 1975) could small rivulets of groundwater discharge be observed over small distances in the upper course for as long as a year.

Hydrogeological studies were carried out in the lower course of Wadi Murwani and in the Khulays basin (ITALCONSULT, 1967, 1976, and H. A. AL-NUJAIDI, 1978). These findings provide data on the wadi filling, the wadi underground and its gradient, and on the tectonics of the area as well.

Using AL-NUJAIDI'S geophysical measurements (vertical electrical sounding and seismic refraction, 1978), which served the purpose of investigating the groundwater situation, an attempt was made to establish the course of the base of the wadi filling (Fig. 52)[1].

In the lower 25 kilometers before the embouchure in the Khulays basin, the floor of Wadi Murwani has an average gradient of 5‰; and that of the alluvial cone is approximately 8‰. Upstream the gradient flattens out over a longer stretch to 1‰, interrupted by smaller steep steps. In the hinterland, both the main wadi and its tributaries turn into steep V-shaped ditches. These are the catchment veins forming the erosion pattern of the large, sudden floods after episodic rainfalls.

The underground of the wadi filling shows a more accentuated change in gradient. Most striking is a large steep step between measuring points 22 and 32 on the longitudinal profile of Fig. 52. If the smaller irregularities in the profile are neglected, the average gradient for this stretch is 8.5‰. As this profile is based only on data from measuring points which in some cases are a number of kilometers apart, the buckle in the gradient between points 26 and 31 indicates that the wadi underground here runs with a less regular gradient than the profile indicates. All indices point to an erosional step between the stretches of the wadi floor continuing up- and downstream with only 0.4 and 0.5‰ gradients, respectively. The course of the steep step is believed to be of an erosional origin rather than of a fault scrap.

Photogeological observation of this area shows four main fault systems (H. A. AL-NUJAIDI, 1978): the predominant NE–SW system; the NW–SE system which plays a major role in the course of the wadis, especially in the Precambrian rocks; and the N–S and E–W fault system, which is of minor importance.

A large fault follows, beyond doubt, the eastern border of Khulays basin. The underground of the Wadi Murwani ceases abruptly at its embouchure in the basin (Fig. 52).

A view of the situation is provided by ITALCONSULT bores (1967) from the eastern edge of Khulays basin and the embouchure of Wadi Murwani.

---

[1] H. A. AL-NUJAIDI used the geophysical measurements only for groundwater investigations; he did not evaluate the data for geomorphological purposes.

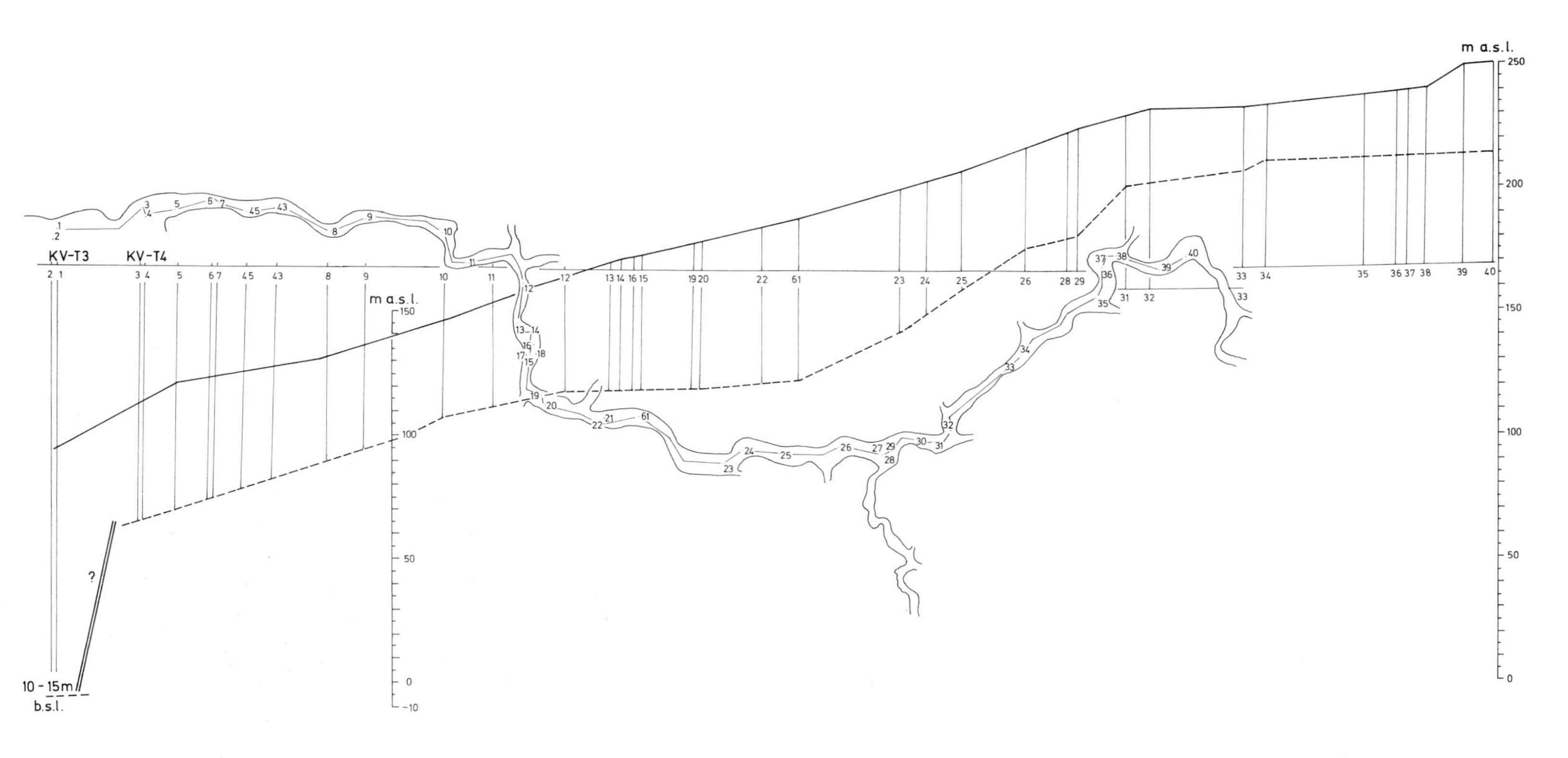

Fig. 52. Wadi Murwani; gradient of the wadi surface and wadi bottom with fault at the margin of Khulays basin. Drawn after seismic data from H. A. AL NUJAIDI (1978).

The ITALCONSULT borehole KV-T 4 at the outlet of the wadi near measuring point 3 of the profile shown in Fig. 52 reached the bedrock underground at a depth of 45 meters (1967, p. 26). This depth is in agreement with the profile.

Just eight kilometers west of this well (KV-T 4), i. e. about five kilometers from the NE edge toward to inner part of the basin, the bore designated as KV-T 3 reached the crystalline bedrock only at a drilling depth of 159 meters.

ITALCONSULT does not give the altitude of KV-T 3 in meters a. s. l. As estimated with Fig. 52, the KV-T 3 bore site is 50–60 meters a. s. l., meaning that the crystalline basement is at least 100 meters under the present sea level.

The KV-T 3 geological log shows only three sediment types: sand and gravel; clay or shale; and sand. The detailed profile given in Fig. 9 and Table 1 in the ITALCONSULT report (1967) indicates sand and gravel at a depth of 0–14.50 meters; sandy clay 14.50–22.00 meters; sand and gravel 22.00–27.00 meters; sand 27.00–29.00 meters; sand and gravel 29.00–33.00 meters; sand 33.00–43.00 meters; sand and gravel 43.00–52.00 meters; clay 52.00–159.00 meters; and crystalline bedrock below 159.00 meters. The first 20 meters are assumed belonging to the Quaternary; from 20.00 to 159.00 meters there is a series of Tertiary sediments, almost exclusively sandy clay with intervals of fine sand and gravels, slightly cemented in places. The lithological logs unfortunately do not mention fossils and it is not possible to make comparisons with the Tertiary sediments on the northern and southwestern edges of Khulays basin.

KV-T 3 was, at the time, the only bore in the basin which reached crystalline bedrock. Bores Nos. KV-T 1 (total depth 22.50 meters, alluvium 16.00 meters), KV-T 2 (total depth 9.80 meters) and KV-T 6 (total depth 56 meters, alluvium 21.00 meters) northwest and north of the basin did not reach crystalline bedrock.

Also of interest are the bores in the lower Wadi Murwani: KV-T 4 (crystalline bedrock at 44 meters depth); KV-T 5 (crystalline bedrock at 28.00 meters); and KV-T 7 (crystalline bedrock at 20.00 meters). Bore KV-T 4 is in the middle of the wadi, KV-T 5 on the southern flank and KV-T 7 on the northern flank of the wadi, showing a short erosion gully at the entrance of the wadi into the Khulays basin. This indicates rapid tectonic sinking of the basin.

As is the case with the wadis in the Shelf of the Arabian platform, most of the wadis in the Jeddah hinterland and their tributaries do not show distinct lower terraces. This is not surprising, as the floods after the episodic but very heavy rains have a much greater effect owing to the greater relief energy in the western part of the Shield than in the Shelf, which has a comparatively flatter gradient. The raging torrents usually cover the entire breadth of the wadi floor.

Of even more interest is a well preserved Quaternary terrace remnant in Wadi Arran, a southern tributary flowing into Khulays basin (Plate II at the end of the book). The surface of this terrace is 2–3 meters above the present wadi floor and has a lightly cemented structure.

Incrusted remnants of tree branches which had only been washed away were recovered from the terrace body; carbon-14 dating showed them to be 5,900 years old (s. Table 7, No. 5903).

This terrace body, with an age of about 6,000 years, is an indication that the heavier precipitation shown for the shelf in the so-called "Neolithic Pluvial" (see Vol. 1, p. 20ff) also affected the western slope of the field.

Usfan catchment area

Coming out of Khulays basin, the Usfan road crosses a saddle in the orographically pronounced southern frame (Plate II, see insertion at back cover). The saddle itself lies in Tertiary sediments: Usfan Formation (marly limestone, shale, sandstone, conquina and conglomerate of Early Eocene), and Shumaysi Formation (sandstone, shales, siltstone of Oligocene Age). Further to the southeast, Shumaysi Formation also forms the flanks of Wadi Fayidah, whereas in the west and south, Precambrian rocks (Samran Series composed of diorite and granite; see Geological Map, I-210 A, G. F. BROWN, and R. O. JACKSON (1963) form the irregular mountainous region. Both Tertiary and Precambrian rocks are covered by the remnants of young basalt flows (Young Tertiary – Quaternary).

Three wadi systems – Wadi Fayidah, Wadi Ash-Shamiya and Wadi As Suqah – run together at Usfan. After a relatively narrow breakthrough through the Precambrian rocks of the Samran Series (green schists, rhyolite), they continue together as the Wadi Ghulah. The Wadi Ghulah disappears some 15 kilometers before the shoreline in the coastal plain. The total area of the Usfan catchment is 2,200 square kilometers. The Usfan catchment also includes structures related to the NW–SE block-faulting system which runs parallel to the Red Sea and was involved in wadi formation. Faults striking NE–SW also occur.

From data obtained by E. A. B. AL-KHATIB (1977) on the Wadi Ghulah and the Wadi Fayidah, and three ITALCONSULT bores (1976), the depth of the wadi channels in the basement is at least approximately known.

Based on 24 geoelectrical measurements, E. A. B. AL-KHATIB (1977, p. 102ff) gives the thickness of the valley filling in Wadi Fayidah (Quaternary and Tertiary sediments). But the measuring points are not given exactly, and the wadi surface has a roughly similar gradient. Therefore, it is not certain that the thicknesses measured in Wadi Fayidah of between 7 and 16 meters, in Wadi Ash-Shamiya, of between 12 and 16 meters, and in Wadi Ghulah, of approximately 50 meters, actually indicate a step in the underground.

The ITALCONSULT bores show that the depth of the crystalline bedrock (at the mouth of Wadi Fayidah going toward Usfan) increases from 35 meters to 44, 49 and 56 meters. This is notable, in that the ITALCONSULT well US-T 1 slightly to the west in the Usfan basin ended at a depth of 208 meters in Tertiary clay, without reaching the crystalline basement.

Also of interest is the observation of the stratification and faults in the sandstone beds (Shumaysi Formation) near the village of Usfan. At first, there are almost horizontal dips of 2°, and then a tilt of up to 30°. This indicates that there is an erosion channel similar to the deeply sunken basin (trench running NW–SE?) at the outlet of Wadi Fayidah, as at the outlet of Wadi Murwani in the Khulays basin.

The US-T 2 well southwest of Usfan (see Plate II) encountered crystalline bedrock at a depth of just 22 meters and thus lies on the western edge of the tectonic depression of Usfan at the beginning of the only outlet to the west which is open today, the so-called Wadi Ghulah.

Absolute altitudes are regrettably scant, even for the ITALCONSULT bores. The only measuring point given on the 1 : 500,000 geographical map shows an altitude of 68 meters a. s. l. west of Usfan. Subtracting the 22-meters thickness of the Quaternary and Tertiary sediments from the nearby US-T 2 well produces an altitude of 46 meters a. s. l. for the crystalline bedrock. This altitude is in agreement with E. A. B. AL-KHATIB'S (1977, fig. IV-37, p. 138) contour lines for the "elevation of the basement rocks a. s. l. as obtained from electrical resistivity in Wadi Ghulah".

These results are significant in two ways. On the one hand, they show that at the outlet of Wadi Ghulah 25 kilometers away from the coast, the bedrock of the wadi floor lies at the present-day sea level. On the other hand there exists a recent erosion step between 20 meters and 45 meters a. s. l. in the crystalline basement (gradient approximately 15‰), preceded upstream by a flattening (gradient 0.25‰).

As regards the depth of the crystalline bedrock in the certainly tectonic Usfan depression, we only know that it is more than 140 meters below the present level of the Red Sea. Considering the small amount of data available, there is good agreement with the situation in the Khulays area.

Unfortunately, only the ITALCONSULT well US-T 3 provides some information on the largest wadi in the Usfan catchment area, Wadi Ash Shamiyah. That bore was just west of Haddat Ash Sham and encountered crystalline bedrock at a depth of 36 meters.

E. A. B. AL-KHATIB (1977) described the sediment sequence in Wadi Fayidah and Wadi Ash Shamiyah as follows. Zone one is 1–3 meters thick and is covered in places by eolian sand sheet. It is made up of fragments of basalts of varying sizes, gravel, quartz grains and fragments of metamorphic and granitic rocks. Zone two is 2–10 meters thick and has a compact horizon of clay mixed with sands and silt, usually red in color. Zone three is the saturated zone, with a thickness averaging 5–10 meters; it is composed of silty sands.

Notable is a description in the lithological log of a layer of fine sand cemented with calcite at a depth of 25 to 30 meters, which is below Zone three. There is reddish silt below this stratum and medium-to-coarse sand and gravels above it. Lack of further data prevents closer interpretation of this accumulation phase, which was probably due to climatic factors.

West of Haddat Ash Sham there are remnants of young basalts arranged discordantly upon the sandstone of the Tertiary Shumaysi Formation at a height of 100 to 120 meters above the present wadi floor. Boulders as large as twice the head size under the basalt can only be explained by heavier runoff under climatic conditions different from those prevailing today.

Information on the genesis of the apparently atypical drainage course of Wadi As Suqah (from SE to NW) is offered by the young breakthrough between Wadi Ghulah and the Usfan depression near the village of Usfan. This modest passage is surely not equal to a catchment area of nearly 2,000 square kilometers and could not suffice for the wadis when they periodically carry water. It may thus be supposed that this passage only developed after the tectonic movements with the NNW–SSE-striking block faults, when the area west of Usfan and Wadi As Suqah was uplifted.

Wadi Fatimah

With a total surface of 4,650 square kilometers, the catchment area of Wadi Fatimah is the second largest wadi system in the Jeddah quadrangle.

The upper courses of the widely branching tributaries of Wadi Fatimah in the north and northeast still extend into the massive basalt flows of the Harrat Rahat, and the remaining catchment area is made up mainly of Precambrian rocks. Tertiary sediments (Shumaysi Formation) are to be found in the middle course, but are not profuse. Extensive alluvial cover occurs in the lower course of the wadi. The Precambrian formations of the Arafat Group (metavolcanic rocks) and the block-faulted Fatimah Formation are less resistant to weathering than the young basalts. The deeply incised gorges are the collecting channels for larger floods after the heavy rains which fall in the higher parts of the Shield, and these chasms reach areas of more than 2,000 meters a. s. l. (Plate II). The most prominent relief in the quadrangle developed in these formations as a result of the weathering and erosion processes combined with tectonic movements.

The high relief energy (in places more than 1,000 meters within a few kilometers) and the proximity to the Red Sea as the deepest drainage level permit the upper courses of the Fatimah system to extend back beyond the wall of the escarpment west and northwest of At Taif as a result of the regressive erosion. This caused a considerable shift to the east of the main Arabian watershed on the Shield (Plate II). The upper courses of the wadis east of the escarpment first run northeast, clearly showing that they previously belonged to the drainage pattern of Wadi Al Aqiq. These wadis are unmistakably diverted and captured by strong headward erosion owing to the steep gradient to the nearby and low-lying erosion level[1].

In its middle course, the Wadi Fatimah is to be considered as a fault-bounded graben, called the Wadi Fatimah graben. The northwest side of Wadi Fatimah was especially affected by block faulting (K. NEBERT et al., 1974). The uplift of the horst-like blocks was accompanied by severe erosion, which produced the extensive fans (up to 30 meters thick) around the present mountains.

The wadi fillings range in thickness from a few meters to 30 meters and probably even more. They consist of sand, gravel and blocks transported mainly by occasional floods after heavy rainfall. The cover of eolian sand within the wadis varies from thin to sporadically thick deposits. Fluvial terraces are well exposed in the middle course of Wadi Fatimah near Al Ashraf, Al Haramis oases, and in the area of Abu Urwah. Morphologically, the terraces in places form low hills, composed of poorly stratified and poorly sorted clastic material including rounded fragments embedded in a loose sandy matrix. There is no doubt about the fluvial origin of these sediments, which were deposited during a time of prevailing humidity (K. NEBERT et al., 1974).

In 1967, ITALCONSULT sank its well FA-2 south of Al Jumum in the lower middle course of the wadi. The geological profile shows an alteration of silt, sand

---

[1] A geological sketch of the middle and lower wadi area may be found in K. NEBERT et al. (1974), p. 3.

and gravels; therefore, it is unclear whether the bore, with a total depth of 35.20 meters, was still in the Quaternary valley filling or had reached Tertiary sediments.

Remarkable is a distinct layer of sand and gravels with pebbles of diameters exceeding 10 centimeters at a depth of 5–16 meters and, below this, a sequence with frequent conglomerates. At a depth of 32–34 meters, another layer of pebbles with diameters greater than 10 centimeters was encountered. The uppermost conglomerate layer is found at a depth of 18 meters, slightly below the groundwater level at the time.

The groundwater of Wadi Fatimah was an early source of water for the city of Jeddah, as the old ruins of a kanat system show. This groundwater is still a part of today's water supply for Jeddah and Mecca. At the time of the survey (1976), the depression of the groundwater body owing to overproduction from the wells was visible in the vegetation of the area.

There are two gravel terrace systems in the middle part of Wadi Fatimah for which an age classification has not yet been possible; it may, however, be said that they date from the Quaternary period. The older 'boulder-gravel terrace' may date back to the Late Pliocene. Preserved rests of these terrace bodies are also developed in the upper Wadi Na'man and its tributaries. Beside this gravel terraces there are remnants of the young, Holocene loess-like terrace (Fig. 53). The profile shows silty calcareous dust material with few intercalated flood sediments and fluviatile gravel at the base.

Fig. 53. Loess-like terrace remnants from Wadi Na'man, SE of Mecca with fluviatile gravel at the base and eolian dust material forming the terrace body only interrupted from some flood sediments. (Photo: H. HÖTZL, 1976.)

2.3.2. Hydrochemistry of the Groundwaters

(P. HACKER, W. KOLLMANN)

2.3.2.1. Introduction

Hydrochemical studies were performed to describe and classify the groundwaters of the Wadi Khulays, Wadi Usfan, Mid-Wadi Fatimah and Wadi Na'man, and to establish relationships between hydrochemistry and geochemistry within and between the catchment areas. Two groups of waters were found with significantly different hydrochemical values. Hydrochemistry formed the basis for the classification of the waters.

The nature of the groundwater in the area depends largely on the petrographical and geochemical texture of the water-bearing rocks and sediments. Other factors which also determine the chemistry of these waters include the primary chemical composition of the precipitate, groundwaters with different chemistry from that of neighboring areas (allochthonous waters), and human influences. Between 1974 and 1978, 162 samples were taken from wells in the catchments of Wadi Murwani (Khulays), Wadi Usfan, Mid-Wadi Fatimah and Wadi Na'man, and analyzed for content of dissolved materials (Table 8). These catchments cover a total area of about 14,400 square kilometers.

Table 8. *Minimum – maximum values of the chemical composition of the groundwaters (in mg/l)*

Catchment (wadis)	Na^+	K^+	Ca^{2+}	Mg^{2+}	Cl^-	SO_4^{2-}	HCO_3^-
Murwani							
'Arran	63–263	4–22	35–132	16–132	131–683	58–269	117–251
n = 29	av. 103	5.5	59	33	199	96	194
Basin of							
Khulays	75–845	5–20	132–407	130–262	683–1897	269–1135	188–312
n = 12	av. 390	9	220	154	822	493	244
Usfan	30–1187	2–6	39–194	11–163	49–2074	32–672	–176
n = 12	av. 608	3	71	46	436	193	60
Fayidah	32–872	1.5–7	50–325	13–211	85–1999	65–777	132–314
n = 27	av. 452	4	105	56	439	212	199
Ash Shamiyah	45–1372	1–8	32–482	27–317	92–1396	26–1980	3–8(?)
n = 23	av. 300	3	176	184	671	708	3
Ghulah	386–1335	3–6	38–357	30–290	649–2808	140–569	3–6(?)
n = 8	av. 740	4	194	135	1686	334	5
Mid Fatimah	30–178	3–14	83–264	2–25	63–1118	48–627	128–319
n = 34	av. 69	6	113	17.5	148	218	192
Na'man	56–315	4–28	54–200	16–95	56–235	42–654	132–343
n = 17	av. 100	5	118	42	126	201	186

av. = average value, n = number of samples.

Ten further samples were taken during an additional sample series in March 1976 by W. KOLLMANN. Ninety-five percent of the samples were analyzed in the laboratory of the Institute of Applied Geology at the King Abdulaziz University in Jeddah using AAS and titration. Individual samples were analyzed for trace elements in the Geologische Bundesanstalt in Vienna.

Plate II shows the entire area from which samples were taken. The mean hydrochemical value was determined for each wadi (Table 9) and entered in the combined triangular diagram (Fig. 54).

Table 9. *Average values of the chemical composition of the groundwaters (in epm %)*

Catchment (wadis)	Na^+	K^+	Ca^{2+}	Mg^{2+}	Cl^-	SO_4^{2-}	HCO_3^-
Murwani 'Arran	43.7	1.0	29.1	26.2	51.9	18.5	29.6
Basin of Khulays	41.6	0.5	27.0	30.9	61.9	27.4	10.7
Usfan	78.1	0.2	10.5	11.2	71.1	23.2	5.7
Fayidah	66.4	0.3	17.7	15.6	61.7	22.0	16.3
Ash Shamiyah	52.1	0.2	17.5	30.2	56.1	43.7	n.m.
mid Fatimah	28.7	1.6	55.6	14.2	33.2	38.8	28.0
Na'man	32.9	0.9	41.0	25.2	31.8	41.0	27.3

The principal lithologies forming these catchments are Precambrian plutonic and metamorphic rocks – gabbro, diorite, quartz diorite, granodiorite, adamellite, granite, schists, amphibolites – and the intervening alluvial valley fillings. These Precambrian rocks are usually well exposed, but in the Khulays, Usfan and Fatimah catchments, the outcroppings of widely spaced blocks and patches are partly covered by Tertiary sandstones, siltstones, limestones, clays and basalts, Quaternary sands and gravels (s. chapter 2.3.1.).

While the metavolcanic rocks outcrop only sporadically in the northern part of the area under study, they do so more frequently in the south, especially in the Arafat Group. These rocks form the most prominent relief in that area.

Apart from the numerous vents, the basalts lie as a discordant cover on the Precambrian strata, or on the hard and soft Tertiary sediments. The surface area covered by basalts decreases from north to south in the area under study.

2.3.2.2. Groundwaters from the Khulays catchment

In the Khulays catchment area (Wadi Murwani, Wadi 'Arran, Khulays Basin), the hydrochemical values of the groundwaters in the tributaries differ significantly from the waters filling the basin.

Cations

Sodium and Potassium

While the average and maximum Na^+ values for the Wadis Murwani and 'Arran (middle) or Wadi Na'man are within normal limits, with 103 and 262 mg/l to 69 and 178 mg/l in Wadi Fatimah and 100 to 315 mg/l in Wadi Na'man, the waters from the Khulays Basin are among the more highly concentrated, with 390 and 845 mg/l. They are comparable to the concentrations in Wadi Fayidah. Although the K^+ value of the tributaries averages 5.5 mg/l, the average value for the basin waters is the highest of all the waters examined. These high K^+ values are presumably the result of the use of potash fertilizer in the relatively intensive cultivation of the oases in the lower course of the tributaries.

Calcium and Magnesium

There are also large differences in the concentrations of alkaline earths between the groundwaters of the tributaries and the Khulays Basin: Ca^{2+}: 59 resp. 220 mg/l; Mg^{2+}: 33 resp. 154 mg/l.

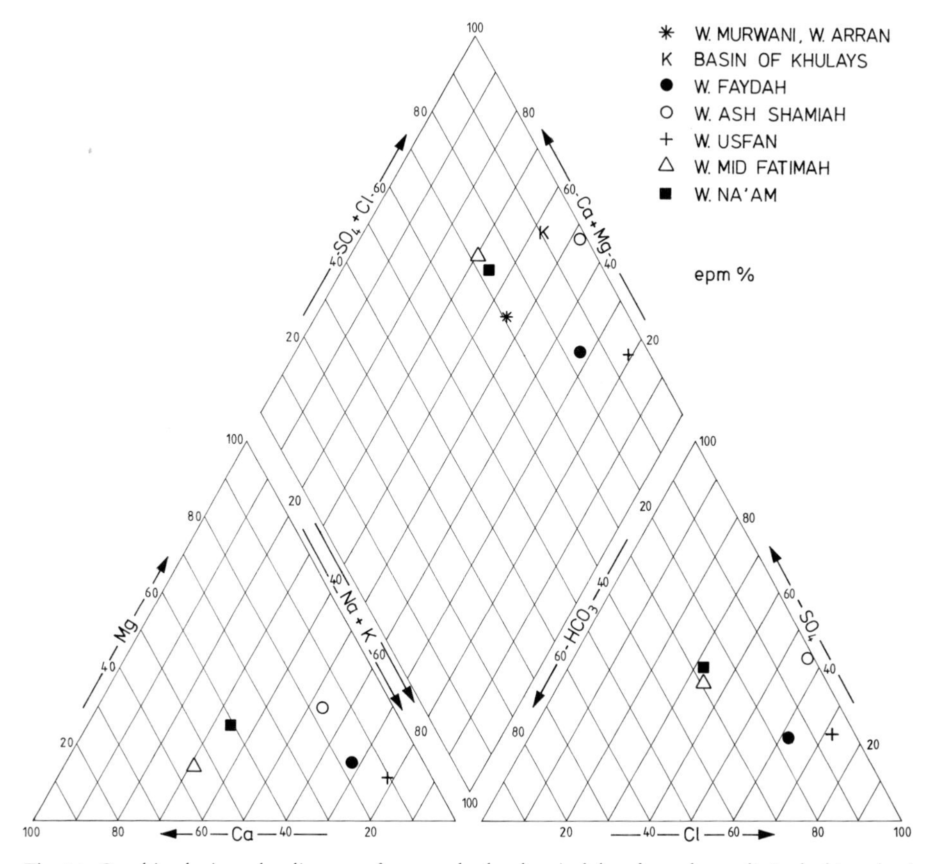

Fig. 54. Combined triangular diagram of average hydrochemical data from the wadis in the hinterland of Jeddah.

In spite of the difference in concentration, the Ca/Mg ratio is about the same: wadis 1.8; basin 1.4. This result, further the facts that the Na/Mg ratio changed to about the same extent in both groundwater types, and that the equivalent sum of Ca^{2+} + Mg^{2+} is greater than the HCO^- ion concentration, are indicative of possible ion-exchange processes.

Anions

Chloride and Sulfate

Both in the two tributaries and in the basin itself, the Cl^- and SO_4^{2-} values of the groundwaters can change considerably over short distances, e. g. the Cl^- from 131–683 mg/l resp. 683–1,897 mg/l, but this range of variation is not as great as, for example, in Wadi Usfan. The observation could be made in this catchment as well that the relatively higher values are found when there is a well with a periodically running pump far to the side of the valley channel and at the center of an intensively cultivated oasis. Evapotranspiration and fertilizing are not alone responsible for the increase in anion concentration; the cone of depression causes most of the irrigation infiltrates to flow back to the pumping well, and the changes in water level caused by the periodic pumping area seems to be also an important factor. This observation is confirmed by comparison of ion concentrations in waters from galleries within irrigation systems, and wells in the same area. A relevant example is provided by a well with a pump and a nearby gallery outlet in the middle course of the Murwani Valley:

	Ca^{2+}	Mg^{2+}	Na^+	K^+	Cl^-	SO_4^2	HCO_3^-	PO_4^{3-}	
Gallery:	42.9	19.9	81.6	5.4	135.3	74.3	190.3	30.7	(in mg/l)
Well:	162.2	132.7	263.3	10.8	683.6	269.0	237.2	76.7	(in mg/l)

The high Cl^- and SO_4^{2-} concentrations in the basin groundwater are not only the results of irrigation in the tributaries and the basin itself; they are certainly also a result of leaching of maritime deposits (Wadi Khulays is a young depression basin) such as evaporation pans or gypsum deposits.

An increase in strontium content could be correlated very significantly with an increase in the proportion of calcium-sulfate salts by solution of gypsum from wadi sabkhah sediments (S. S. AL-SAYARI and J. G. ZÖTL, 1978. K. WIRTH (1974) mentioned this fact in the course of trace-element studies in spring waters from gypsum-bearing Keuper, and interpreted this as a differentiated precipitation process during evaporation with enrichment by further ions, such as magnesium, sodium, potassium, lithium and rubidium. But the firm relationship found in the strontium concentrations could not be found with any other ion except lithium, owing to the additional influence of geological and human factors.

Table 10. *Sr^{2+}, Mg^{2+}, Na^{+}, K^{+} and Li^{+} concentrations corresponding to $CaSO_4$ salt content (further alkaline earths and alkali sulfates are also present) in samples collected 1976 (cf. Fig. 55)*

1976	$CaSO_4$	Sr^{2+}	Mg^{2+}	Na^{+}	K^{+}	Li^{+}
		milliequivalents per kg				
1	21.5	$51 \cdot 10^{-3}$	3.5	37.1	0.25	$5.5 \cdot 10^{-3}$
2	2.2	$3 \cdot 10^{-3}$	3.1	7.1	0.08	$0.3 \cdot 10^{-3}$
3	2.8	$7 \cdot 10^{-3}$	2.1	3.7	0.16	$0.6 \cdot 10^{-3}$
4	2.5	$5 \cdot 10^{-3}$	1.6	3.3	0.14	$0.6 \cdot 10^{-3}$
5	2.5	$10 \cdot 10^{-3}$	2.0	4.2	0.17	$0.6 \cdot 10^{-3}$
6	3.7	$6 \cdot 10^{-3}$	2.1	4.2	0.16	$0.6 \cdot 10^{-3}$
7	5.5	$14 \cdot 10^{-3}$	4.8	24.8	0.26	$2.0 \cdot 10^{-3}$
8	4.3	$15 \cdot 10^{-3}$	2.7	6.3	0.11	$0.6 \cdot 10^{-3}$
9	4.1	$9 \cdot 10^{-3}$	2.9	9.4	0.20	$0.9 \cdot 10^{-3}$
10	21.5	$66 \cdot 10^{-3}$	5.3	39.6	0.65	$4.9 \cdot 10^{-3}$
Correlation coefficient (r)		+0.98	+0.66	+0.93	+0.77	+0.98

Bicarbonate (carbonate hardness)

The HCO_3^- values are highest in comparison to those from the southern wadis with 194 mg/l (tributaries) and 244 mg/l (basin), corresponding to a carbonate hardness of 9.1 and 11.5° dH, respectively. This is probably due to the large percentage of surface area covered by basalt as compared to the neighboring southern region.

2.3.2.3. Groundwaters from the Usfan catchment

Cations

Sodium and potasium

The average and maximum Na^{+} concentrations are higher here than in the wadis discussed above (Wadis Fatimah, Na'man and Murwani/Khulays): Usfan 608 and 1,189 mg/l; Fayidah 452 and 872 mg/l; Ash Shamiyah 600 and 1,372 mg/l; Ghulah 740 and 1,335 mg/l. The K^{+} concentrations, however, are lower throughout the Usfan Basin than in the Khulays catchment.

Calcium and magnesium

The Ca^{2+} and Mg^{2+} concentrations differ in the groundwaters of the tributaries of this catchment (Table 8). In the Fayidah and Usfan areas the Ca^{2+} values are lower than in upper Fatimah and Na'man, while the Mg^{2+} values in the former are slightly higher. The Ash Shamiyah and Ghulah waters, however, are remarkably high in Ca^{2+} (average 176 and 194 mg/l) and much higher in Mg^{2+} (average 184 and 135 mg/l). This difference in Ca^{2+} and Mg^{2+} concentrations is obvious in the Ca/Mg-ratio: Usfan 0.9, Fayidah 1.1 and Ash Shamiyah 0.6, as compared to Fatimah 3.9 and Na'man 1.6. The comparison of cation concentrations (especially

Na^+ and Mg^{2+}) in the Wadi Fayidah and Wadi Ash Shamiyah waters strongly suggests an ion exchange, based on the increase of Mg^{2+} in contrast to Na^+. This is possible because shales of the Shumaysi Formation outcrop only in Wadi Ash Shamiyah. Its bedrock components and weathering products in the aquifer are most suitable for absorbing Na^+ in exchange for Mg^{2+}.

Generally, when the positive base-exchange index is negated by negative values, it is most often a matter of exchange of alkaline earths in the water for alkalis in the aquifers. With the exception of the samples from the Khulays Basin and Wadi Ghulah, all the samples tested show an equivalent sum with ($Na^+ + K^+$) $> Cl^-$, suggesting that the remaining alkali sulfates and hydrobicarbonates, respectively, are due to ion-exchange processes, or man-made pollution.

Anions

Chloride and sulfate

These ions vary considerably in concentration from one well to another in this catchment, especially in the downstream portions. As in all other wadis, the concentration depends on the relative position of the well with respect to the valley channel. The range of Cl^- in Usfan is 49–2,074 mg/l (average 436 mg/l), and of SO_4^{2-} 32–672 mg/l (average 193 mg/l), and differs only slightly from that of Wadi Fayidah (Cl^- 85–1,999 mg/l, average 439 mg/l; SO_4^{2-} 65–777 mg/l, average 212 mg/l), but differs greatly from Wadi Ash Shamiyah (Cl^- 92–1,396 mg/l, average 671 mg/l; SO_4^{2-} 26–1,980 mg/l, average 708 mg/l). Otherwise, the Ca^{2+} and Mg^{2+} concentrations in these three wadis show the same uniformity on the one side and deviation on the other (Table 8).

The high Cl^- content in the groundwater of Wadi Ghulah (649–2 808 mg/l, average 1,686 mg/l) is very probably the result of intensive irrigation, with the inflow of highly saline water from wadisabkhah upstream combined with low flow velocity and high evapotranspiration. The SO_4^{2-}-content ranges from 140 to 569 mg/l, average 334 mg/l, and is thus intermediate between that of Ash Shamiyah and Usfan. This is logical because Wadi Ghulah is the outlet for these two tributaries.

Bicarbonate (carbonate hardness)

The HCO_3^- concentrations again display two distinct groupings. One group includes Wadi Fayidah and Wadi Usfan (132–314 mg/l, average 199 mg/l; and 3–176 mg/l, average 60 mg/l), and the other Wadi Ash Shamiyah and Wadi Ghulah (3–8 mg/l, averages of 3 and 5 mg/l).

2.3.2.4. Groundwaters from the catchments of mid-Wadi Fatimah and Wadi Na'man

Cations

Sodium and potassium

The concentrations of Na^+ and K^+ are relatively low. The Na^+ content varies in the upper Wadi Fatimah from 30–178 mg/l, and in Wadi Na'man from

56–315 mg/l. The K^+ contents range from 3–14 mg/l and from 4–10 mg/l respectively.

Calcium and magnesium

The Ca^{2+} content varies from 83–264 mg/l in the upper Wadi Fatimah, and from 54–182 mg/l in Wadi Na'man. Mg^{2+} ranges from 0.2–25 mg/l and from

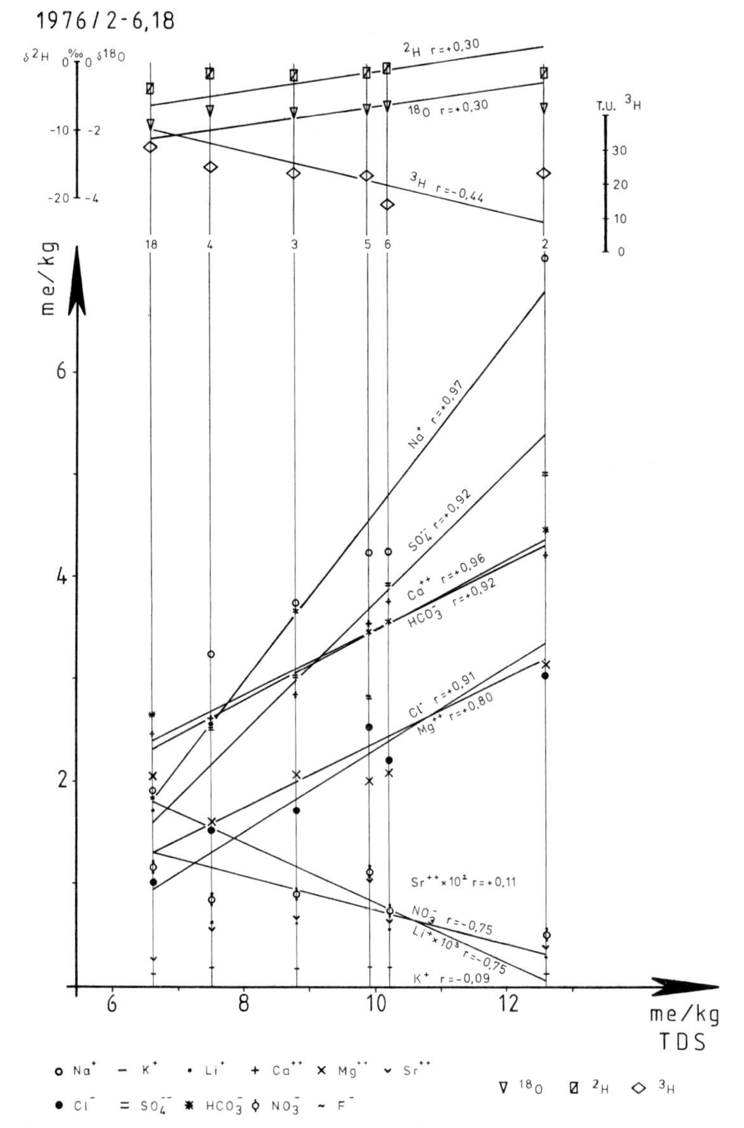

Fig. 55. Chemical composition and isotope content of the groundwater in the upstream part of the Wadi Fatimah (1976/3–6) and the possible mixing components between the waters from mainly Quaternary deposits (Type 1976/2) and joint-waters from basement granite-gneiss-catchment area (Type 1976/18).

16–95 mg/l, respectively. The average Ca/Mg ratio (in eq) for Wadi Fatimah is 3.9, and that of Wadi Na'man is 1.6. This low Ca/Mg-ratio may be attributed to mica containing Mg^{2+} in the storage unit.

Anions

Chloride and sulfate

In the upper Wadi Fatimah, the Cl^- values are between 63 and 1,118 mg/l, with SO_4^{2-} between 48 and 627 mg/l. The range in Wadi Na'man is 56–235 mg/l for Cl^- and 52–654 mg/l for SO_4^{2-}.

The chemical composition and isotope content of the groundwater in the upstream part of Wadi Fatimah is shown in Fig. 55.

Bicarbonate (carbonate hardness)

Bicarbonate is present in these two wadis in a relatively high concentration. The respective averages for the two wadis are 192 and 186 mg/l, corresponding to carbonate hardness of 9° and 8.8° dH.

2.3.2.5. Groundwaters within Jeddah

The following list includes the maximum and minimum values of the most important cations and anions in groundwater samples from boreholes within the Jeddah city limits. The number of samples, unfortunately, is not known.

Table 11. *Samples from boreholes within the Jeddah city limits (from various private reports)*

	in mg/l	
	min.	max.
Ca^{2+}	300	10,400
Mg^{2+}	400	6,000
Na^+	32,000	180,000
K^+	100	700
SO_4^{2-}	51	3,708
Cl^-	212	17,809
HCO_3^-	55	260

These values indicate that some of these waters are highly concentrated NaCl-waters (Red Sea water: Ca^{2+} 411, Mg^{2+} 1,290, Na^+ 10,800, K^+ 392, Cl^- 19,400 mg/l; Ca/Mg = 0.19) owing mainly to the proximity to the coast (marine intrusion) and the geochemical composition of the sediments (evaporation pans). The potassium content, even in its minimum value, is very high; this is indicative of human contamination, which could often be observed by eye at open excavations.

2.3.2.6. Conclusions

The total concentration of dissolved ionized material in the groundwater samples in the area studied ranges between 300 and 7,500 mg/l. Waters with a total concentration above 1,000 mg/l can be classified as significantly mineralised. The waters are mainly in the neutral pH range of 6.8–7.4.

The mineral content of the groundwaters studied depends mainly on the mineral composition of the rocks and sediments in the catchment area. Evaporation processes favored by natural conditions, e. g. more limited discharge on the sides of the wadis, or by artificial influences, e. g. irrigation, are responsible for an increase of mineralization, especially as regards Na^+ and Cl^-. Salinity always increases toward the sides of the valley. Disregarding human influence, hydrochemical features can therefore essentially be attributed to the differences in the underlying rocks of each catchment and intensity of evaporation processes.

There are prominent differences between the chemistries of the groundwaters from the Na'man and Fatimah catchments as compared to those from Khulays and Usfan.

NaCl-waters may originate from leaching sediments with diffuse salt storage (from evapotranspiration intensified by irrigation) or massive evaporites. As a rule, these waters contain more Cl^- than Na^+ ions (as in the Wadi Ghulah or the Khulays Basin), and so differ from the Cl^--containing waters of the bedrocks. This is exemplified by the waters from Wadi Usfan and Wadi Fayidah.

Waters containing sulfate very often come from gypsum deposits, although bacterial oxidation of organic material in tidal mud flats can also be a source. Water that has leached gypsum always contains an abundance of alkaline earths (Table 8), especially Ca^{2+} (mostly exemplified by the waters from Wadi Fatimah and Wadi Na'man). Sulfates in crystalline bedrock mainly come from oxidation of sulfide ores, through bacterial reduction from sulfates.

Among the cations (Table 8), sodium ions predominate in the Usfan (about 66 epm%) and Khulays catchments (about 42 epm%). Chloride is the main anion with about 63 epm% in the Usfan catchment, 57 epm% in the Khulays catchment, and 32 epm% in the upper Wadi Fatimah and Wadi Na'man. Sulfate is present with about 29 and 39 epm% respectively.

The chemical composition of the water samples in the area studied leads to a division which agrees closely with the specific geology of the catchment:

1. **Khulays catchment:**
   Wadi Murwani, Wadi 'Arran: $\underline{Na}$ – Ca – Mg – $\underline{Cl}$ – HCO_3 waters
   Khulays Basin: $\underline{Na}$ – Mg – Ca – $\underline{Cl}$ – SO_4 waters

   In this catchment the groundwater is dominated by Na^+ and Cl^-. Alkaline earths are represented with over 20 eq-%.

2. **Usfan catchment:**
   Wadi Fayidah: $\underline{Na}$ – $\underline{Cl}$ – SO_4 waters
   Wadi Ash Shamiyah: $\underline{Na}$ – Mg – $\underline{Cl}$ – SO_4 waters
   Usfan area: $\underline{Na}$ – $\underline{Cl}$ – SO_4 waters

Here, the elements Na^{+}, Mg^{2+}, and Cl^{-} dominate, followed by SO_4^{2-}. The increase in Mg^{2+} concentration may not be related to an ion exchange, but rather to the somewhat different geological situation in the catchment.

3. Mid Wadi Fatimah catchment: Ca – Na – SO_4 – Cl – HCO_3 waters
Na'man catchment: Ca – Na – Mg – SO_4 – Cl – HCO_3 waters

Here, Ca^{2+} and SO_4^{2-} dominate, followed by Cl^{-}, Na^{+}, HCO_3^{-} and Mg^{2+} (>20%). Lenses of gypsum in the wadi filling determine the chemistry of these groundwaters.

Aluminum and zinc were demonstrable only as very scant traces in some samples. The concentration range for Al^{3+} was from $< 1 \cdot 10^{-3}$ to a maximum of $9 \cdot 10^{-3}$ mg/l (or 0.1 to 1.0 meq/kg); for Zn^{2+} the concentration was $< 1 \cdot 10^{-3}$ to $40 \cdot 10^{-3}$ mg/l (or <0.03 to 1.2 meq/kg). A relationship to total solution content or to the hydrogeological circumstances in the catchment area (proximity to the basement, sand-gravel body of the wadi filling or wadi-sabkha development) cannot be determined from the samples available.

The fluoride anion attained higher concentrations; it appears to have a low correlation with phosphate-nitrate content (probability approximately 90–99%), suggesting that fluorine is mainly present owing to pollution (Table 12).

Table 12. *Fluoride and phosphate contents*

1976	F^{-}	PO_4^{3-}
1	1.05	1.1
2	0.35	0.7
3	0.35	0.4
4	0.39	0.6
5	0.52	0.9
6	0.48	0.4
7	0.54	0.6
8	0.24	0.3
9	0.58	0.4
10	0.66	0.6
Correlation coefficient r_F:		+0.71

2.3.3. Isotopic Composition of Water Samples

(P. Hacker, H. Moser, W. Stichler, W. Rauert, J. G. Zötl)

2.3.3.1. General discussion

In the course of the field studies made in 1976 in the vicinity of Jeddah, water samples were taken from chosen wells in the wadis (Fig. 56 and Table 13, Nos. 1–10), and examined for their content of the environmental isotopes deuterium (^{2}H), oxygen-18 (^{18}O) and tritum (^{3}H)[1].

[1] Isotopic contents were measured at the GSF-Institut für Radiohydrometrie, Neuherberg/Munich, FRG. For details on measurements and interpretation of environmental isotope contents see S. S. Al-Sayari and J. G. Zötl (1978), and H. Moser and W. Rauert (1980).

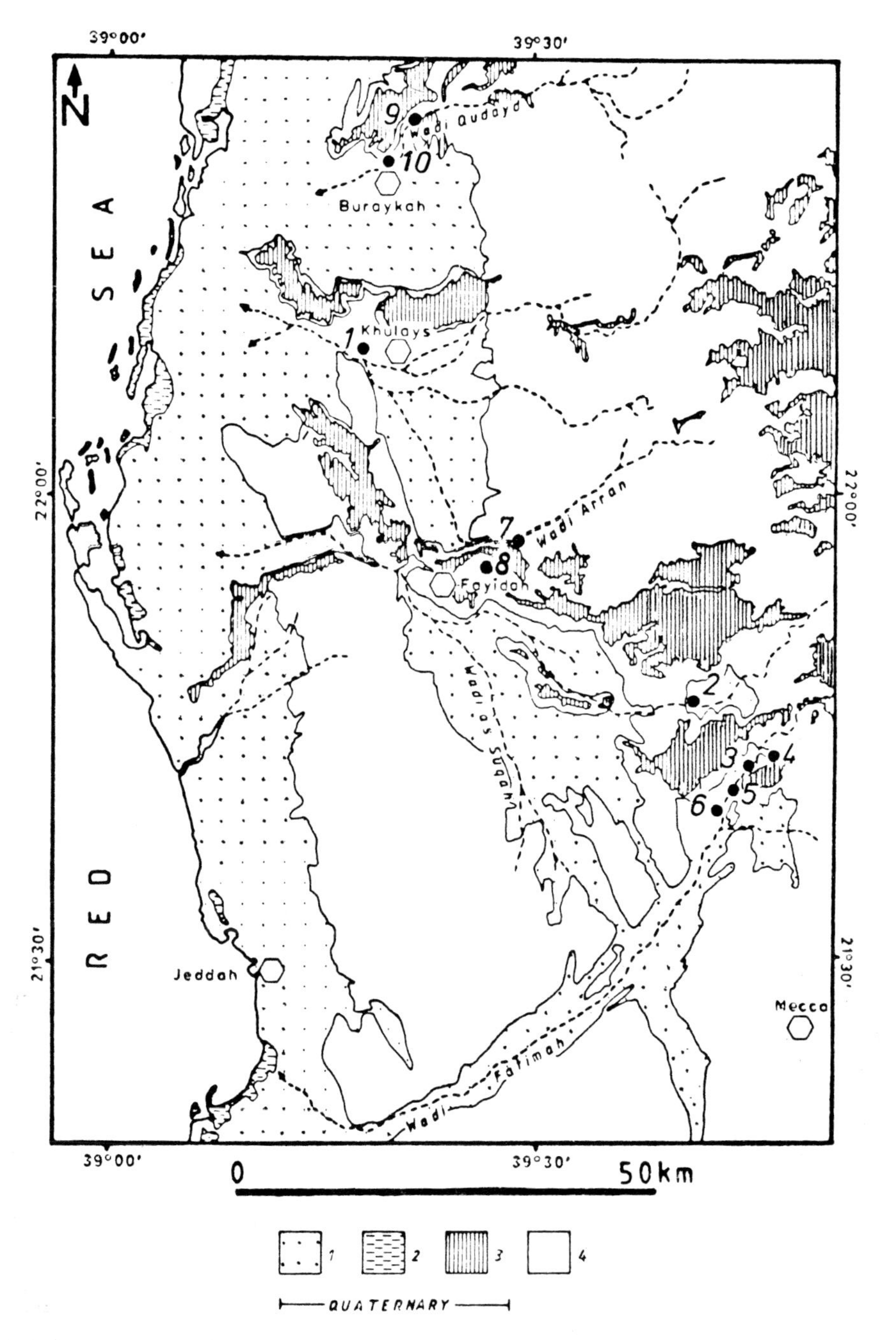

Fig. 56. Location map of water samples for isotopic measurements; sampling March 1976 (drawn by W. KOLLMANN). Table 13, Nos. 1–10 *1* mainly sands and gravels; *2* sabkhahs; *3* Young Tertiary basalt; *4* mainly Precambrian rocks.

Table 13 shows the results of isotope analysis for samples taken in March 1976. For the purpose of comparison, results from Al Madinah area are listed in addition to the results from the Jeddah area (Nos. 1–13). Fig. 57 shows the δ^2H–$\delta^{18}O$ relationship for the sample series 1976 from the Jeddah area. In all, the values are scattered around the line $\delta^2H = 8\ \delta^{18}O + 10$, whereby the values are rather narrowly limited to the area $-5‰ < \delta^2H < 0‰$, or $-1.8‰ < \delta^{18}O < -1.2‰$ for

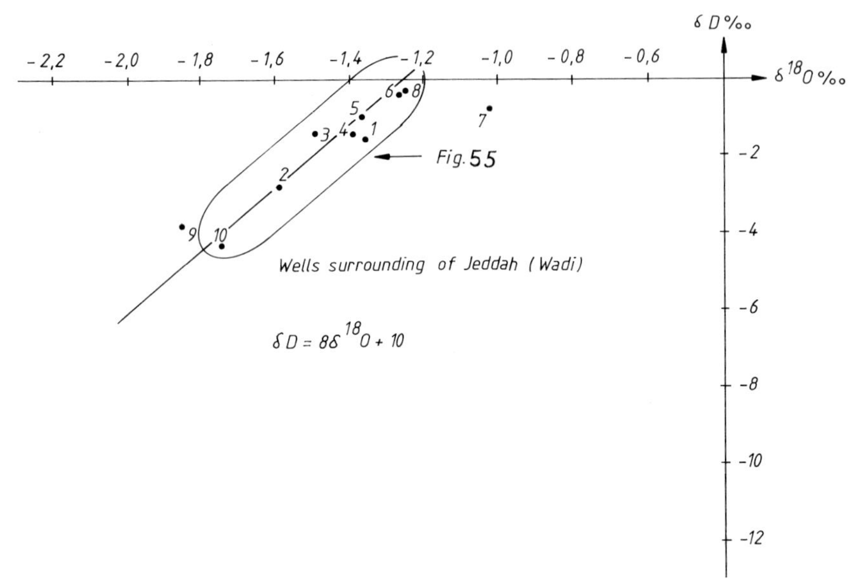

Fig. 57. Environmental isotopes δ^2H-$\delta^{18}O$ relationship for 10 samples of field reconnaissance 1976 (instead of Fig. 55 read 56).

their ^2H and ^{18}O contents. The samples from the country around Al Madinah (Table 13, Nos. 14–20) show very scattered isotope contents, sometimes with δ values considerably lower than those for the waters from the Jeddah area. Pronounced deviations from the precipitation lines were seen in samples 7, 12 and 17 are surely due to the effect of evaporation. Nos. 11–13 have been collected two years later.

The variation in isotope contents in the waters from the Jeddah area is probably due to the mixing of waters with different underground residence times. This is indicated by the relatively high and variable ^3H contents (13.8–25.7 TU), which in turn suggest a relatively large proportion of rainwater from recent decades. Precipitation in this area occurs as more or less excessive showers which rapidly soak into the underground.

In contrast, for the waters from Al Madinah area (Table 13, Nos. 14–20), the relatively low and highly variable ^2H and ^{18}O contents, must be attributed to different catchment areas at higher altitudes (isotopic altitude effect). As the excessive precipitation occurs mainly in the winter and early spring. there are probably, generally speaking, no additional seasonal variations in the isotope content of the precipitation superimposed upon this altitudinal effect, and thus

Table 13. *Environmental isotopes 1976 in deep wells of Jeddah area (Nos. 1–13, Fig. 55) and Al Madinah region (Nos. 14–20)*

Name	No.	δ^2H (‰)	$\delta^{18}O$ (‰)	TU
Khulays	1	−1.8	−1.36	16.4±1.1
Haddat Ash Sham	2	−2.4	−1.59	25.7±2.1
ENE Jeddah	3	−1.4	−1.49	23.7±1.7
ENE Jeddah	4	−1.4	−1.38	25.2±1.9
ENE Jeddah	5	−1.0	−1.37	22.9±1.9
ENE Jeddah	6	−0.2	−1.27	13.9±1.0
Wadi 'Arran	7	−0.8	−1.13	24.1±2.0
E Usfan	8	−0.3	−1.25	13.8±1.0
Buraykah	9	−3.8	−1.86	23.2±1.7
Buraykah	10	−4.3	−1.75	19.5±1.6
E Mecca	11	−3.6	−1.84	32.3±2.4
E Mecca	12	−3.1	−1.35	27.5±2.0
E Mecca	13	−2.8	−1.68	33.7±2.4
Madinah	14	−8.6	−2.29	1.2±0.4
Hulayfah	15	−6.5	−2.48	41.8±2.9
NE Al Madinah	16	−5.2	−1.80	25.1±1.7
NNW Al Madinah	17	−12.2	−2.35	3.2±0.6
NW Al Madinah	18	−8.1	−2.17	1.2±0.4
S Al Madinah	19	−7.7	−2.60	5.8±0.8
S Al Madinah	20	−8.2	−2.28	8.1±0.9

obscuring it. The influence of an isotopic mass effect with the highly variable precipitation intensity has, however, not as yet been determined.

Further isotope analyses are available for samples taken by P. HACKER in 1977 and 1978 from groundwaters in the Jeddah area (Table 14; sample location see numbers on Plate II, see insertion at back cover), and from surface waters and local rain showers in Jeddah and At Ta'if (Table 15). Compared with the results of the sample series taken in 1976 from groundwaters in the same area, these values show considerably greater variation ($0.0‰ > \delta^2H > -10.2‰$, or $-1.1‰ > \delta^{18}O > -2.6‰$), but most of them are located around the line $\delta^2H = 8\ \delta^{18}O + 10$. Here as well, the explanation for the variation is also to be sought in the different components of rapidly percolating rainwater, as is also indicated by the correspondingly greater variation in ^3H content (1.1–20.6 TU) as compared to the 1976 sample series. In March 1972 one of the authors experienced a severe hailstorm in this area; the ground was covered with ice, and then there was a raging surface runoff in the otherwise dry wadis. This experience makes it easier to understand the highly variable components of fast runoffs in the shallow groundwater manifested in the isotope contents (see Table 14). The fact that these variations were less pronounced in the sample series taken in 1976 is probably due to the fact that considerably deeper wells were chosen for that series, and that the influence of short-term infiltration processes was thus evened out over a longer period of time.

Table 14. *Content of environmental isotopes of water samples of the surroundings of Jeddah collected May 1978 (see Plate II, see insertion at back cover)*

No.	Location	δ^2H (‰)	$\delta^{18}O$ (‰)	^3H (TU)
38	Usfan	−4.5	−2.23	17.0±1.4
60	Wadi Ash Shamiyah	−0.8	−1.11	2.8±0.6
65	Wadi Ash Shamiyah	−1.2	−1.15	1.1±0.6
27	Wadi Fayidah	−7.1	−2.21	10.6±1.0
68	Wadi Ghulah	−10.2	−2.60	11.2±1.1
67	Wadi Ghulah	−8.4	−2.08	10.3±1.0
69	Wadi Ghulah	−5.0	−1.90	8.8±0.9
70	Wadi Fatimah (wells)	0.0	−1.40	1.4±0.6
71	Wadi Fatimah (gallery)	−2.3	−1.44	20.6±1.7
72	Wadi Na'man	−2.4	−1.65	17.3±1.4
73	Wadi Na'man (one of the wells for Mecca)	+1.6	−1.09	14.7±1.2
61	Wadi As Suqah (surface water)	+55.7	+11.95	13.0±1.1

Table 15. *Environmental isotopes in precipitation events and in one sea water sample*

Location	Date	δ^2H (‰)	$\delta^{18}O$ (‰)	^3H (TU)
Precipitation Jeddah	December 13, 1977	+3.6	−1.37	6.8 ± 0.7
Precipitation Jeddah	February 02, 1978	−16.6	−3.46	7.3 ± 0.8
Precipitation At Ta'if	April 19, 1978	−4.9	−2.34	17.8 ± 1.3
Sea Water Jeddah, depth 20 m	April 23, 1978	+6.0	+1.55	
Precipitation IAEA Jeddah	1970	—		19
	1971	—	—	15

It must also be taken into consideration that the samples were taken from different aquifers. This is seen, for example, in the groundwaters of Usfan, Wadi Fatimah and Wadi Na'man catchment areas; these waters can be divided into two groups according to their ^3H content, and these two groups come from different aquifers. The groundwaters with ^3H contents between 10 and 20 TU flow in the upper, mostly clastic layers, while the waters with ^3H contents between 1 and 3 TU are found in lower groundwater levels which are covered by a layer of finely clastic clayey materials of different thickness. This material probably comes from a period of lesser average precipitation or precipitation intensity than prevails today; its thickness in the individual wadis ranges to tens of meters and is related to the thickness of the entire wadi filling or the size of the wadi channel. This division by aquifer may sometimes also be seen in the δ^2H and $\delta^{18}O$ values. The groundwaters with low ^3H contents (1–3 TU) thus usually have higher δ values than those of the group with ^3H contents between 10 and 20 TU.

Finally, it should be noted that these samples were taken at different times (different years and seasons), and rapid infiltration of local precipitation with varying isotope content can result in changes in the δ value of the groundwater.

For example, samples Nos. 27, 61 and 68 showed differences of approximately 5 δ^2H ‰ as compared to samples from May 1978 and April 1977, while other groundwaters show very constant isotope contents over a number of years (see also Table 16).

Fig. 58 shows the $\delta^2H - \delta^{18}O$ relationship of the sampled rainwaters, the groundwaters from different depths, and the surface waters (see Table 16). The deviation from the line $\delta^2H = 8\ \delta^{18}O + 10$ clearly shows that the groundwaters of

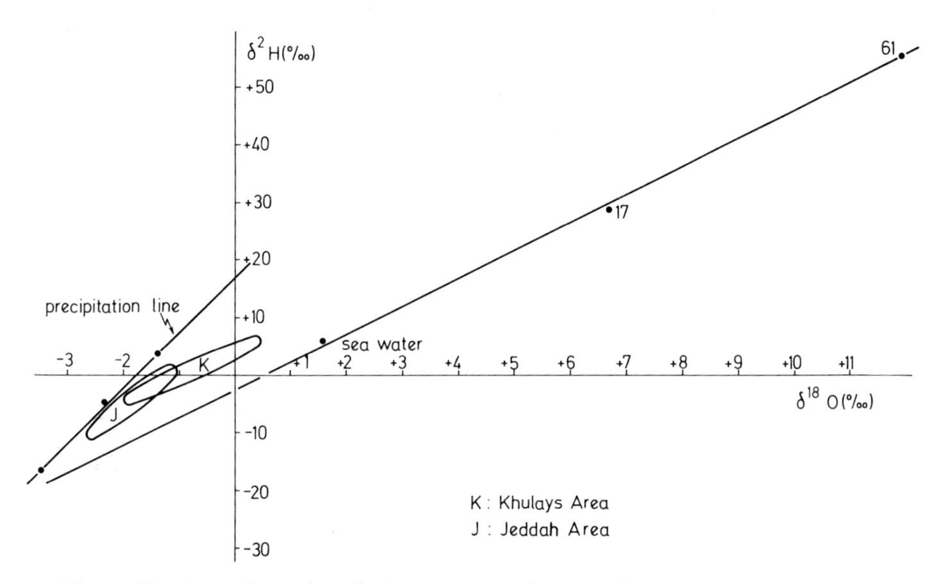

Fig. 58. δ^2H-$\delta^{18}O$ relationship of rain water, groundwater and sea water 1977 (P. HACKER).

the individual wadis were exposed to different degrees of evaporation as compared to the rainwaters. The groundwaters from the Wadis Na'man, Fatimah and 'Arran, and from Usfan with a value of 8 for the slope of their $\delta^2H - \delta^{18}O$ relation thus differ distinctly from the groundwaters from Wadi Murwani with a slope of 5. Extremely high evaporation effects may be seen in the δ values of surface waters (small artificial groundwater exposures in the Khulays Basin and in the lower course of the Wadi Ash Shamiyah, where the slopes in the $\delta^2H - \delta^{18}O$ relation have decreased to a value of 2. Otherwise, the seawater sample fulfills the $\delta^2H - \delta^{18}O$ relation of the surface waters.

2.3.3.2. Special studies in the Khulays catchment area

Table 16 summarizes the ^2H and ^{18}O contents of waters sampled repeatedly between 1976 and 1978 along the Wadi Murwani and Wadi 'Arran down to the Khulays Basin (see Plate II). The $\delta^2H - \delta^{18}O$ relation of these values is shown in Figure 58.

Figure 59 shows, first, that the waters of the Wadi 'Arran, a small tributary to the Khulays Basin, vary distinctly in their δ values from those of the Wadi

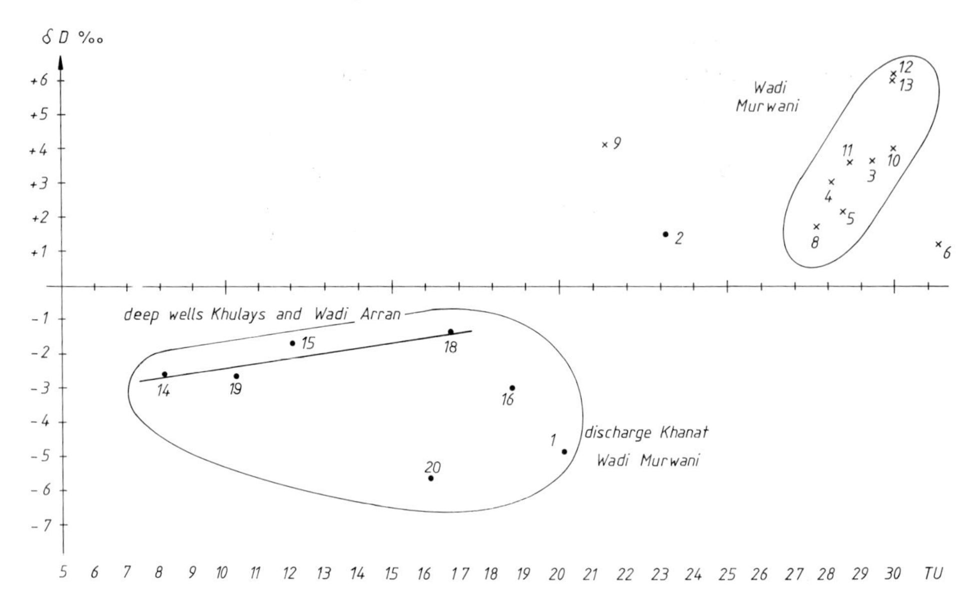

Fig. 59. δ^2H-$\delta^{18}O$ relationship of groundwater samples from Wadi Murwani and basin of Khulays (sample location see Plate II).

Fig. 60. Location No. 1 Wadi Murwani (Plate II). Outflow of the distroyed kanat continuously enriching the wadi groundwater. Evaporation influences strongly the isotope composition of the groundwater downstream. (Photo: J. G. ZÖTL, 1972.)

Murwani; this must be due to a different catchment area, together with the local nature of the precipitation events mentioned previously. In this context it must be noted that the deep wells in the Khulays Basin resemble those of the groundwaters in Wadi 'Arran in their δ values, and not those of Wadi Murwani except sampling location No. 1. This can mean that the groundwater in the Khulays Basin is supplied for the most part by influx from the Wadi 'Arran which, unlike the Wadi Murwani, has not as yet been extensively exploited for irrigation wells.

In Wadi Murwani, the sampling location No. 1 is the upper water outlet (flow approximately 50 l/sec) from a terrace slope, so that the δ values from this sampling location should represent those of the groundwater unaffected by evaporation. The water flows some 500 meters before it soaks back into the ground (Figs. 60 and 61). The water is then enriched in ^{2}H and ^{18}O owing to evaporation, but farther down-stream the δ values again decrease owing to admixture of groundwater which is not enriched by evaporation.

Groundwater fed to an oasis also seeps into the wadi floor through defective pipes. Here it forms a groundwater stream near the surface which also mixes in shallow wells with shallow groundwater the ^{2}H and ^{18}O contents of which are uninfluenced by evaporation; this is then seen as upwelling groundwater in the orographically irregular wadi floor.

The δ values of samples Nos. 12 and 13, taken from a relatively mineral-poor groundwater flow (560 μS, 80 l/sec) and some 2.5 kilometers apart in a flat wadi bed exposed to sunlight, surprisingly show a relatively slight evaporation effect. Despite the high temperature, conditions favorable to evaporation (air stagnation, high relative humidity) must not have prevailed here at the time the samples were taken.

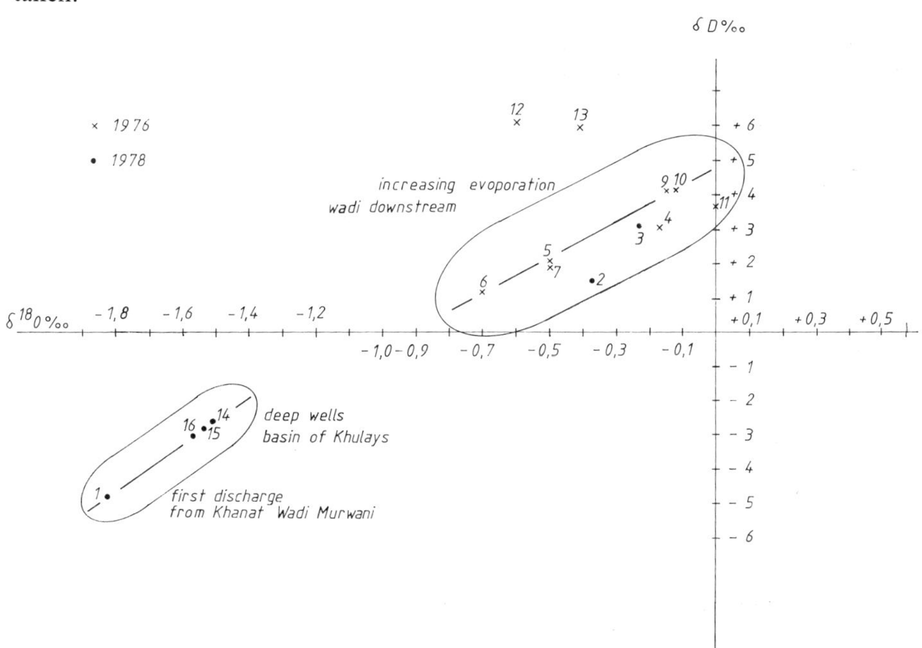

Fig. 61. δD-$\delta^{18}O$ relationship groundwater downstream of Wadi Murwani.

In the deep groundwaters, an evaporation effect is scarcely to be seen in the δ values, as they contain more groundwater from the upper course of the wadi and infiltrates from brief, severe floods, both of which have never been exposed to evaporation, than seepage water from near the surface which has been exposed to evaporation.

Table 17 also contains the corresponding ^3H values for the water samples. Here it is remarkable that the ^3H contents of the groundwaters in Wadi 'Arran and the Khulays Basin are roughly the same as the yearly averages in 1970/71 for ^3H in the rainfall in Jeddah, but the values for Wadi Murwani are considerably higher. Apparently rainwater with ^3H contents around 30 TU infiltrated this wadi during severe local floods in the winter of 1974/75. Similar ^3H values were seen at the same time in East African rainwater, and the same air masses could have reached the Jeddah area.

Table 16. *Stable isotope data of shallow groundwater and surface water of Wadi Murwani, Basin of Khulays and Wadi 'Arran (1976–1978)*

No.	Water type	02. 06. 1976		25. 09. 1976		04. 04. 1977		22. 05. 1978	
		δ^2H (‰)	$\delta^{18}O$ (‰)	δ^2H (‰)	$\delta^{18}O$ (‰)	δ^2H (‰)	$\delta^{18}O$ (‰)	δ^2H (‰)	$\delta^{18}O$ (‰)
	Wadi Murwani								
1	groundwater discharge	—	—	—	—	—	—	−4.8	−1.82
2	gallery (upper part)	—	—	—	—	—	—	+1.5	−0.37
3	gallery (lower part)	+3.7	−0.09	+3.3	−0.02	+3.6	+0.12	+3.1	−0.23
4	shallow well	+3.0	−0.18	+3.3	−0.06	+2.2	−0.22	—	—
5	shallow well	+2.1	−0.50	+1.6	−0.28	+1.8	−0.31	+2.7	−0.54
6	shallow well	+1.2	−0.70	+0.2	−0.67	−0.3	−0.67	—	—
7	groundwater discharge	+1.9	−0.50	+0.3	−0.48	—	—	—	—
8	groundwater discharge	—	—	—	—	+1.7	−0.18	—	—
9	shallow well	+4.1	−0.15	+2.6	−0.04	—	—	+1.6	−0.62
10	groundwater discharge	+4.1	−0.12	+2.5	−0.06	—	—	—	—
11	shallow well	+3.7	+0.02	+1.0	−0.54	+0.5	−0.68	+1.1	−0.84
12	surface runoff	+6.2	+0.60	—	—	—	—	—	—
13	surface runoff	+6.0	+0.41	—	—	—	—	—	—
	Basin of Khulays								
14	deeper well	—	—	—	—	—	—	−2.6	−1.51
15	deeper well	−1.7	−1.21	−1.5	−1.35	−3.9	−1.45	−2.7	−1.53
16	deeper well	—	—	—	—	—	—	−3.0	−1.57
17	surface waters	—	—	—	—	—	—	+29.0	+6.64
	Wadi 'Arran								
18	well (downstream)	—	—	—	—	—	—	−1.4	−1.21
19	well (midstream)	—	—	—	—	—	—	−2.7	−1.63
20	well (upstream)	—	—	—	—	—	—	−5.6	−1.89

The slight decrease in ^3H values along the Wadi Murwani can be explained not only by radioactive decay during the period of flow, but above all by mixing processes in the groundwater aquifer.

All in all, the relatively fast reaction of the isotope contents in the groundwaters, and in most cases the reaction as well of ^3H to precipitation events, suggest a short time of residence for these groundwaters, and thus a small storage volume below ground.

Table 17. *Tritium content (TU) of shallow groundwater and surface water of Wadi Murwani, Basin of Khulays and Wadi 'Arran (1976–1978)*

No. Plate II	Date of sampling 02. 06. 1976	25. 09. 1976	04. 04. 1977	22. 05. 1978	Remarks
			Wadi Murwani		
1	—	—	—	20.2±1.5	high values express fast influence of precipitation excess
2	—	—	—	23.3±1.8	
3	29.4±2.1	28.1±3.0	25.3±2.2	—	
4	28.1±2.1	28.9±2.3	26.2±2.7	—	
5	28.5±2.0	27.0±2.2	27.5±2.3	24.4±1.8	
6	31.3±2.0	29.0±4.8	26.6±2.9	—	
7	43.6±3.0	30.4±2.7	—	—	
8	—	—	27.7±3.1	—	
9	21.3±1.5	28.4±2.4	—	26.8±2.1	
10	30.0±2.1	28.3±2.3	—	—	
11	28.7±2.2	31.6±2.8	30.5±2.7	25.8±1.8	
12	30.3±2.1	—	—	—	
13	30.0±2.4	—	—	—	
			Basin of Khulays		
14	—	—	—	8.1 0.8	almost the same values for different horizons
15	12.1±1.0	12.0±1.4	11.5±1.6	12.1±1.2	
16	—	—	—	18.7±1.4	
17	—	—	—	11.5±1.0	
			Wadi 'Arran		
18	—	—	—	16.8±1.3	
19	—	—	—	10.4±1.1	
20	—	—	—	16.2±1.3	

2.4. South Tihamah and Farasan Islands

2.4.1 Geology of the Tihamat Asir[1]

(E. MÜLLER)

2.4.1.1. General morphologic and geologic features

In the southwestern part of Saudi Arabia, between the escarpment of the Asir Mountains in the east and the Red Sea coast in the west, the Plain of Jizan or

[1] Investigations financed by the Saudi Arabian Ministry of Agriculture and Water, Ar Ryadh.

Tihamat Asir (south) forms a narrow NW-trending coastal strip which is part of the almost 2,000 kilometers long marginal corridor following the border mountains of the Arabian Peninsula. Along a length of some 190 kilometers, this coastal plain is limited by an area of low hills rising from 100 to 200 meters on the average and reaching 550 meters at Jabal At Tirf. These hills, most of which are of volcanic origin, correspond to the eastern end of the Red Sea basin (Plate III, see insertion at back cover). They are dominated by a second parallel foothill band consisting of metamorphic schists and reaching elevations between 200 and 700 meters. The strongly dissected Asir Mountains with culminations of more than 2,000 meters are formed by greenstone and granite and overlook the whole area.

The Plain of Jizan itself which averages about 40 kilometers in width is entirely composed of alluvial deposits, and slopes down gently and regularly from 100 meters to the sea level. The only relief of the Jizan Plain is caused by four little volcanic cones near Sabya and Abu Arish in the east and the salt dome of Jizan in the west. To the north, the Jizan Plain is limited by metamorphic rocks of the basement complex and the huge basaltic eruptions of Al Birk Plateau. Southward, the Jizan Plain continues into Yemen to the southern end of the Red Sea.

Several important intermittent wadis transverse the Plain of Jizan. They are supplied by the summer and spring rain falling especially in the area of the Asir Mountains, and distribute their floods over the plain in numerous branches. Only the upper courses of the wadis in the mountainous area are perennial, forming narrow and deep valleys with locks in the volcanic zones.

The sediments brought by the wadis are very coarse-grained in the eastern part of the plain. They become thinner and very fine-grained in the vast deltas at the end of the wadis.

The boundary between the plain and the sea is formed by a typical sandy rectilinear shoreline which follows the structural lines of the Red Sea depression. The coast line is imprecise and variable because of tides and seasons. It forms periodically-flooded sabkhahs with salt and gypsum originating from evaporation.

The Tihamah belongs to the eastern margin of the Red Sea rift valley. Its evolution started in Oligocene and continued to recent geologic time. The geologic history of this area, however, began a long time before the development of the recent tectonic pattern with the deposition of Precambrian eugeosynclinal sedimentary and volcanic rocks. These series, now metamorphosed and heavily denudated, form the truncated upland of today's Arabian Nubian shield complex, and had been affected by several Precambrian orogenic and plutonic events before the cratonisation occurred. The Precambrian series are exposed all along the northeastern flank of the Red Sea.

Many volcanic intrusions, dyke swarms and flows which occurred during the development of the Red Sea rift are characteristic in the Jizan Plain. Several small volcanic cinder cones, which are still in a good state of preservation, confirm that volcanic activity has continued to recent geologic time.

The Jizan Plain has been the subject of several geological studies. Geologists of the U. S. Geological Survey have drawn a geological map on a scale of 1 : 500,000 using aerial photo interpretation. The map was published in 1959 by the Saudi Arabian Ministry of Finance and National Economy (G. F. Brown,

R. O. JACKSON, 1958 and 1959). It covers the whole Asir quadrangle (Map I-216 A and I-217 A of the official Index map of the Arabian Peninsula). Aeromagnetic surveys, performed in 1962 by the Directorate General of Mineral Resources, and a reflection seismic survey, executed by Petromin in 1963 in the western coastal area of Tihamat Asir, have followed. In 1966, detailed seismic studies were performed in the Jizan Plain by Auxerap to determine the location of the oil exploratory well of Mansiyah I. The well is about 40 kilometers north of Jizan and is 3,932 meters deep.

Some information about the Jizan Plain has been published in generalized reports of the tectonic and sedimentological aspects of the Red Sea Valley by M. GILLMAN, et al. (1966), G. F. BROWN (1970), F. K. KABBANI (1970), L. DUBERTRET (1970), J. D. LOWELL, and D. J. GENIK (1972), and others. In the frame of a development program, detailed geologic and hydrogeologic studies in the area of Wadi Jizan were carried out by ITALCONSULT in 1965. Another report on these investigations was published by V. COTECCHIA, et al. (1970). However, one of the most important reports on this study was published by M. GILLMAN in 1968 and titled "Primary results of a geological and geophysical reconnaissance of the Jizan coastal plain in Saudi Arabia." Many of his results will be reflected in the following geologic report.

The geology of the Jizan Plain is shown in the geologic map (Plate III, see insertion at back cover). This map was drawn up on the basis of interpretation of aerial photographs on a scale of 1 : 30,000, as well as through field investigation. At the beginning, it should be mentioned, that the geologic investigations of this study have been carried out within the limits posed by the frame of groundwater exploration. Thus, as also mentioned above, there has been no need to go into more detailed investigation.

2.4.1.2. Stratigraphy

Basement complex (Precambrian)

The basement of the area consists of metamorphics and mainly acid plutonites. According to the official map (G. F. BROWN and R. J. JACKSON, 1958), these are all pre-Permian. However, in more recent publications (R. G. COLEMAN et al., 1972, J. D. LOWELL et al., 1973), these rocks have been described as of Precambrian age.

The basement rocks outcrop all over the eastern margin of the area and form the hilly pediment of the Asir Mountains. In the northern part of the Jizan Plain there are exclusively highly altered metamorphic schists. The most common lithotypes are microcrystalline sericite schists and coarse-grained quartz schists. Chlorite schists and phyllites occur less commonly. Farther east, these series pass into greenish colored amphibolitic greenstones (an area which has not been mapped during this study).

The sericite schists have a finely schistose and crenulated texture with alternating thin bands of phyllosilicates and very fine-grained feldspar-quartz-associations. The lithological character of this strata which contain albite, chlorite, sericite and epidote indicates a low metamorphic grade (Epizone) everywhere.

In the southern part of the Jizan Plain (south of Wadi Jizan) the basement consists of granites and granodiorites. These are slightly metamorphosed and highly altered. Sometimes quartz monzonites and quartz diorites are associated. The rock is intersected by numerous pegmatitic veins of quartz and orthoclase.

Khums Formation (Early Jurassic)

In the Jizan Plain a narrow faulted strip of various sandstones, stretching NW–SE, crops out along the western margin of the basement complex. The strata are described as of Jurassic age and superpose the basement with angular unconformity. This so-called Khums Formation is composed of a dark tan quartzitic sandstone with beds of red and grey-green shale and siltstone. The average dip is about 30–40° W.

Three different series have been determined by the USGS and by Auxerap. The type section at Wadi Khums has been described in detail by M. GILLMAN (1968).

Unit 1 "Lower Sandstone" – 480 meters

This has medium- to coarse-grained, heterogeneous, grey sandstone with cross-bedding, rounded pebbles of basement material, and very coarse conglomeratic material at the base. The cement is argillaceous and siliceous near the bottom, and feldspathic near the top.

Unit 2 – 110 meters

a) fine- to medium-grained, feldspathic sandstone with phosphatic inclusions;

b) argillaceous chlorite or siliceous silt with calcareous inclusions;

c) fine-grained, quartzitic sandstone;

d) conglomerate with well-rounded, metamorphic pebbles and argillaceous chloritic cement; and

e) intercalations of violet, ferruginous, silicified shale. Volcanic material is abundant throughout the unit.

Unit 3 "Upper Sandstone" – 290 meters

Fine-grained, well-bedded, siliceous sandstone (rarely argillaceous) with ferricrusts in the top layers.

The thickness of the Khums Formation is greatly variable. From 880 meters at the type section in Wadi Khums in the south, to only 10 meters in Wadi Jizan to the north. The thickness gradually increases to 100 meters near Qadhab and to 200 meters in an isolated outcrop north of Gaim.

The outcrop north of Gaim is composed of medium- to coarse-grained, heterogenous sandstone with angular grains and kaolinitic cement. The top 20 meters of the outcrop is ferruginous.

Hanifa Formation (Late Jurassic)

In the southern part of the Jizan Plain, between Wadi Fija and the Yemen border, the Khums sandstones are partly overlain by massive layers of limestone which are named "Hanifa Formation" (USGS terminology). The rocks are blue-grey, silicified and brecciated, fossiliferous limestone and are considered to be part of upper Jurassic. The series has a maximum thickness of 720 meters at Um Araj in the following sequence from bottom to top.

"Lower limestone" – 500 meters

a) 70 meters of dark grey, crystalline or sublithographic limestone, and dark-grey, dense, bituminous dolomite with chert nodules at the top;
b) 30 meters of grey, argillaceous and very fossiliferous limestone; and
c) 400 meters of dark-blue, sublithographic limestone in thick massive layers with some intercalated beds of brown, shelly limestone.

Shelly limestone or "Upper Limestone" – 200 meters

Purple, fragmental limestone, brown, shelly limestone, and marl and shale (with chlorite).

The Hanifa Formation contains fossils, which indicate an age from Oxfordian to Kimmeridgian.

Northward, the Hanifa Formation becomes very thin – and with a single exception east of Masliyah – does not exist in the North of Wadi Fija. East of Masliyah, at the eastern bank of Wadi Baysh, a small outcrop consists of grey, dense, slightly silicified limestone which contains shell fragments of different sizes. At the base, conglomerates with rounded pebbles of basement materials occur. The layers form a slight uparching fold with a 20° NE trending fold axis.

Tertiary deposits

Due to downwarping and flexuring which developed with the subsidence of the Red Sea graben and which affected the pre-Miocene land surface, the facial character of the Tertiary sediments is greatly varying in the Tihamah Plain (Fig. 62). To simplify the complicated depositional factors in the area of deposition it can be said that two sedimentary environments predominated during the late Miocene (penesaline to saline sedimentary conditions) which resulted in thick saliferous deposits in the western part of the area (salt dome of Jizan) and littoral conditions in the east resulting in the deposition of the so-called Baydh Formation (G. F. Brown and R. O. Jackson, 1959). The latter contains large volumes of tuffaceous material originating from volcanic activity which accompanied the Tertiary tectonic movement. M. Gillman (1968) believes that the Jizan flexure (see Plate III, insertion at the back cover) has been the limit between these two zones. Reactivations of more intensive tectonic movements since the end of Miocene have produced large quantities of clastic material in the rifted zones which covered the Jizan Plain from the foothills to the coast. This process went on intermittently until Pleistocene times.

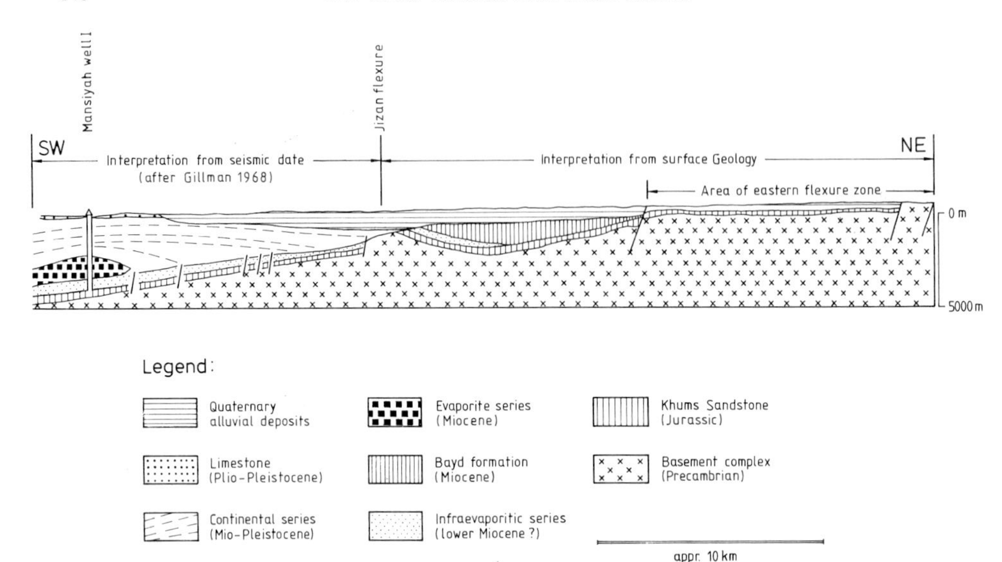

Fig. 62. Generalized Geologic cross section of the Jizan Plain (Southern Tihamah).

The Baydh Formation (Miocene) lies unconformably on the Jurassic and on the basement. Outcrops of Baydh Formation are scarce, though these strata underly the Quaternary to a great extent. Two sections, however, are well exposed.

The section in the Jizan Valley (which was not mapped in this study) from the base upwards is as follows after (M. GILLMAN, 1968):

a) 60 meters of breccia and volcanic tuff with blocks of basement rocks;

b) 400 meters of greenish silicite with layers of greenish shale, rare limestone, volcanic tuff, and at the base, interstratified flows of dacite, dolerite and obsidian; and

c) 600 meters of coarse volcanic tuffs and dacite flows with a few siliceous layers and numerous andesite and quartz porphyry sills.

The succession continues in the Ad Darb – Wadi Baydh area in the northern part of the Jizan Plain. Between the basement and the outcrops of the Baydh Formation, the first 1,200 meters of succession are concealed by recent deposits, after which it continues.

d) 1,200 meters of chiefly grey, buff, red and green sericite with intercalations of green or pinkish shale, volcanic tuffs and diabase sills.

West of the line of the foothill outcrops, the Baydh Formation, which dips approximately 20° W, disappears beneath the Quaternary of the Plain.

Evaporitic Series and Continental Series of Clastic Material (Miocene to Pleistocene) have been encountered by the Mansiyah Well I some 40 kilometers north of Jizan. According to M. GILLMAN (1968), the lithological description is as follows:

0 to 171 meters	sandy limestone with minor sandstone or clay layers;
171 to 999 meters	"Upper continental member" red shale with metamorphic and volcanic rock pebbles;

999 to 1250 meters	"Middle continental mmember" red shale with thin dolomite and anhydrite beds;
1250 to 2230 meters	"Lower continental member" red shale with sandstone and conglomerates;
2230 to 3472 meters	Evaporite series: salt (halite) with rare anhydrite layers; and
3472 to 3992 meters	Infraevaporitic series composed of gray shale with conglomerate, silt and grey, fine-grained sandstone.

Due to the upward pushing of the salt in the Miocene these formations are forming an anticline (Fig. 62).

The strata series is partly exposed in the area of the town of Jizan forming salt plug which is covered by caprock of mostly red shale, dolomite and anhydrite, and of well-rounded fragments of basement and volcanic rocks.

According to G. F. BROWN (1970) and M. GILLMAN (1968), this strata is believed to be partly the lateral eastern equivalent of the Baydh Formation. Therefore, this term has usually been applied to these deposits. However, since the facial character of these sediments is much different from the type Baydh Formation, the author prefers to use the term "Continental series of clastic material."

Quaternary deposits

Almost all of Jizan Plain is covered by young clastic sediments. In the east, the cover sediment is discordant with the Tertiary Dyke-Formation and the Baydh-Formation. In the west, however, they follow without a sharp caesur the Tertiary 'Continental series of clastic material' in an equal lithologic development. Following the literature (G. F. BROWN and R. O. JACKSON, 1959), later researchers have mapped these young sediments as Quaternary deposits. Concerning the problem of stratigraphic correlation, see Chapter 2.4.1.2. The young series are differentiated on the geologic map (Plate III) as follows:

Boulder Gravel deposits (Pleistocene – Late Pliocene?)

Dissected fan-like terrace gravel deposits, about 10 meters thick, lie on the basement along the major wadis northeast of the plain. The deposits consist of coarse pebbles, gravels, and sands, and have been described by W. R. GREENWOOD (1976) in the Al Qunfudah region (North Tihamah). This author believes that these deposits indicate a period of wadi aggradation, which may be related to the onset of dessication at the end of a wetter phase of the Pleistocene. The regolith was transported by storm runoff to form alluvial fans where the wadis broadened at the base of the scrap. Continued desiccation and the absence of soil cover caused the stream bed to change from mixed mud, gravel, and boulders to mostly gravel and boulders. Reduction of velocity of the flow caused boulders to be dropped in the upper courses, a fact which led to an increased erodibility in the lower courses. This erodibility caused dissection of the fans. Several stages can be seen in this dissection.

Old Terrace Alluvial deposits (Pleistocene)

They are exposed along the western margin of the basement complex where they form a NW-stretching band, approximately five kilometers wide.

In the northern part between Wadis Wasi and Atwad, the gravels are composed of poorly sorted, predominantly very coarse pebbles in a sandy groundmass. The well-rounded pebbles are derived from dike and basement materials. The sands are often contaminated by reddish and brown argillaceous materials, which provoke the reddish colour of wide outcrop areas. Basal conglomerates within the strata are common. In the area of Wadi As Sirr – Wadi Baydh (drillings 6J82, 6J88) the strata are 40–50 meters thick, and contain two thick layers (10–12 meters) of clay, and lie directly on the Miocene dike rocks. Westward this strata seem to disappear along a NW-trending fault line. The fault causes a thickening of younger Quaternary deposits farther west.

South of Wadi Wasi between Jabal Akwah and the Yemen border, the Pleistocene sediments have different character. They are predominantly composed of fine sands and silts, sometimes well consolidated and slightly clayey. Lenses and layers of fine to medium-grained pebbles are abundantly intercalated. Southeast of the volcanic cones in Wadi Sabya area, interstratified consolidated tuff, with a maximum thickness of three meters, is widely exposed. In several areas of this wadi, the tuff is underlain by a thick (5–7 meters) layer of loess which contains various freshwater gastropods.

All the strata are believed to be of Pleistocene age because they are overlain by the lava sheets mentioned above.

Holocene Alluvial deposits

In the western and central parts of the investigated area, the Pleistocene alluvials are covered by a younger series of alluvial deposits. The outcropping material, which forms the actual surface of the plain, consists of fine quartz, feldspar, mafic minerals, mica, and sericite sands. The grains are predominantly angular and rarely rounded. To the west, the sands become more silty and are weakly consolidated. Small lenses of gravels of lava materials are found in the sands.

Recent Alluvial deposits

These alluvials form vast terriform lands along the wadis. The sandy groundmass contains well-rounded gravel of different sizes and origins.

2.4.1.3. Tectonic features and volcanism

Particularly during the deposition of the Baydh Formation, large flexure-like zones of westward dipping faults have been formed in the area of the Jizan Plain. The most important one stretches parallel to the outcropping basement in the east and corresponds to the eastern edge of the Red Sea graben. Affecting the basement and also the Khums and Baydh formations, the fault forms a set of fractures which provoke a rapid deepening of the sedimentary basin. A very important volcanism marks this flexured zone. Its direction within the investigated area is NW–SE. The geoelectric measurements performed within this study have located several of these faults in the Wadi Shahdan area east of Wadi Baysh and west of the outcropping Jurassic layers. This system is affected by another system of faults which cuts the flexure area in an almost northeast direction and which provokes lateral displacements of the flexure. One of these faults (Wadi As Sirr displacement) is perhaps responsible for the change in direction of the Wadi Baysh northeast of Masliyah after this wadi has left the foothills. Another fault south of Wadi Wasi causes the retreating of the flexure zone to the northeast. West of this flexured area aeromagnetic and seismic surveys have revealed a homocline which is about 25 to 35 kilometers wide with a northwest–southeast strike and low westward dip (M. GILLMAN, 1968). Some northwest-trending faults affect this homocline, which is believed to be the deposition area of the Baydh Formation (see Plate III, insertion at back cover). Also, by reflection seismic surveys, a second flexure zone (Jizan-Flexure) has been located parallel to the first one. It is the limit of a second homocline in the west and provokes a westward thickening of the clastic "Continental series". This zone is believed to be the limit between two sedimentary environments during the Miocene.

Although no earthquake epicenter has been recorded along the Red Sea coast, the fault zones are still in an active state as is shown by raised Pleistocene coral reefs, faultings of Pleistocene sediments, recent lava effusions and the presence of hot springs (Wadi Khulab, Wadi Jizan).

In 1962, a precision systematic airborne magnetometer – scintillation counter survey was carried out over the Jizan area. The data indicate an excessively faulted sedimentary section that is more than 5,000 meters thick in some parts and which reaches (to the west of the coast line) a maximum thickness of about 8,000 meters (Fig. 62).

The tectonic movements during the Tertiary have been accompanied by a widespread volcanic activity which has produced large amounts of diabase, dike swarms, layered gabbros, granophyres and rhyolitic dikes occupying the NW-trending main tectonic lines.

The magmatic rocks form huge irregular masses (Jabal At Tirf, Jabal Hathah, Wadi As Sirr area) and numerous swarms of dikes which have injected the basement, the Khums, and the Baydh formations as well.

The dike swarms range in width from one kilometer to approximately five kilometers. The thickness of the dikes varies between ½ meter and 20 meters. Dips of the dikes are very steep (50° NE to vertical). In some areas vesicular lavas with poorly developed pillow structures dip steeply southwest.

The dike rocks are predominantly fine-grained diabase primarily consisting of calcic plagioclase and clinopyroxene. Textures are primarily ophitic.

These dark green and black rocks appear compact and massive, but due to the presence of preferential cleavage plains, they very easily break into irregular pieces. Near the contacts to the basement the diabase contains xenoliths of quartzites and basement materials. East of Jizan, in the upper Wadi Amlah-Maqab area, a large crescent-shaped layered gabbro body is associated with granophyre intrusives and forms the important rise of Jabal At Tirf (R. G.

COLEMAN et al., 1972). The layered gabbro is approximately eight kilometers long, and 2.5 kilometers wide; the layers dip steeply (40–60 to SW).

A second layered gabbro body (Jabal Hathah), 65 kilometers northwest of Jabal At Tirf, has a semicircular outcrop and is located near Masliyah.

Younger Pleistocene to Holocene basaltic lava flows occur west of the main eastern flexure zone and form several isolated cinder cones, which are still in a good state of preservation. Three superimposed lava sheets at Jabal Akwah, overlapping Pleistocene alluvials, confirm at least three eruptions during the Quaternary time. The eruption fissure seems to belong to the set of faults, some of which have been revealed by the geoelectric survey west of the outcropping Jurassic in the Wadi Baysh area. The main rock is composed of a very hard vesicular block lava with platy fracture and contains large amounts of olivine. Flow structures are very common within the lava sheets, whereas real pillow lava has not been observed. Onion-skin exfoliation is a characteristic weathering process in these rocks.

2.4.2. The Development of Tihamat Asir During the Quaternary

(A. DABBAGH, R. EMMERMANN, H. HÖTZL, A. R. JADO, H. J. LIPPOLT, W. KOLLMANN, H. MOSER, W. RAUERT, J. G. ZÖTL)

2.4.2.1. General considerations

Various studies on the geology of southern Tihamah are currently available. However, these are mainly concerned with the economic-geology aspects and give only little consideration to the Quaternary Period. The geological maps with a scale of 1 : 500,000 published by the Saudi Arabian Ministry of Finance and National Economy show the occurrence of Quaternary deposits. This area is to be found on maps I-216 Tihamat Ash Sham (G. F. BROWN, and R. O. JACKSON, 1958) and I-217 Asir (G. F. BROWN, and R. O. JACKSON, 1959). The Quaternary deposits are adjusted to the scale, with no further differentiation in the presentation.

Some detailed results of specialized studies are available. Drilling holes and geological studies undertaken for petroleum exploration provide preliminary information on the total thickness of the Quaternary layers. In the borehole Mansiyah 1, nearly 40 kilometers north of Jizan, M. GILLMAN (1968) reported sandy limestone and thin layers of clayey sandstone in the upper 171 meters. The next 828 meters is the so-called "upper continental series" containing red shale and isolated layers of pebbles. In comparison to the Durvaraz borehole in Sudan, W. GILMAN suggests a possible classification of this series as Quaternary deposits. Other authors, however, have already considered this series as being Plio-Pleistocene.

The geophysical survey made in the course of hydrogeological studies provides most of the information on the thickness of Quaternary deposits in the coastal-plain area. A detailed geodetic survey with geoelectric profiles in the southern Tihamah was made for feasibility studies and irrigation and water-supply projects (ITALCONSULT 1965, V. COTECCHIA et al., 1970, HALCROW and PARTNERS, 1972, GERMAN CONSULT, 1978). Further details on the geoelectric measurements made by a GERMAN CONSULT subcontractor are given by E. MÜLLER in this volume (sect. 2.4.1. and 2.4.3.).

The seismic and geoelectrical profiles first show varying thickness of the Quaternary layers (Fig. 62). A closure in the immediate coastal region (Mansiyah anticline) with its upper continental deposits being overlain by marine, calcareous sediment, is followed eastward by a wide, slightly asymmetric syncline. The synclinal axis is placed approximately in the middle of the Tihamah (Jizan flexure). The young alluvial deposits within this flexure have a thickness of several hundred meters. On the synclinal limb rising toward the west, the alluvial deposits are superimposed upon the "upper continental series" and interfinger with the marine calcareous sediments. Eastward, however, the deposits overlie the Miocene Baydh Formation, and further east overlie the Dyke Formation. In a faulted area (flexure 1 according to M. GILLMAN, 1968), these young alluvial deposits overlie the Jurassic Khums Sandstone. This zone is approximately 10 kilometers wide. Eastward of the flexure, the crystalline basement outcrops into the surface after further step faults.

The alluvial filling of this wide syncline shows a differentiated sequence, which is described below according to the drill logs (ITALCONSULT, 1965):

1. a layer of silty sand (1–3 meters);
2. unconsolidated sand mixed with gravel and pebbles (up to 20 meters);
3. lenses of clayey material;
4. unconsolidated sand and gravel; and
5. consolidated alternate bedding of clay and sand.

The upper boundary of this last layer of consolidated clay and sand lies between 40 and 150 meters below ground surface in the individual bore holes. The upper layers of loose sand and gravel are thickest (up to 150 meters, including the upper sand horizon in the absence of the separating clay lenses) where they are connected to an old channel system. This is the old covered drainage pattern, which corresponds in part to the present courses of the wadis, but in places varies from them and runs at a slant to the present direction of outflow. These old channels are particularly important sources of ground water with low mineral content (compare section 2.4.3.).

It is remarkable that no indication of a marine transgression could be found in any of the holes drilled in these alluvial deposits. Intercalated beds of marine sediments are not found until the west edge of the Tihamah where the coast is today. This means that the sedimentation during the Quaternary kept pace with the slow sinking of the coastal plain and was compensated for by the clastic sedimentation from the hinterland. With its slope down to the coast, most parts of the 40 kilometer-wide Tihamah area still give the impression of a morphologically uniform alluvial plain. The sediment grain size, however, decrease continuously from the hilly hinterland down to the coast. This is true for the entire Quaternary sequence.

Morphologically, a slight accentuation of the relief, owing to the young wadis, is seen only on the eastern edge of the Tihamah, in a 10 kilometer-wide block which is less down-faulted. Several meters are more deeply eroded and show recognizable terraces composed of sandy to gravelly material, and of coarse gravel with intercalated thick beds of clay north of Wadi Malahah. In Wadi Sabya and Wadi Dhamad, a layer of tuffa (see geological map, Plate III) forms a prominent key horizon within this terrace formation. Downward from the wadi, these terraces apparently dip down to a flexure parallel to the edge of the graben and are increasingly overlain with young sediments (first main flexure according to M. GILLMAN, 1968). The terrace slope decreases to the west and disappears completely in the increasingly drier and diverging wadi courses.

In addition to the terraces, remnants of an old aggradation plain have been preserved in the transitional area to the hilly hinterland. This older accumulation is especially remarkable owing to the presence of boulders up to 70 centimeters in diameter which W. R. GREENWOOD (1976) designated as storm-tide accumulations.

E. MÜLLER'S geological map (Plate III) shows the occurrence of the Quaternary sediments mentioned in the area of southern Tihamah between the Yemen border in the south and Wadi Atwad in the north. Field studies lasting several weeks were carried out by the authors to examine these Quaternary deposits in more detail, with special attention to the connections among the main wadis and the sequences of sedimentary strata. The most important details concerning the individual wadis and sections of the coast are described below.[1]

2.4.2.2. *Quaternary sediments in the wadis*

Wadi Khulab

Wadi Khulab was followed from Suq Al Ahad upward to the village of Kuba on the Yemen border. Suq Al Ahad is located in the center of Tihamah. Downward toward the west, the wadi gradually loses its identity in the accumulation plain which is partially covered by dunes. Going up the wadi, a distinct channel is observed with terraces up to 10 meters high. This is basically the main terrace, made up of coarse gravel. Isolated remnants of a younger terrace made up of finely clastic sediments extend from it. The wadi bed itself is formed of reworked alluvial particles of sand and gravel, and in part of fine-grained sabkhah sediments.

An example of this sort of development may be seen at a point nearly 20 kilometers above Suq Al Ahad where the tributary Wadi Khitam flows into the main wadi. There the main wadi turns upstream toward the southeast in the area of the Tertiary Dyke Formation, whereas Wadi Khitam flows into it straight from ENE. Here, the wadi channel is about 120 meters wide. The recent wadi bed lies eight meters below the main terrace which has very thin accumulations in places.

---

[1] Because MÜLLER's map was finished 1978 it was not possible to include the results of our detailed studies. The reader who is refered to this map (Plate III) from our description concerning the areal distribution of the different Quaternary sediments should notice that there are differences in the designation and adjoining. Our "boulder gravel" plane is nearly identical with the fanlike boulder gravel deposits; our main terrace essentially includes the old terrace alluvial deposits. The middle terrace is mapped mailny as terrace alluvial deposits and partly includes the terraces alluvial deposits. The lower terrace suits partly the terraced alluvial deposits and the reworked alluvium.

Rocks of the Tertiary Dyke Formation thus outcrop on the floor of the tributary Wadi Khitam. Individual dikes in the form of long, thin ridges in some places rise above the surface of the terrace. The main terrace is composed of poorly sorted, coarse clastic material which includes subrounded boulders with diameters up to 50 centimeters.

On the orographic left side, the younger terrace extending from the main terrace appears as a break in the dip of the slope. On the right side it forms an erosion surface 4 meters deeper than the main terrace. At the embouchure of Wadi Khitam, the terrace is 50 meters wide. The conditions of exposure allow a good view into the structure of this young terrace.

This terrace consists of a sequence of fine-grained clastic sediments, mainly in the form of an alternate bedding of fluvial sands, wadi-sabkhah sediments and eolian inclusions.

A few meters above the embouchure, Wadi Khitam is partially dammed off from Wadi Khulab by an intermediary passage of the Dyke Formation running at right angle to it. This passage appears as a sill approximately 1 meter in height. The occasional surface runoff thus forms a pool which persists for most of the year. The sill is diorite-porphyry and is covered in the terrace profile by an irregular calcareous sinter sequence up to 1.2 meters thick. The sinter encloses large pebbles from the dyke stone and redeposited pebbles from the main terrace. This sinter layer dips down to the main wadi and disappears uner the wadi bed where it is covered by the finely clastic sequence of strata described earlier for the younger terrace. ^{14}C dating of the sinter shows an age older than 32,000 years (see Table 18, sample No IRM 7612). The sedimentological characteristics indicate that the younger body of the terrace was formed under drier climatic conditions than presently prevail. Sedimentation has almost completely filled the channel in the main terrace. Unfortunately, no borings are available here to provide information on the depth of the channel and its filling. The dissection of the young terrace indicates later erosion under moist climatic conditions (or perhaps only higher seasonal run-off). Currently, sedimentation predominates in this section of the wadi, mainly in the form of reassorted sands and gravels, but also as silt in the form of wadi-sabkhah sediments.

Farther up the wadi, in the broad valley within the foothills, the planation of the main terrace as a "Reg" is particularly pronounced, and covered by isolated dunes. The spreading of this terrace surface gives the impression that the wadi extended farther to the south before it was filled with gravel. This impression is supported by the electrical profiles P and Q (see section 2.4.4.2.), which show a channel nearly 100 meters deep running toward the southwest.

Today's Wadi Khulab with its twist above the embouchure of Wadi Khitam was thus formed on the northern edge of the old accumulation channel as a young epigenetic breakthrough during erosive dissection of the main terrace. Occasional terrace steps, visible only in certain parts of the main terrace, are apparently erosion steps, or remnants of the middle terrace described below. They indicate that the erosion of the surface of the main terrace had occurred continuously, and not in one phase.

The fine clastic material of the younger terrace is completely eroded in some sections of the wadi. The interfingering of the two terraces can, however, be seen

Table 18. *Results of ^{13}C and ^{14}C measurements on samples from the Southern Tihamah and Farasan Islands. Analysis performed by GSF – Institut für Radiohydrometrie, Munich–Neuherberg, Federal Republic of Germany*

Locality	Geographic Coordinates	Material	IRM Lab. No.	δ ^{13}C (‰ PDB)	^{14}C content (% mod.)	Assumed initial ^{14}C content (% mod.)	^{14}C age uncorr. (years B. P.)
Harrat Al Birk small wadi 4 km S Al Birk	lat 18°12′ long 41°33′	sinter limestone from the terrace surface in the wadi	7621	− 7.9	7.0±0.4	85 100	20,000±500 21,300±500
Harrat Al Birk small wadi 4 km S Al Birk	lat 18°12′ long 41°33′	sinter limestone from the wadi floor	7622	−11.6	47.6±2.5	85 100	4,700±400 6,000±400
Harrat Al Birk small wadi 4 km S Al Birk	lat 18°12′ long 41°33′	sinter limestone with marine components from young wadi channel	7623	−11.8	43.2±1.4	85 100	5,400±200 6,700±200
Harrat Al Birk mouth of Wadi Dahin	lat 18°08′ long 41°35′	corals from the raised reef, 4 m-terrace	7624	− 0.6	1.3±0.4	100	$34,800^{+2,800}_{-2,000}$
Harrat Al Birk mouth of Wadi Dahin	lat 18°08′ long 41°35′	shells from the raised reef, 4 m-terrace	7625	− 0.1	7.0±0.4	100	$39,700^{+7,200}_{-3,700}$
Harrat Al Birk southwest part of the coast	lat 17°50′ long 41°50′	calcified tuffs, from a cone near to the coast	7626	−10.9	4.5±0.6	100	$24,900^{+1,200}_{-1,100}$
Harrat Al Birk 7 km N road deviation to Muqabil	lat 17°49′ long 41°54′	corals from a raised reef	7620	+ 1.3	2.2±1.0	100	$30,700^{+4,900}_{-3,000}$
Wadi Jizan margin of basalt flow E of Al Bayd	lat 17°06′ long 42°51′	sinter limestone from the margin of the basalt flow	7610	−10.1	39.2±2.0	85 100	6,200±400 7,500±400
Wadi Jizan tributary wadi downstream Maloki bridge	lat 17°01′ long 42°55′	sinter limestone below loess-like sediments (s. sample No. 7608)	7609	− 5.9	31.1±1.1	85 100	8,100±300 9,400±300

Wadi Jizan tributary wadi downstream Maloki bridge	lat long	17°01′ 42°55′	snail shell from loess-like sediments above sinter limestone (s. sample No. 7609)	7608	− 9.1	67.6±3.1	100	3,000±400
Wadi Jizan right bank of Maloki dam	lat long	17°03′ 45°57′	calcified roots from a tuff layer	7629	− 6.9	37.1±1.6	100	8,000±300
Wadi Jizan 5 km upstream Maloki storage lake	lat long	17°00′ 43°01′	carbonate cement from terrace gravels	7611	− 8.2	9.4±1.1	85 100	$17,700^{+1,000}_{-900}$ $19,000^{+1,000}_{-900}$
Wadi Jizan Maloki-crater	lat long	17°04′ 42°55′	snail shell from the eolian sediments in the crater	—	− 7.5	91.9±3.8	100	700±300
Jabal Umm Al Qumam Wadi Maqab	lat long	16°52′ 42°58′	calcified roots from the older eolian complex	7614	− 9.6	41.0±1.8	100	7,200±300
Jabal Umm Al Qumam Wadi Maqab	lat long	16°52′ 42°58′	calcrete layers between the older and younger eolian complex	7615	− 9.1	36.6±1.6	100	8,100±400
Wadi Khulab mouth of Wadi Khitam	lat long	16°45′ 43°01′	sinter limestone from the basis of young terrace accumulation above Dyke Formation	7612	− 7.8	<1.8	85 100	>31,100 >32,400
Farasan Islands Farasan Al Kabir NE coast, opposite Sajid	lat long	16°46′ 42°00′	shells from a beach ridge 5 m above re- cent surf notch	7616	+ 3.3	55.5±2.8	100	4,700±400
Farasan Islands Farasan Al Kabir NW part near Hussein	lat long	16°47′ 41°58′	coral from the top of the upper reef layer	7618	− 2.7	11.0±0.7	100	17,700±500
Farasan Islands Farasan Al Kabir NW part near Hussein	lat long	16°47′ 41°58′	coral from the conglo- meratic lower reef layer	7619	− 0.4	1.5±0.5	100	$33,500^{+3,500}_{-2,400}$

in some places. The hot springs of Ayn Al Hara (Kote 158 on Map I-217) are one such place (Fig. 63). The hot springs emerge from the edge of erosion remnants of a NW-striking rock vein with acid volcanic rocks and a pegmatite vein, which approaches the recent wadi channel.

In the spring area, the main terrace on the left side of the wadi is an erosion surface in the outcropping solid rock. On the right side of the wadi, the accumulated sediments are coarse gravels. The planation of the main terrace is nearly 13 meters higher than the present wadi bed. The young terrace body lies

Fig. 63. Hot spring of Ayn Al Hara, upper part of Wadi Khulab. The hot water emerges from the left bank on the margin of a basement spur. Lower terrace (with two steps) built up by silts and fine sands. The higher terrace is formed by a basement plain corresponding with the main gravel terrace on the right bank. (Photo: H. HÖTZL, 1978.)

before the main terrace and is about 5 meters high. It is made up mainly of loess-like, silty, fine sands.

The recent sedimentation, which is gravelly in this area, increases in thickness within the wadi channel going up to Kuba, which is overshadowed by steep mountain chains. This sedimentation reduces the height of the main terrace to merely 6 meters. In the vicinity of Kuba, where thermal waters re-emerge in the wadi bed, the body of the main terrace is made up of boulders up to 0.5 cubic meter in size.

Wadi Fija and Wadi Maqab

Wadi Fija and Wadi Maqab, with their tributaries, drain the accumulation plain bounded in the west by the young volcanoes of Jabal Umm Al Qumam and in the east by Jabal At Tirf foothills formed by the Dyke Formation. This plain was originally a uniform talus fan slanting from the foothills toward the west to the coast. Because this talus fan corresponds very closely to the main body of the terrace, these surfaces have been dropped from MÜLLER'S geological map (Plate III, see insertion at back cover) with uniform conventional signs. In comparison with the actual gravelly terrace accumulation, which was sedimented in the drainage channels from the mountains, the talus fan shows poorer grain-size distribution and poorer grain rounding. The grains of this material are, in general, considerably finer than the gravels and pebbles from the main terrace.

The old talus fan bears in the upper courses of Wadi Fija and Wadi Maqab the two volcanic cones of Jabal Umm Al Qumam with their volcanic flows which mainly follow the gradient to the west. A sample was taken from the bottom flow directly on the east edge of Jabal Umm Al Qumam; K/Ar dating showed an age of 0.9±0.3 million years (see Table 19, sample SA 77/19), confirming the Quaternary age of these young basalts in the area east of Jizan, which the geomorphological situation had already indicated.

The old accumulation plain was already eroded when these basalts erupted, as must be concluded from the deep location of individual basalt flows on the northern edge of the Jabal Umm Al Qumam, where some of them are at the level of the present-day wadi channel. After eruption of the basalts, eolian accumulations formed in the lee of the two volcanoes; slight surface run-off alone sufficed to erode and deposit them. Today, between the volcanoes in the west and the foothills in the east, there are occurrences of thick sequences of eolian sediment that are deeply carved by a branching pattern of erosion channels.

These channel profiles show within the eolian sediment sequence an older soil horizon with a calcareous crust and a layer with root tubes. It appears to divide older, consolidated sands from the younger, looser deposits. The soil horizon shows a surface divided by individual channels, which in part was then covered over by the loose sediments. There is recent dune-sand accumulation in the lee of the basalts, but today erosion is again predominant.

Samples from both the calcareous crust and the completely calcified roots were used to determine the age of this soil horizon (Table 18, IRM No 7614 and 7615). ^{14}C dating showed an age of 8,100 years for the calcareous crust, and 7,200 years for the roots.

Downstream, both wadis are shallow channels cut into the young valley fills. Owing to the small drainage area, they very soon flatten out, first with sandy – then with increasingly silty-recent sediment, and finally disappear halfway to the coast.

Wadi Jizan

Wadi Jizan is narrowed by young basalt eruptions in the transitional area between mountains and coastal plain. The wadi originally was uniformly wide but in this section was forced to create a new bed, in places canyon-like. This

Table 19. *Potassium/Argon ages of young basalts from the southern Tihamah. Analysis performed by the Laboratory for Geochronology, University Heidelberg*

Sample No.	Sample locality	Geographic Coordinates	Material	L. f. G. Run-No.	K(%)*	^{40}Ar (rad) $\cdot 10^7$ cm^3 STP	Ar (atm)%	K/Ar-age** in mill. years
SA 77/52 A	Harrat Al Birk small wadi 4 km S of Al Birk	lat 18°12′ long 41°33′	weathered basalt from the top of the flow cut by the wadi	3922 3953	1.07	0.57	89	1.42±0.41
SA 77/52 B	Harrat Al Birk small wadi 4 km S of Al Birk	lat 18°12′ long 41°33′	fresh/weathered basalt few meters below the top of the flow	3906 3955	1.04	0.56	89	1.35±0.41
SA 77/21	Jabal Akwah southern margin, coastal plain, 30 km NE of Jizan	lat 17°10′ long 42°44′	weathered basalt from the basal flow directly above the tuff layer	3908 3965	1.38	0.24	94	0.44±0.26
SA 81/66	Wadi Jizan left bank 6 km below Maloki dam	lat 17°02′ long 42°55′	fresh/weathered basalt from main terrace	4720	0.78	0.25	91	0.8±0.3
SA 77/19	Jabal Al Umm Al Qumam, E margin of the basalt	lat 16°53′ long 42°58′	fresh basalt from the basal flow near eruption center	3891 3957	1.58	0.56	91	0.90±0.31

* Potassium content of the K-Ar-rock powder.
** Constants from 1977 (R. H. STEIGER, E. JÄGER, 1977).

caused repeated shifting of individual small channels, as may be seen in a tributary south of the Maloki Bridge. A wadi which originally directed northwest was deflected there by a basalt flow and then carried on along the edge of the basalt in a southwest curve. This tributary wadi is brought back into the old drainage direction to the northwest by the still regressing erosion from the new breakthrough section of the main wadi.

No absolute datings were available till now to give the exact age of the basalts there. Particularly, the stones from the basalt flow from the Maloki dam in the old wadi channel proved to be too weathered to provide a reliable age value for this relatively young basalt. Extensive sampling within the context of the Quaternary Project turned up suitable material for dating about one kilometer below the Maloki Bridge. It produced a value of 0.8±0.3 million years (Table 19, sample No. SA 81/66). This value agrees nicely with the basalt dating from Jabal Umm Al Qumam (Table 19, sample No SA 77/19), or the data from R. G. COLEMAN et al. 1977 for Wadi Jizan basalts. Together they give an age of about one million years for the young lava filling in the middle section of Wadi Jizan.

This classification is also supported by the fact that in the middle and lower course of Wadi Jizan, gravels of this young volcanic flow are only found in the uppermost meters of the alluvial deposits (V. COTECCHIA et al., 1970). The young volcanism may be divided into three phases:

1. explosive phase with eruption of tuffa;
2. outflow of lava with the formation of the basaltic layers and flows extending into the valleys; and
3. explosions with the formation of the volcanic cones made up of tuffa and lava layers.

During the preliminary studies for the Maloki storage dam in the upper part of this narrowing of the wadi, geoelectrical measurements and exploratory drillings were made (ITALCONSULT 1965, V. COTECCHIA et al., 1970). They provide a picture of the original form and position of the wadi channel. The geoelectrical profiles (cf. chapter 2.4.4.2) show first an old channel in the present wadi area with a base at about 145 meters above sea level, or about 15 meters higher than the base of the recent wadi bed. An even deeper channel is shown on a profile from ITALCONSULT (1965). Directly under the Maloki crater, the Quaternary basalt is in a broad depression upon the basement at about 110 meters above sea level (Fig. 64 a).

These old channels had already been filled with coarse to fine clastic sediments before the basalt eruption. In the area of the Maloki storage dam, this fill-up extended to a height of 160 meters above sea level. This is in good agreement with the main terrace in the storage area, and with the area adjacent to the upper end of the reservoir. There, the planation of the main terrace is at 165 to 170 meters above sea level.

The stratification of this main terrace is rather variable. Primarily fluvial, gravelly fine sands made up of gneiss and granite detritus are described from the exploratory drillings in the vicinity of the reservoir dam. In some drillings, the basis of this terrace accumulation consisted of very fine, lightly cemented, red sands, with sands and coarse pebbles beneath. The upper layer of this main terrace is made up of a number of meters of sand with boulders (diameter up to 50 centimeters) both in the upper part of the reservoir and in the transverse valley below the dam. Farther down, finely granular sediments prevail. An outcrop at the upper end of the reservoir showed only inclusions of silt and clay (thickness 4 meters) between the coarse gravels and the basement beneath.

In places, the boulders are cemented gompholite with a calcareous material. The calcareous cement was at first considered to be young, and related to the washing out of the loess (see below), but ^{14}C dating showed an age of about 17,700 years (Table 18, sample No. IRM 7611). This value is not, however, definite, because precipitations occurred on an apparently older cement during Holocene wet periods.

The accumulation plain of the main terrace was again eroded to terraces even before the volcanic eruption of young basalt. Today, this main terrace dips below the young sediments eastward from Abu Arish (Fig. 64c). As early as 1970, V. COTECCHIA et al. related this to a downwarp on a flexure or fault. A further terrace is very nicely developed in the narrow transverse valley of Wadi Jizan through the basalt sheet (Fig. 64 a). It consists mainly of gravels and boulders up to 50 centimeters in diameter. This is mainly basement material that at least in part was rearranged from the older main terrace or coarse boulder gravel. The terrace is some six meters higher than the recent wadi bed. It counts as a part of the

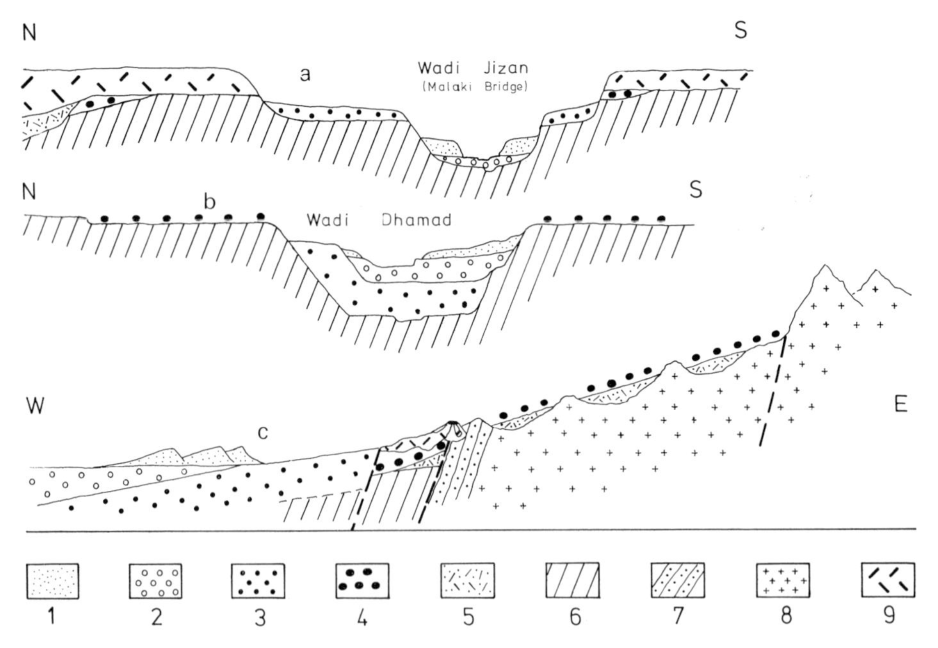

Fig. 64. Schematic geologic cross sections from Wadi Jizan and Wadi Dhamad. a) Cross section of Wadi Jizan, near Maloki bridge, showing the stratigraphic position of the basalt flow (about 1 mill. years) and of the down cutting of wadi channels. b) Cross section of Wadi Dhamad west of the basement block, c) Terrace plains in a section from the basement to the coast (*1* eolian dust and sand accumulation; *2* gravel and sands of the lower terrace; *3* gravel of the middle terrace; *4* main terrace and coarse gravel and boulder of old planation; *5* sandy silty accumulation; *6* Bayd- and Dyke-Formation, Tertiary; *7* Jurassic Khums Sandstone; *8* Precambrian basement; *9* young basalts).

middle terrace (see below) which separated out in the following sections in the Tihamah area.

Considerable surface remnants of this main terrace are preserved in the widened valley of Wadi Jizan and its many branches above the narrow passage at Maloki. Both the terrace surfaces and the channels eroded into them are still covered in places by a young loess-like dust deposit. These sediments can accumulate to a considerable thickness (2–3 meters) in the terrace channels. The subsequent erosion then removed a large part of these finely clastic sediments, so that a young terrace body with a height up to 2.5 meters developed.

Parts of this dust deposit are Holocene, as could be documented in the narrow transverse valley in the basalt area. These sediments persist there in the form of small inclusions of terrace remnants. A calcareous sinter deposit was found in the tributary wadi under the Maloki Bridge on a base of finely clastic sediments with occasional land snails directly above the rock of the Dyke Formation. ^{14}C dating of this sinter gave an age of about 8,100 years (Table 18, sample No. IRM 7609). Dating of the snail shells collected from the overlying loess-like deposit produced an age of about 3,100 years (Table 18, sample No. IRM 7608). A further sinter dating of a sample from the western edge of the basalt sheet is comparable to these two values; this date was established as 6,200 years (Table 18, sample No. IRM 7610).

West of the Quaternary basalt, where the main terrace has a flexure-like dip, the highest geomorphological element is another alluvial terrace body. In the following discussion, it will be called the middle terrace. It may be traced to within some 5 kilometers of the coast, where the terrace step gradually decreases and is covered by recent sediments. Because of the parallel courses of the neighboring wadis (Jizan, Dhamad and Maqab), the terrace is divided into individual long, flat crests. Between Wadi Dhamad and Wadi Jizan, the terrace persists only as a narrow strip which is interrupted in places. At El Ragan, halfway between Abu Arish and the coast, where the recent wadi channel approaches this terrace body directly, the terrace step still has a height of three meters. South of Wadi Jizan to Wadi Maqab, this terrace forms a broad crest some 10 kilometers wide. There are often dune deposits on this accumulation body, which is mainly built up of sands.

Wadi Jizan is cut into this middle terrace in the form of a 5-to-10 kilometer wide wadi bed, the floor of which today consists mainly of fine-sandy to silty sediments. This is the lowest terrace, a sort of flood terrace, engraved with recent drainage courses 0.5 to 1.5 meters deep which diverge sharply and gradually disappear on their way to the coast. In contrast to the finely granular sediments of the lowest terrace, sand and gravel are deposited in the recent wadi channels in the eastern area, while fine sand and silt occur only toward the coast.

Wadi Dhamad

In its middle section, Wadi Dhamad runs parallel to Wadi Jizan and only some 10 kilometers north of it. Its particularly straight southwest course from the mountains to the coast is noteworthy. This course shows very clearly the old drainage direction perpendicular to the strike of the graben faults.

The wadi was traced into the area of the outcropping of the Tertiary Dyke Formation in the transitional area between the mountains and the Tihamah. The young basalts of the Maloki area are bordered there by Wadi Dhamad in the north. Starting at the southeastern eruption centers, a considerable part of the catchment area, dipping northwest in the direction of the wadi, is taken up by basalts. They cover a channeled relief just inside the boundary area between the basement and the Dyke Formation. Following the path of an older channel of Wadi Dhamad, a basalt flow streamed for more than 20 kilometers to the southwest. The present wadi then was eroded into the northern edge of this basalt flow, laying it free in places in a sort of relief inversion as a narrow basalt crest.

There are still extensive occurrences of the old boulder gravels on the edge of the crystalline basement east of the Dyke Formation (Fig. 64 b). Their distribution and position give the impression of an originally irregular aggradation. The original relief of the surface upon the basement is difficult to evaluate, owing to repeated erosion and dissection and the resultant roll down of the large boulders on the younger cuts (Fig. 64 c). The few cuts show clearly that, in this area, the base of the boulder gravel lies above the younger main terrace. The boulder gravel is apparently seldom thicker than five meters; often it is only a layer of pebbles one meter thick.

The recent wadi channel itself is usually limited by the more or less broadly developed main terrace. This terrace consists apparently of quite a considerable extent of redeposited boulder gravels. In places the gravel is solidified with calcareous cement.

As neither borings nor geoelectrical profiles are available, nothing can be determined about the depth of the channel, which is filled with accumulation from the main terrace. The minimum thickness of the aggradation is given by the terrace slope, which is up to four meters high. The individual sections with the Dyke Formation cut onto the terrace base appear to be younger epignetic water gaps, and do not provide information about the depth of this old filled-in channel.

The young channel cut into the main terrace shows flatter terrace steps next to the planation of the main terrace. Because of the limited conditions of exposure, no conclusions could be drawn as to the extent to which these intermediate steps made up of gravel and sand are only erosion phases, or equivalents of the middle terrace formed in the middle and lower courses.

Loose sediments, mainly of eolian dust, once again make up the youngest terrace formations. In places they lie in front of the main terrace as fully developed terrace steps, whereas in other places they constitute a layer only 0.5 to 1.5 meters thick on the intermediate steps mentioned above. Farther up the wadi – in the basement area – these loose sediments extend increasingly to the main terrace. The material there is mainly redeposited weathered detritus.

The heavy runoffs subsequent to this sedimentation caused the removal of much of these finely clastic loose deposits. At present, mainly sands and gravels are deposited in this part of the wadi bed, owing to the ephemeral runoffs.

Wadi Sabya

The middle section of Wadi Sabya from the margin of the mountains to the vicinity of the coast below the locality of Sabya is accompanied by an extensive terrace complex. The young volcanic eruptions of the Jabal Akwah further accentuate the geological structure (Fig. 65, 66).

The oldest accumulations are the old boulder gravels in the foothills and farther eastward. Extensive old plain remnants persist from just north of Wadi Sabya to Wadi Baysh. The boulder gravels are usually very well rounded and sometimes reach a diameter of 70 centimeters. There are no more such surfaces west of the main marginal graben dislocation, which is touched by a dip in the Tertiary Dyke Formation (Fig. 64 c). The aggradation surface of the main terrace is the determinative morphological element in the adjacent eastern part of the Tihamah. It has a reasonably uniform width of 15 kilometers between Wadi Dhamad in the south and Wadi Baysh in the north. It extends to the edge of the mountains and has a considerably steeper slope to the west than the present wadi bed. West of the volcanic cone of Jabal Akwah, it submerges under younger, flatter aggradations.

The conditions of exposure show the structure of this main terrace to be very heterogeneous. Such lower sections as are exposed show reddish, silty to fine-sandy sabkhah-like or eolian sediments, which are generally consolidated and

Fig. 65. Wadi Sabya with Jabal Akwah. In the foreground terrace step of the right bank: gravels with the intercalated tuff layer. Background stratovulcano of Jabal Akwah with short lava flows on the left side. (Photo: H. HÖTZL, 1981.)

cemented in the dessicated slight breaks. There are also some conglomerate sands and gravels. The coarse clastic sediments increase noticeably in the upper bed of the terrace body; in some places they are distributed over surfaces, and in others they clearly serve to fill channels in the deeper, finely clastic layers (Fig. 66).

The planation of the main terrace is characterized by a top bed made up of coarse gravels. Its pebbles are as large as 0.3 meter and are usually covered with

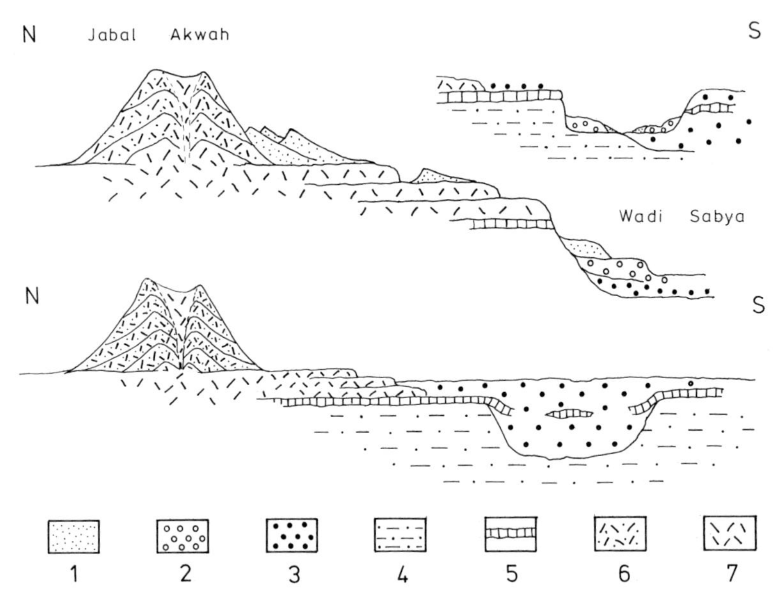

Fig. 66. Schematic cross sections from Wadi Sabya. Lower profile: former sequence of channel filling showing tuff eruption parallel to gravel accumulation of middle terrace. Upper profiles: terrace sequence caused by repeated down cutting and sedimentation (*1* eolian dust and sand accumulation; *2* gravel and sand of lower terrace; *3* gravel and sand of middle terrace; *4* silty sandy accumulation of main terrace; *5* tuff layer; *6* ash and basalt layer; *7* basalt).

desert varnish. As the layer is generally not very thick, it appears to have resulted from beginning erosion of the main terrace; the finely granular components were washed out and the coarse fraction remained. Extensive old surfaces of the main terrace are covered with large pebbles from north of Jabal Akwah to Wadi Baysh.

In the vicinity of Wadi Sabya, the planation of the main terrace is partially covered with a layer of tuff with a thickness of 0.3 to 1.5 meters. This may be taken to be the initial phase of the volcanic eruptions of Jabal Akwah. The appearance of this tuff on the main terrace, as well as on various intermediate levels – in some places in individual channels – indicates that the main terrace was dissected before the eruption of this loose volcanic material.

On the southern edge of Jabal Akwah, where the basalts extend as far as the wadi, the tuff lies directly over the reddish basal silt and fine sand complex. There it is directly covered by the oldest basalt flow. In the area farther up the wadi, the tuff layer is lapped by a gravel sequence (Fig. 67). The height and extent of the tuff clearly show that at the time of its eruption, the surface was dissected by

Fig. 67. Gravel terrace (middle terrace) Wadi Sabya south of Jabal Akwah; intercalated tuff layer with thickness up to 2 m. (Photo: H. HÖTZL, 1977.)

Fig. 68. Main terrace body of Wadi Sabya SE of Jabal Akwah. The outcrop shows the change of fine and coarse grained sediments; gravel layer with channel filling structure. (Photo: H. HÖTZL, 1977.)

channels more than 15 meters deep. These channels extended down to the sand-silt complex, the surface of which was thus sculptured, in places, like a terrace. These channels had begun to be filled with gravel before the tuff eruption, as is shown by the finding of tuff embedded in these channels (Fig. 68). The later coarse clastic sedimentation then filled the entire channels, especially in the basement mountains, and also overlaid some parts of the tuff-covered terrace surface.

It was hoped that absolute dating of the Jabal Akwah tuff and basalts would provide indications about the age classification of the sediments of the main terrace. The decomposition of the tuff was, unfortunately, too extensive, apparently because of the effects of groundwater. The age value (more than seven m. y.) obtained from this irregular K/Ar relationship is obviously too high, compared to the basalts, and is, therefore, misleading. For example, the oldest flow of the three basalt sheets of Jabal Akwah was dated. The sample (Table 19 sample No. SA 77/21) taken on the edge of the wadi directly above the tuff showed an age of 0.44 ± 0.26 million years.

A new sedimentation phase deposited gravels upon the tuff. This is viewed in connection with the formation of the middle terrace. The situation east of Jabal Akwah, however, is not entirely clear, owing to the overlapping of these gravels onto the main terrace.

During the course of this formation of the middle terrace, erosion predominated in the basement area immediately to the east. The planation there corresponding to the main terrace was slightly dissected, and the sediments completely removed in places. Sedimentation occurred in the western piedmont. The existing channels in the cropping area of the Dyke Formation filled up progressively, whereas farther to the west, the sedimentation even extended in some sections to the surface of the main terrace. This occurred only in a narrow strip along the edge of the mountains; farther to the west, the sedimentation remained limited to the more deeply incised channels.

The covering of the tuff shows this clearly. From the west edge of the dyke zone the tuff strikes out as the top bed of the main terrace surface. Toward the west, it is then covered by fan-like deposits of gravel. East of Jabal Akwah these gravels on the surface of the main terrace thin out, and only the tuff remnants in the deeper channels are covered by this younger sedimentation. There, the typical deeper-lying middle-terrace body is formed by the incomplete filling of the channels.

Above Jabal Akwah, the recent wadi channel lies as much as 15 meters below the level of the tuff-covered main terrace. Owing to the flatter gradient of the wadi bed, which is presently undergoing aggradation, the height of the terrace decreases to some eight meters below Jabal Akwah. Individual flattish remnants of terraces, appearing in some places as erosion steps and in others as accumulation steps, may be seen in the narrow, deeply incised wadi section. A terrace body (middle terrace) thus developed in the widening wadi, which because of its flatter gradient again overlaps the old remnants of the main terrace only a few kilometers below Jabal Akwah. This remarkable down-dipping of these two old surface elements in the section between Wadi Jizan and Wadi Baysh on a northwest line parallel to the edge of the graben indicates a tectonic subsidence or thickening.

The middle terrace consists of redeposited sands and gravels, as well as a considerable amount of silty-to-clayey sediments. As a result of the flatter gradient of the wadi channel in comparison to the surface of the terrace, the height of the terrace decreases continuously down the wadi. The step of the middle terrace is substantially levelled off a few kilometers below the locality of Sabya. At this point, the wadi has fanned out extensively and the mainly finely granular sedimentation extends to the terrace surface, so that toward the coast there is a relatively uniform sedimentation surface, with extensive eolian accumulation.

Below Jabal Akwah, an additional flat terrace body appears in the widening wadi (Fig. 66). It consits mainly of silt and fine sand. The intensive cultivation of this surface has erased much of the terrace-like nature of these sediments. Sand deposits dominate in the wide channel itself, which occasionally carries ephemeral runoffs.

Wadi Baysh

Wadi Baysh is one of the most important drainage systems for the southern Asir highlands sloping away to the Red Sea, and for the coastal plain. Worthy of consideration is its course, which deviates markedly from the other wadis and apparently belongs to an older drainage system. Its upper course is directed toward SSW, its middle course runs NS between Masliyah and Al Qashah, and only in its lower course does it follow the other wadis SW, perpendicular to the edge of the graben.

The survey permitted exploration of only the middle section between Baysh and Masliyah, as well as of individual tributary wadis between the localities of Al Haqu and Qura. Between Baysh and Masliyah, the wadi bed is made of sand and is rather wide. The middle terrace follows it on the left side, flanked by the main terrace surface, which is somewhat set off and widens to the south. On the right side of the wadi there are broad remnants of the lowest terrace with eolian and fluvial sands and silts.

At Masliyah, the recent wadi passes very close to a narrow projecting ridge of the Dyke Formation at a gorge. Slightly farther north, however, the old wadi, marked by the terrace planation of the middle and main terrace, crosses the outcrop of the Dyke Formation in a relatively broad cross-section of the valley.

In the basement area above Masliyah, the wadi forms a wide, meandering valley in places. There is a well-developed meander core at the embouchure of the tributary Wadi Akkas. The southern, inactive meander bed is occupied there by the middle terrace. Otherwise, the terrace steps appear as erosion edge in rock, occasionally with a scattering of gravel.

In this part of the basement at the edge of the mountains, the old, coarse boulder gravel plain which today is cut by shallow channels, is determinative. Occasional rock crests project from this surface. Its steeper gradient in comparison to the wadi bed is shown by the differing heights of the respective terrace steps. At the meander core mentioned above, it is some 15 meters above the present-day wadi bed, but eight kilometers farther up the wadi, it is already 35 meters above it.

In the vicinity of Ghamr, Al Haqu and Qura, individual wadi courses have thoroughly excavated these coarse-gravel surfaces. Isolated erosion remnants of the basement with coarse gravels have been conserved. Remarkable in this area are young terrace bodies; below, they consist of sharp-edged gravel, while their upper layers are made up almost exclusively of eolian dust sediments along with reassorted silts and weathered material.

Wadi Baydh and Wadi Samra

Between Wadi Baysh and Wadi Atwad, the main terrace is a uniform strip some 12 kilometers broad (it is erroneously designated as a Baysh Formation on geological map I-217 A); its western portion joins the narrow outcropping of the Dyke Formation going SE–NW. In the stratigraphical profile, the clastic sediments of the terrace body overlie the rocks of the Dyke and Baydh Formations discordantly. The entire block, including the terrace, dips slightly to the west. Wadi Baydh and Wadi Samra cut through this stratigraphic nonconformity in their subsequent course, revealing the profile, in the deep erosion channels, of the stratigraphic structure of the terraces (Fig. 69, 70).

The base is formed, so far as is visible, by the steep-faced stones of the Dyke Formation. Only in the western part of Wadi Atwad are the rocks of the Baysh Formation also exposed extensively. Directly west of the Baysh – Ad Darb road,

Fig. 69. Main terrace step of Wadi Baydh. The terrace is built up by changing clastic sequence, in the lower part sandy silty in the upper part gravel with pebbles. Recent wadi sediments: sands, silts and wadi sabkhah. (Photo: H. Hötzl, 1977.)

Fig. 70. The gravel terraces of the Wadi Samra. On the top and slopes enrichment of pebbles and boulders. On the left bank the main terrace with probably erosive notch of middle terrace. (Photo: H. HÖTZL, 1977.)

the oldest terrace sediments are exposed to a thickness of eight meters. These are silty Sabkhah deposits with lenses of eolian and fluvial sands. They are cemented with gypsum or calcareous material.

In places, root horizons and dune structures may be seen clearly. East of the road, for example, these finely granular sediments thin out toward the east. There, also, the younger deposits of the Dyke Formation are seen. Farther to the west, the deeper, finely clastic sediments dip under the wadi bed with the flattening terrace slope. The layer overlying the finely clastic sequence mainly shows conglomerate gravel layers with sand inclusions. The more resistant gravels are dissected as ledges in the steeper terrace slopes. This overlying sandy-to-coarse gravelly sequence has a total thickness of some eight meters. The sedimentation of the main terrace is then completed with a layer of coarse pebbles some two meters thick. This is also the terrace surface, developed as a broad "reg" field.

West of the Baysh – Ad Darb road, the rise of the wadi bed to the planation of the main terrace is clearly marked by the step of the middle terrace. There, the latter is some eight meters above the wadi bed and about six meters below the planation of the main terrace. The middle terrace first appears as an almost purely erosional terrace. Farther down the wadi, however, its sand and gravel accumulations increase. Because of the flatter gradient, these sediments extend over the down-dipping planation of the main terrace about 10 kilometers west of the road. Farther to the west, the middle terrace forms a uniform aggradation plain in which the wadis are only shallowly incised, or disappear in places completely.

Only Wadi Atwad in the north is more deeply incised; at the locality of Atwad, it has produced a terrace step five meters high, made up of sands and occasional layers of gravel. The surface of the middle terrace is covered by extensive dunes throughout the entire area.

2.4.2.3. *The coast*

The coast in the Jizan area

The Tihamat Asir fronts upon a pronouncedly flat coast broken only by the slight 50 meters elevation of the Jizan salt diapir. The nature of the coast line is determined by the terrestrial foreset beds of detrital accumulation from the mountains rising in the hinterland. Its rather straight course over its entire length of nearly 100 kilometers is astonishing for a coast formed by material washed down from the hinterland. Its direction toward NW and parallel to the graben suggests a tectonic pattern. This is supported by the temporal constancy of the position of the coast. As mentioned in section 2.4.1.2., there was no substantial marine transgression beyond the coast after the deposition of the Baydh Formation. The present-day coast appeared after deposition of the "upper continental series" at the beginning of the development of the Plio-Pleistocene reef flat (Mansiyah boring, W. GILMAN, 1968). The borderline between continental and marine sedimentation has remained much the same since then. The coast is still also preceded by a reef over almost its entire length.

The area immediately adjacent to the coast is taken up by a sabkhah zone virtually as long as the coast, and with a width up to some kilometers. It consists of silty, clayey sediments with high salt content owing to the high groundwater level and the intensive evaporation, or alternating layers including such evaporation products. In some places, such as south of Jizan, part of this sabkhah zone is covered with sand dunes.

This sabkhah area must have been largely flooded over at the high-water mark 4,000 to 6,000 years ago (Flandrian Transgression). This sabkhah area may be the transgression surface, or it may mark the short period of advancement of the sea at that time. Going inland, the sabkhah zone in the area of the wadis gradually changes, following a slight but distinct rise in the sections between the wadis, into the alluvial accumulation plain of the middle terrace. The occasionally distinct step between the middle terrace and the sabkhah area could be a further indication of the extent of the transgression of the sea.

A young fossil soil horizon was found, in a sample taken from this accumulation plain, south of Jizan on the Al Madhaya road. It demonstrated a green-gray silty sabkhah sequence with corresponding oxidative discoloration and shrinkage cracks. Old dune sand was on top.

South of Al Madhaya, the wandering dunes exposed a comparable soil-formation horizon. Here, beneath the dunes, clay shards and pieces of *Tridacna*, shells were found which were apparently used in a pottery to fire lime. A more exact dating has not yet been made.

The coast in the Al Birk section

North of Ad Darb in Wadi Atwad, the coastal plain narrows and is then interrupted for a length of 100 kilometers. This is due to the projection of the crystalline basement to the west; at Qahmah, it reaches all the way to the coast. Additionally, the coastal area is taken up by extensive Plio-Pleistocene basalt eruptions (Jabal Hayil, Harrat Al Birk, and basalts from Kiyat). Thus, except for some parts, this area has the character of a steep coast. Because of the young basalts and possible marine terraces as well, this section was of particular interest with regard to the temporal classification of the Quaternary. For closer investigation, an expedition was undertaken along the new shore road from the south to the locality of Al Birk.

The extensive plateau basalt erupted from numerous eruption centers extending to 25 kilometers inland. The basalt generally flowed along the gradient down to the coast. The basalt sheet is thus largely made up of a series of basalt flows. It is not possible to classify or integrate into stages of the individual flows starting at the coast. The geological survey was thus limited to a selection of samples which, it was hoped, would provide indications of possible geochemical changes in the course of time. The various flows visible in the deeply incised Wadi Dhahaban are especially impressive.

An absolute dating with the K/Ar method was obtained for a basalt flow south of Al Birk. There, a small tributary wadi had formed terraces in the basalt sheet. To date the sinter on this terrace, a sample was taken of the basalt from the upper edge to determine the beginning of the dissection. The samples produced an age of 1.4±0.4, and 1.3±0.4 million years, respectively (Table 19, samples No. 77/52 A and 77/52 B). These samples can by no means determine all of the volcanic activity in this area, which extended over a long period of time, but do confirm that basalt eruptions continued into the Pleistocene. The numerous, well-preserved volcanic cones belong to an especially young eruptive phase. Some such cones, as the Jabal Ar Raqabah or the Jabal Wasm, are directly on the coast. R. G. COLEMAN et al. (1975) cite ages of 0.3 to 0.5 million years for these young volcanoes.

The often steep slopes down to the coast show surprisingly few indications of older high-water marks. Most remarkable are the remnants of an accreted reef flat which has been preserved in a number of places. On such occurrence was uncovered during the road building some seven kilometers north of the turnoff for Muqabil. ^{14}C dating of a piecc of coral from the reef gave a value of about 30,700 years (Table 18, sample No. IRM 7620). Even though this value can be related to the last major interstadial rise in the sea level, it must nonetheless be interpreted very cautiously. Recrystallizations took place in the coral moss which probably involved carbon uptake. The value obtained, therefore, is probably to be understood to be older than 30,000 years.

A similarly located reef is also preserved in a bay by the volcanic island some 15 kilometers south of Al Birk. Its upper edge is some 4.5 meters a. s. l. Dating of a coral sample gave some 34,800 years, and of mussel shells, about 39,700 years (Table 18, sample No. IRM 7624 and 7625).

The expedition also provided unequivocal indication of the high sea water mark some 6,000 years ago. Poorly developed sea caverns and beaches are preserved. In the bay near the volcanic island mentioned above, 15 kilometers south of Al Birk, there is an older beach formation in the form of a small reef 2.5 meters a. s. l.

Mussel and gastropod shells were found in a small tributary wadi just 4 kilometers south of Al Birk, two meters a. s. l., in a young switchback channel trench. The shells were cemented in sinter which gave a ^{14}C age of some 5,400 years (Table 18, sample No. IRM 7623). The same age was shown for sinter formations farther up the wadi on the lowest terrace in the form of large sinter basins (diameter up to 3 meters, Fig. 71) (^{14}C age some 4,700 years. Table 18, No. IRM 7622).

Sinter found there, in the vicinity the erosion terrace cut higher up into the basalt, showed a ^{14}C age of some 20,000 years (Table 18, sample No. IRM 7621). Here, it can also be postulated that older sinter secretion (see above) within these old, porous materials led to new secretions, making this age a mixed value. The actual development of the calcareous deposits on the older terrace probably took place at a time which cannot be determined by radiocarbon. The lower value limit would be that of the above-mentioned basalts, which would also be the lower limit for the beginning of the wadi erosion.

Fig. 71. Large sinter basins in the small wadi 4 km south of Al Birk (scale: hammer). (Photo: H. HÖTZL, 1977.)

2.4.2.4. Absolute dating of the young clastic sediment sequences

The difficulties in regard to the dating of the young clastic series begin with the establishment of a lower age limit. One of the first reference points is provided by the Baydh Formation which, as the youngest relatively dateable Tertiary formation on the eastern edge of the Tihamah, is located directly below the young clastic series. It can be classed as Miocene, and probably Lower Miocene. A further reference point is provided by the evaporites in the western part of the Tihamah on the coast; studies in the Sudan and Egypt show that their upper portions belong to the Middle-Upper Miocene. In the Mansiyah 1 borehole these evaporites are covered by a 2,000-meter-thick "clastic continental series". The youngest deposit above that is a calcareous sequence 170 meters thick.

A period for the deposition of the clastic series from the Late Miocene into the Holocene must therefore be considered.

The "boulder gravels" belong to the oldest above-ground formations of these young clastic deposits. They are remnants of an old plain of gravel deposits and are found only east of the main graben fault in the vicinity of the crystalline basement. This tectonic limitation of their occurrence shows that they were deposited before the last major movements of this main border fault. Yet this does not secure their position, on the basis of the morphological circumstances, to the next youngest the main terrace.

The extent of the main terrace is also determined mainly by tectonics. It overlays the block faulted Tertiary Dyke and Baydh Formations; this overlaying proceeds from west to east, and with increasingly younger series (see Wadi Baydh and Wadi Samra). This is bordered in the east by the crystalline basement (main border fault), as it dips in the west to a parallel fault or flexure (flexure 2, according to W. GILMAN, 1968). The whole block apparently tipped slightly to the west after the deposition of the main terrace sediments.

If a corresponding change in the tectonic position of the dyke block in comparison to the basement is considered in relation to this dumping, the planation of the coarse boulder gravel and main terrace could be viewed as a uniform accumulation plain. The coarse boulders then correspond to the upper level of the main terrace consisting of boulder gravels. These would have reached farther to the east over the basement as the youngest element of this terrace, as mentioned above. The difference in grain size between the coarse boulder gravels and the gravels of the dyke block would here be due simply to the increasing distance from the mountains.

As the young tectonic shift had less effect in the southern part of the Tihamah, this could explain why, in this region, the main terrace can be followed morphologically down to the basement (see section on Wadi Jizan and Wadi Khulab), and why there are no areas of coarse boulder gravel.

The basalts provide an additional opportunity for narrowing down the age of the sediments of the main terrace. The oldest previously known absolute age value for these overlying basalts of 0.9±0.23 million years (R. G. COLEMAN, et al., 1975, compare also sample No. 81/66 and 77/19, Table 19) came from one of the younger volcanic cones near Wadi Jizan. It must be borne in mind that there the even older basalt flows passed over the planation of the main terrace, which had already been formed.

The deposition of the uppermost coarse gravel and the subsequent deep erosion of the planation of the main terrace would have required enormous runoffs and a much moister climate than what prevail today. The depth of the channel, as shown by the geoelectrical profiles in the Tihamah area, extended to more than 50 meters below the accumulation plain of the time. The buckle in the gradient in the area of the flexure west of the dyke block indicates that tectonic activity resumed after this channel was carved.

The entire Red Sea area shows that after a period of relative quiet in the Miocene, tectonics were reactivated in the Pliocene with the opening of the central graben. This indicates, along with the observations mentioned above, that the main terrace was already dissected in the Early Pleistocene (?) or better Late Pliocene.

Sedimentation and erosion events in the Pleistocene and Holocene are thus limited to the aggradation of the middle terrace, covering wide areas mainly in the central part of the Tihamah; to its slight dissection with the deepening of the present-day wadi channels; and to the deposition of lower-terrace sediments within these channels, including the various eolian accumulations. The total thickness of these Plio-Pleistocene sediments is as much as about 150 meters in the area of the Tihamah which sank before the dyke block. This is in good agreement with the thickness of the "Plio-Pleistocene" calcareous coastal sediments (170 meters) in the Mansiyah 2 borehole. This means that the clastic sedimentation in the central and western Tihamah area kept pace with the sinking of this block. This offers an explanation and a confirmation of the temporal constancy of the course of the shore mentioned in previous sections. With the exception of the few meters' height of the step of the terraces, the Quaternary sediments may, to all intents and purposes, be examined only by means of the bore profiles.

2.4.3. Hydrogeology of Tihamat Asir[1]

(E. MÜLLER, K. FREDRICH, H. KLINGE)

2.4.3.1. General Remarks

Quaternary alluvial deposits form the only important groundwater bearing substratum within the Jizan Plain. Due to very limited amounts of precipitation and very high evaporation rates in the plain, the groundwater in the alluvial deposits is exclusively recharged by percolation of surface water which is brought down by the wadis from the mountain areas. The quality of the groundwater is very heterogenous within the aquifer. This is caused by the fact that the freshwater reservoir is underlain everywhere by a complex of groundwater which is highly saline, the interface being quite undulated.

[1] Investigations financed by the Saudi Arabian Ministry of Agriculture and Water, Ar Riyadh.

Within the last few years, two organizations have undertaken groundwater investigations in the Jizan Plain area. A detailed study was made by ITALCONSULT in 1962–1963 resulting in the FAO-report of 1965 on the „Land and Water Surveys on the Wadi Jizan" (ITALCONSULT, 1965). The investigations of the alluvial deposits and the groundwater of the Jizan Valley area included the performance of six drillings with pumping tests and a geoelectric survey. A more general report of the groundwater resources in the Jizan Plain was undertaken by SOGREAH in 1970 within their "Water and Agricultural Studies" of area VI which covers the whole Tihamah Plain between the Yemeni and the Jordanian borders.

In November, 1975, the Ministry of Agriculture and Water Resources contracted GERMAN CONSULT to carry out the consulting services for the South Tihamah Agricultural Development Project. The objectives of the project are to assist the farmers in irrigating more lands from floods and to improve the productivity of the lands by flood control, distribution of irrigation water, wells, drainage and onfarm investment. Within the frame of this project, the following hydrogeologic work has been carried out between Wadi Atwad in the north and Wadi Haradh near the Saudi-Yemeni border in the south (excluding the areas of Wadi Dhamad and Wadi Jizan, which were investigated by ITALCONSULT, in 1965):

a) Investigations on the properties of the aquifer including:
the geologic mapping
the performance of 18 exploratory drillings and 22 piezometer wells which supplied information about the stratigraphic properties of the aquifer
the performance of 17 pumping tests which supplied information about the hydraulic properties of the aquifer
the geoelectric survey (2000 geoelectric soundings) which especially revealed the shape of the fresh – saline water interface

b) Inventory of the groundwater resources including:
a comprehensive well inventory which supplied information about surface and surface fluctuation of sub-soil water, actual tapping of groundwater and average water quality (some 330 wells were inspected)
an estimation of groundwater recharge by means of hydrologic, climatologic, and hydraulic data, which give an idea of available groundwater resources
isotopic analyses of ^{14}C, ^{2}H, ^{3}H and ^{18}O which supply information about age, origin of saltwater and freshwater, and consequently, recharge of groundwater

c) Investigations on the groundwater quality including:
chemical analyses Na^+, K^+, Mg^{++}, Ca^{++}, Cl^-, NO_3^-, SO_4^{--}, HCO_3^-, Eh, pH-value, hardness of water
chemical classification of groundwaters

d) hydro-geological model calculations by means of data processing
The following report is an outline of the main results of those investigations carried out in the years 1975–1978 and summarized by GERMAN CONSULT (1978).

2.4.3.2. The aquifer

Lithologic sequence

In the northeast the Quaternary aquifer is, more or less, limited by the outcrops of the basement and the dike complex respectively. In principle it forms a flat trough-shaped body with a longitudinal axis trending from NW to SE (see Plate III). The axis line at which the alluvials have developed a maximum thickness of at least several hundred meters seems to be located along the Jizan flexure in the center of the plain. East of the axis line the aquifer generally overlaps the Baydh Formation. To the west the aquifer overlaps the continental series of the Mansiyah-anticline. In the northeast between Jabal Hathah and Wadi Atwad the aquifer lies directly on the Miocene dike rocks. In this area and east of a long north-west-trending Pleistocene fault, the aquifer diminishes rapidly due to the As Sirr displacement. Therefore, beyond Wadi Atwad, groundwater occurrence is very limited.

In the eastern area of Wadi Shahdan the base of the aquifer is the Jurassic Khums Sandstone and the thickness of the aquifer is very small. This is caused by the fact that in this area the most effective fault line of the eastern flexure zone is displaced to the southwest.

The 18 exploratory wells drilled by GERMAN CONSULT revealed the existence of an aquifer which is very heterogeneous in its stratigraphic sequence. The sequence consists of various layers and lenses of sands, silts, clays, gravels, and boulders. It seems that the areal/extension of all layers which have been encountered is very limited. Therefore, it is not possible to correlate specific layers between wells.

Some 70 percent of the penetrated water-bearing aquifer material consists of more or less impermeable silts and clays. Most of the silts contain calcium-carbonate; therefore, the CONSULTANT believes that many of these silts actually are reworked loess. The clays, most of which are sandy and silty, are reddish in color in the lower parts and in places also in the upper parts of the strata. This leads to the conclusion that most of the Quaternary aquifer is of Pleistocene age. Holocene sediments form only a thin cover on top of the Pleistocene aquifer.

Layers and lenses of sand and gravel are interstratified everywhere in the aquifer. The gravels are poorly sorted with maximum grain sizes exceeding five centimeters. Important intercalations of angular boulders have been encountered by drilling hole 6 J 87 adjacent to the As Sirr displacement, hole 6 J 77, and hole 6 J 86 southwest of the volcanic cones in the Wadi Sabya area.

There is no evidence that the underlying beds of the Baydh Formation have been penetrated by any drilling. However, red argillaceous beds alternating with fine and coarse gravels, which SOGREAH (1970) describes as belonging to the Baydh Formation, have also been encountered by the CONSULTANT (Drillings 6 J 79, 6 J 83, 6 J 84 and 6 J 86). On the other hand, the borings 6 J 82 and 6 J 88 in the north have encountered the basaltic bedrock of the dike complex at depths of 39 and 53 meters. Due to the As Sirr displacement, this formation forms the bottom of almost dry alluvium in this area.

Geoelectric results

In addition to the exploratory drilling, 2,000 geoelectric soundings have been performed using the Schlumberger-electrodes-sequence to identify the sequences of the aquifer. The geoelectric graphs, being of the two-step descending type, have been used to identify the following sequence:

- an upper layer of relatively high resistivities (100 ohm.m);
- a middle layer of medium resistivities (10–30 ohm.m); and
- a conductive lower layer of low resistivities (5 ohm.m).

By means of calibration measurings at drilling sites, it was possible to correlate the sequence of conductivity to the following hydrogeological units:

- The upper layer with high resistivities corresponds to dry alluvium.
- The medium resistivities layer corresponds to the freshwater-bearing alluvials, except in the eastern area of section D, where resistivities of 20–30 ohm.m. are caused by the displaced Jurassic sandstones. High resistivities (50–500 ohm.m) within the medium resistivities layer occur in the area of drilling site 6 J 77 (see section D). It is evident that these are caused by the thick intersections of basaltic sheets which were penetrated in that drilling (Figs. 72, 73).

Chemical analyses of groundwater, especially made for geophysical interpretation, show that the mean resistivities recorded usually correspond with the following hydrochemical data:

Conductivity: 0,5–2 mmhos/cm
Chlorides: 50–400 mg
Dissolved solids: 400–1600 mg

- The low resistivities generally correspond with alluvials intruded by saline water. This has been proved by chemical analyses of well samples all over the area. However, in some eastern parts of the plain a coherence with argillaceous strata cannot be excluded.

Over most of the area the limit between medium resistivities and low resistivities is absolutely definite. In areas where the definite limit is replaced by an indistinct transition, a limit has been established below 10 ohm.m, which corresponds with an aquifer (gravel, sand, silt) containing water of a conductivity of roughly 5000 mmhos/cm.

Due to the high and strongly varying concentration rates of chloride in the groundwater it is impossible to reveal any lithologic stratification in the subsoil of the Jizan Plain. This is caused by the fact that the resistivities of clays and sands are in the same low order since they are bearing salt water of high chloride concentration.

The hydrogeological evaluation of the geoelectric survey has yielded the following results in detail:

a) The main groundwater body of the Jizan Plain is highly saline.

b) The complex of saline water is overlain by a body of fresh groundwater.

c) The freshwater reservoir actually is a subterranean channel-like system analogous to the present drainage pattern and to former wadi systems. Wadi migrations have been caused by tectonic events and by basaltic volcanism which has altered the ancient landsurface.

d) The maximum thickness of the freshwater body is more than 100 meters and has been located in the Wadi Baysh and Wadi Sabya areas.

Without regard to the groundwater flows connected with the present system, six subterranean freshwater flows have been located (Figs. 72, 73). Very similar graphs are plotted by ITALCONSULT (1965) for the Wadi Jizan portion lying between Wadi Sabya (Fig. 72) and Wadi Maqab (Fig. 73).

The 6 subterranean freshwater flows are:

– The ancient drainage channels of Wadi Baysh with a direction from NE to SW.

The change in direction of this wadi has obviously been caused by the As Sirr displacement. The extremely undulated interface to the underlying salt water west of the river can be explained by a chronological succession of several movements from west to east. This wadi deviation is considerable and – in the area of section D – amounts to more than 10 kilometers, forming a pattern of several drainage channels. The maximum thickness of the freshwater-bearing layer is about 100 meters and has developed in the area of section C, two kilometers northwest of Baysh. The freshwater-bearing layer has been penetrated by the drilling 6 J 77. At the drilling site, it has a thickness of approximately 60 meters and consists of layers of clays, sands and gravels. The gravel layer has a total thickness of 20 meters. Possibly, these groundwater channels are connected with

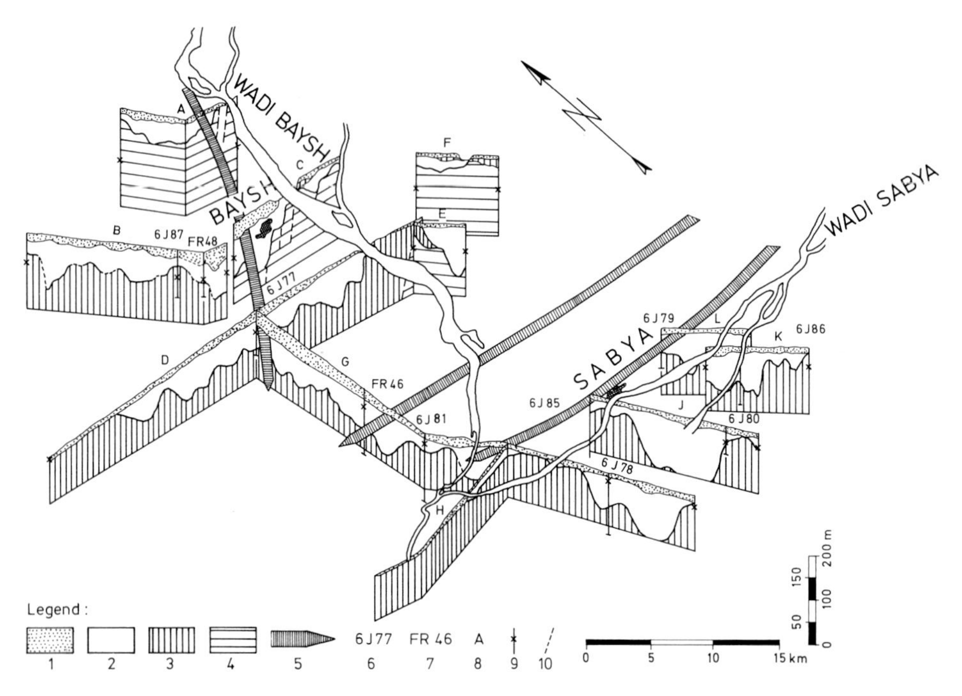

Fig. 72. Fence diagram of the freshwater bearing alluvium (northern area, Jizan region: *1* dry alluvium; *2* freshwater bearing alluvium; *3* conductive substratum [mostly alluvium intruded by saline water]; *4* bedrock; *5* ancient subterranean freshwater flow; *6* drilling number [e.g. 6J85]; *7* number of SOGREAH drilling; *8* section; *9* sea level; *10* fault).

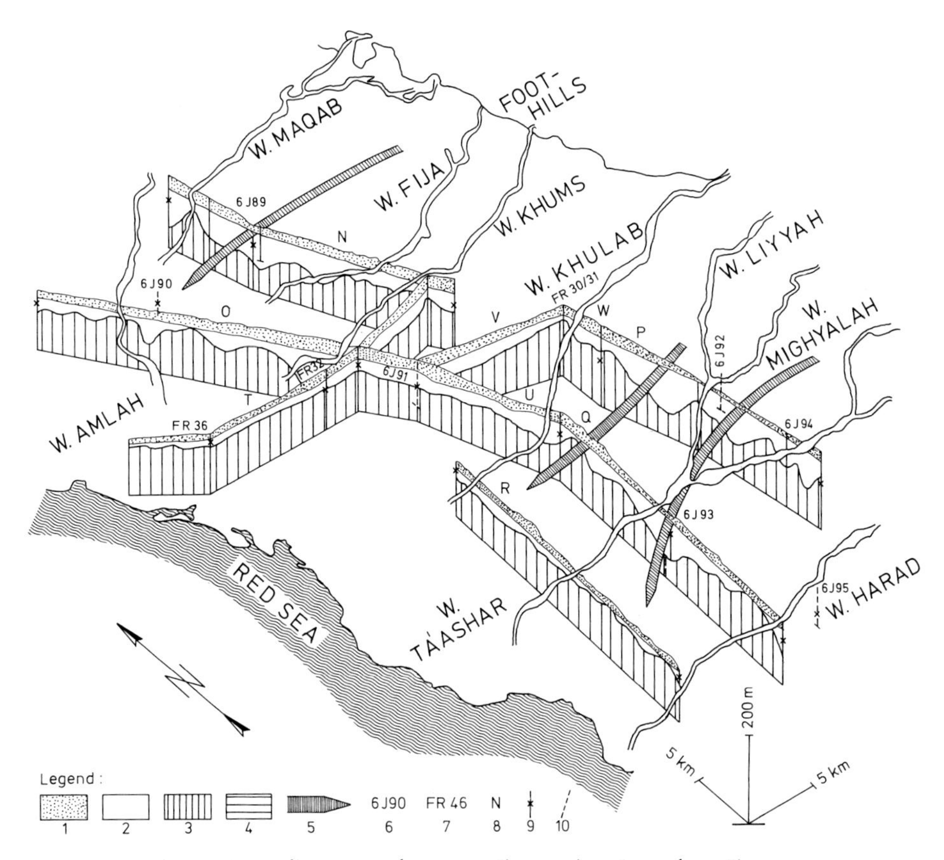

Fig. 73. Fence diagram southern area, Jizan region. Legend see Fig. 72.

diversion channels, through which flood-water from Wadi Baysh is diverted to flood-irrigated areas to the west of Wadi Baysh. Actually, the freshwater channels which have been revealed by the geoelectric survey underlie these diversion channels. Therefore, it would seem to be evident that the groundwater channels are, at times, fed by percolation from the overlying surface flood channels.

– Two ancient drainage channels west of Jabal Akwah with an E–W-direction.

It is apparent that this former wadi system has been cut off the ancient surface drainage pattern due to a fairly rapid morphologic modelling of the former landsurface connected with the volcanic eruption of Jabal Akwah.

The northern drainage channel is detected by the geoelectric logs only in its lower course. Therefore, its direction cannot be determined. However, a chemical analysis of well water 6 J 84 shows that the direction of flow could possibly be as stated above. In the area of well 6 J 84, the freshwater-bearing layer has a minimum thickness of 55 meters (the top of saline water has not been encountered) and consists of an interbedding of clays, silts, and gravels. The gravels, divided into four layers, have a thickness of 15 meters.

The southern drainage channel has been delineated by means of sections G, I, and L. It passes the vicinity of Sabya and provides the drinking water supply of this village. Crossing section M, the freshwater-bearing layer thickness increases to 100 meters. Northwest of Sabya, however, the depth decreases (well 6 J 85) to 30 meters. In its lower course, the freshwater flow joins the drainage channel of the recent Wadi Baysh.

– The ancient drainage channel between Wadi Maqab and Wadi Fija with a direction from E to W. This former wadi bed lies beneath the volcanic cones in the eastern section of this region. The geologic history seems to be similar to the hydrogeological development in the Jabal Akwah area. The freshwater was located by means of the geophysical sections N and O. The thickness of the freshwater-bearing layer is approximately 75 meters in the area of section N and gradually increases to the west. It has been penetrated by the drillings 6 J 89 and 6 J 90, the two logs being very different. The freshwater beds in the vicinity of 6 J 90 are formed by a group of gravels and sands (20 meters) which are overlain by two thick layers of clay and silt.

Drilling 6 J 89 has revealed a strongly silty aquifer which has four layers of sands and fine gravels totaling 10 meters in thickness.

The freshwater channels are possibly fed at times by Wadi Maqab tributaries draining the area to the west of a volcanic cone situated south of Wadi Maqab. Drilling site 6 J 89, where the freshwater-bearing aquifer has its maximum thickness, is located adjacent to a tributary.

– Two ancient drainage channels in the Wadi Khulab-Ta'ashshar area with a direction from E to W. The origin is not known.

They have been located by means of geophysical sections P and Q. The freshwater-bearing layer has developed a maximum thickness of at least 100 meters (Section P) in its upper course. While flowing to the west the depth is rapidly decreasing. Passing section R the flows are present only by way of intimation. The southern freshwater beds have been penetrated by drilling 6 J 93. At the drilling site they have a depth of approximately 70 meters and are formed by alternating stratifications of coarse and fine layers. The coarse layers have a total thickness of approximately 18 meters.

Hydraulic properties and piezometric surface

For the evaluation of the pumping tests, which have been performed in connection with the exploratory drillings, only nonequilibrium well formulae and graphical methods have been used. Preference was given to methods of THEIS and HANTUSH-JACOB (WALTON). The JACOB'S straight line method could be applied only with great care, as the drawdown- and recovery curves have two major slopes. The results of the pumping tests are shown in Tables 20, 21.

All pumping tests indicate a semi-confined aquifer with an influence of leakage. The leakage is indicated by fast stabilization of the dynamic level and by fast recovery after pumping has been stopped.

It is caused by vertical movement between the layers of different transmissivities, especially at those wells where the uppermost layers have not been screened.

Table 20. *Summary of results of pumping tests (Wadi Groups 1 and 2)*

No.	Location	Depth m	T m^2/s	Storage Coeff.	r/L	L	c	Q/s $m^3/h.m.$	Remarks
6 J 77	SW of Baysh	92.50	$4 \cdot 10^{-3}$	$4 \cdot 10^{-4}$	0.4	190	104	4.3	t/t' = 13
6 J 78	Jedayen W. Sabya	114.00	$1.5 \cdot 10^{-2}$	$5 \cdot 10^{-4}$	0.06	515	204	18	t/t' = 10
6 J 79	W. Nakhlan-Sabya	54.00	$2.0 \cdot 10^{-3}$	$2 \cdot 10^{-4}$	0.8	65	25	5	t/t' = 40
6 J 80	SE El Gara	44.70	$4.0 \cdot 10^{-3}$	$4 \cdot 10^{-4}$	0.4	106	33	7	t/t' = 200
6 J 81	Jumeima, W. Baysh	68.75	$1.2 \cdot 10^{-2}$	$2 \cdot 10^{-4}$	0.08	370	132	10	t/t' = 26
6 J 82	N of Abul Gaid	44.50	$3 \cdot 10^{-3}$	$3 \cdot 10^{-5}$	0.1	432	720	4	t/t' = 16
6 J 83	SE of Abu Sala	66.00	$3 \cdot 10^{-3}$	$1.3 \cdot 10^{-4}$	0.1	410	628	2.2	t/t' = 14
6 J 84	N of Sabya	124.00	$1.3 \cdot 10^{-2}$	$5 \cdot 10^{-4}$	0.25	277	80	14	t/t' = 9
6 J 85	Matherith, W. Sabya	120.70	$3.5 \cdot 10^{-3}$ $1.2 \cdot 10^{-3}$	$9 \cdot 10^{-5}$ $2 \cdot 10^{-4}$	– 0.08	– 640	– 4000	– 5	t · 1.5 min barrier in 50 meters recharge in 250 meters
6 J 86	NW of Baysh	73.00	$2.7 \cdot 10^{-3}$ ($1.5 \cdot 10^{-3}$	$3 \cdot 10^{-4}$ $3.3 \cdot 10^{-4}$ for t 4 min)	–	–	–	1.1	reaching boundary in 140 m, 310 m, and appr. 500 m

Table 21. *Summary of results of pumping tests (Wadi Groups 3, 4 and 5)*

No.	Location	Depth of installation in m	T m^2/s	Storage Coeff.	r/L	L	c	Q/s $m^3/h.\ m.$	Remarks
6 J 89	appr. 11 km south of Abu Arish on dirt road to Sug Al Ahad	75.00	$3.5 \cdot 10^{-3}$	$1.4 \cdot 10^{-3}$	0.15	370	453	5.5	(t/t') = 3, max. capacity (short term) 38 m^3/h
6 J 90	between Muhaddin and Khuzna, west of Qaim	58.00	$1.8 \cdot 10^{-2}$	$4 \cdot 10^{-4}$	0.1	406	106	25	(t/t') = 42
6 J 91	at Ghowaidiya, east of As Sirr	97.00	$4 \cdot 10^{-3}$	$5 \cdot 10^{-5}$	0.2	246	175	10	(t/t') = 70
6 J 92	at Major, Wadi Mighiliya	73.00	$3.5 \cdot 10^{-3}$	$1.8 \cdot 10^{-4}$	0.4	84	23	10	(t/t') = 87
6 J 93	east of Sadiyah and Huthrur	80.00	$2 \cdot 10^{-2}$	$1 \cdot 10^{-3}$	0.2	185	20	12.3	(t/t') = 11
6 J 94	at Al Zeeb, Wadi Ta'ashshar	59.70	$5 \cdot 10^{-4}$	$2.5 \cdot 10^{-4}$	0.8	52	63	2.0	(t/t') = 76
6 J 95	near Hathira and Aswadiyah Wadi Haradh	70.70	$1.2 \cdot 10^{-2}$	$2 \cdot 10^{-4}$	0.08	390	147	19.2	(t/t') = 2.5

Apart from two wells with boundary conditions and one well with very low transmissibilities, the other wells can be divided into two groups:

A. $T = 1.2 - 2 \cdot 10^{-2}\ m^2/s$

Group A represents the wells with full wadi connections.

B. $T = 2 - 4 \cdot 10^{-4}\ m^2/s$

Group B comprises wells with diminished wadi connections.

The storage coefficient is between 0.9 and $5 \cdot 10^{-4}$. Two wells have much smaller coefficients (3 to $5 \cdot 10^{-5}$), and two wells have larger storage.

On the basis of the specific capacity (ratio of rate of discharge to the stabilized drawdown of one day pumping) the wells can be divided into 3 groups:

A. wells near wadis

$$Q/s = 25 - 10\ m^3/h \cdot m.$$

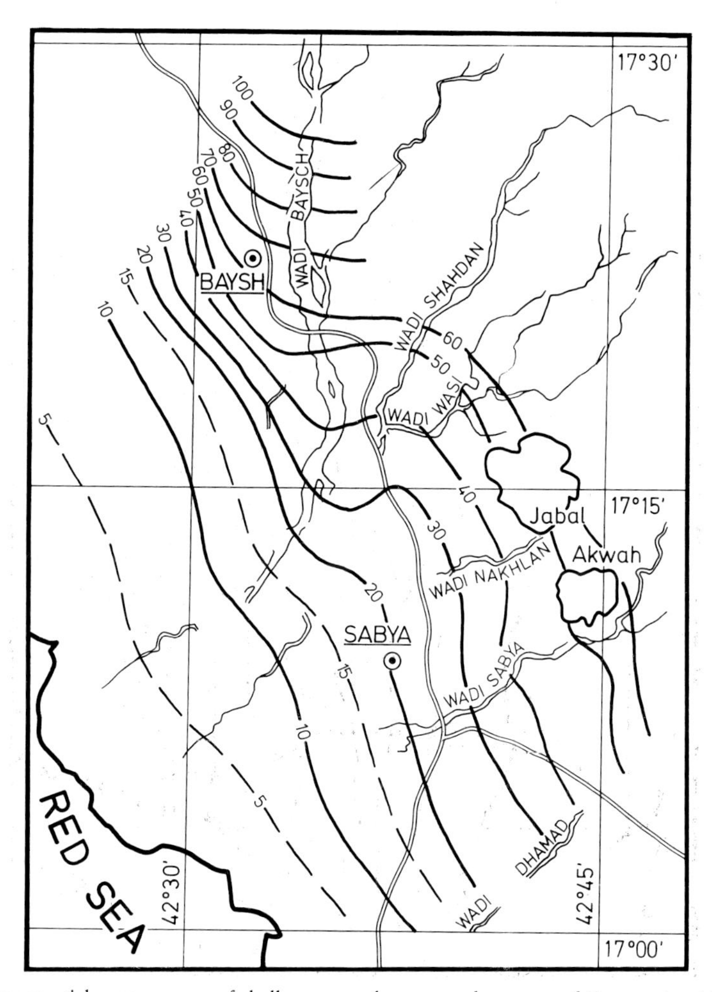

Fig. 74. Isopotential contour map of shallow groundwater northern area of Jizan region (April–May 1976). Contours in meters above sea level.

B. good wells between wadis

$$Q/s = 10 - 4\ m^3/h \cdot m.$$

C. wells with limitations

$$Q/s = 4 - 2\ m^3/h \cdot m.$$

All results ought to be regarded as a unit only. It is not feasible to take transmissibility and storage alone and to neglect leakage, for the error would be considerable.

Due to the influence of leakage, the extrapolated drawdown after one month of continuous operation of wells is only one-half or even less compared with the early time properties of the aquifer.

The comprehensive well inventory included a registration of some 330 wells.

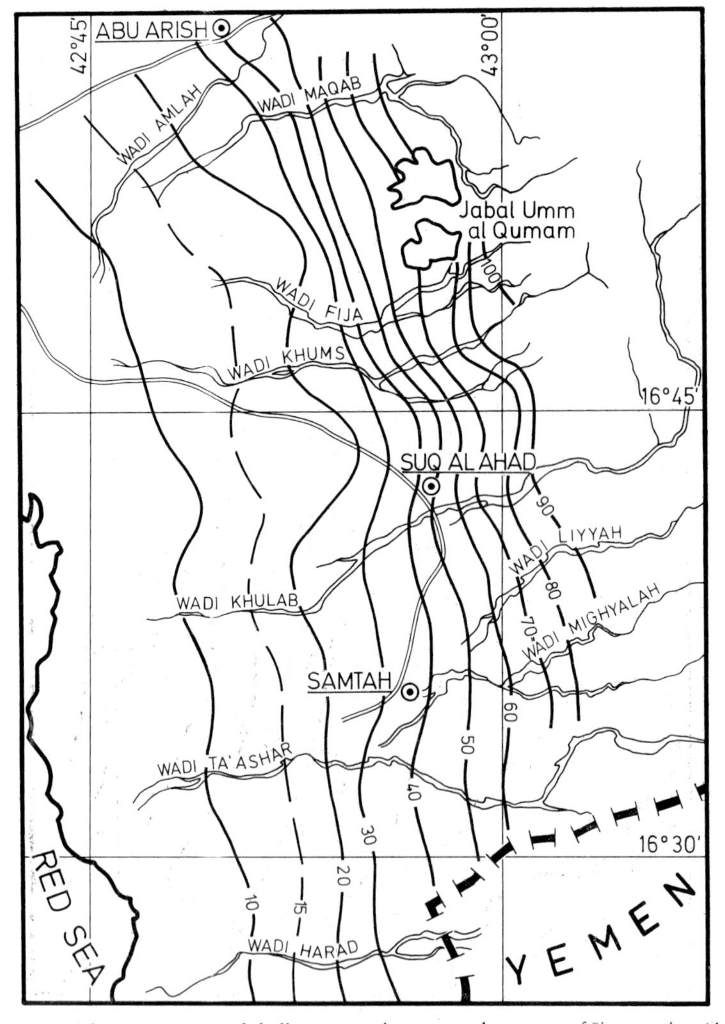

Fig. 75. Isopotential contour maps of shallow groundwater southern area of Jizan region (Aug.–Sept. 1977).

Table 22. *Groundwater extraction of the project area*

Location	Domestic Water supply						Pumping irrigation wells			Total		
	Hand-dug wells			Pumping wells								
	l/s	MCM/a	%	l/s	MCM/a	%	l/s	MCM/a	%	l/s	MCM/a	%
Wadi Group 1	3.8	0.12	18	7.3	0.2	32	11.6	0.41	50	22.7	0.7	100
Wadi Group 2	2.6	0.19	2	1.1	1.0	26	87.0	2.7	72	120.9	3.8	100
Wadi Group 3	2.2	0.07	9	5.1	0.2	20	17.6	0.6	71	24.9	0.8	100
Wadi Group 4	0.7	0.02	3	6.0	0.2	22	20.3	0.6	75	27.0	0.9	100
Wadi Group 5	1.1	0.03	1	12.1	0.4	17	58.6	1.9	82	71.8	2.3	100
Wadi Group 6	1.1	0.03	/	/	/	/	/	/	/	1.0	0.03	/
Projekt Area	11.6	0.36	4	61.6	1.9	23	195.1	6.1	73	268.3	8.5	100

Most of the wells in the investigation area are hand-dug. The water is raised according to the traditional method using a rubber bucket fixed on a rope. Drilled pumping wells for domestic water supply are located predominantly in bigger villages and towns. Drilled pumping wells for irrigation purposes are concentrated in the areas of Wadi Sabya, Wadi Amlah and Wadi Ta'ashshar.

The drilled wells have an average depth of 40 to 60 meters; they are equipped with a 16 or 26 HP-engine. Due to the unsatisfactory construction of the wells the yield is relatively low (8 to 14 l/s depending on the installed engine). In total, some 70 tube wells have been registrated.

The isopotential contour line maps are drawn on the basis of a levelling of water tables of the 330 wells observed (Figs. 74, 75).

Table 23. *Groundwater recharge of the project area*

	Surface runoff	Groundwater recharge		
Location	RCM/a	l/s	MCM/a	% of runoff
Wadi Group 1*	105	305	9.6	9.1
Wadi Group 2	19.1	231	7.4	36
Wadi Group 3	15.4	123	3.9	25
Wadi Group 4	29.4	156	4.9	17
Wadi Group 5	68.2	428	13.5	10
Wadi Group 6	10.1	146	4.6	45
Project Area	251.2	1389	43.8	17

* Including Wadi Shahdan, Wadi Nakhlan.

The direction of the underground flow is roughly parallel to the wadi beds; that means a general direction from the NE to the SW, e. g. from the foothills toward the sea.

The contour map indicates that the slope of the water table is influenced by the wadis. The contour lines in this range are diverted to the sea; that means the groundwater is fed by infiltration through the wadi beds.

The piezometric gradient varies from 0.5 percent in the upper parts of the wadis to 0.07 percent at the beginning of the coastal deltas.

Under natural conditions the depth of the water table below surface is low within the wadi beds, and increases toward the area between the wadis (Fig. 76, 77). In the wadi beds there are depths of less than five meters; and sometimes water pits of one meter, depending on the annual season, can be found, whereas in the areas between the wadis the groundwater table decreases to a depth of more than 30 meters.

The seasonal variation of the groundwater table in wells depends on the location of the well. In wells which are directly influenced by the wadi the variation is between 1.5 and 2.0 meters. In wells between the wadis the fluctuation is less than 0.5 meter.

Table 24. *Groundwater resources of the South Tihamah project area*

	Natural recharge		Present extraction			Proposed extraction			Free capacity	
	l/s	MCM/a	l/s	MCM/a	% of Rech.	l/s	MCM/a	% of Rech.	l/s	MCM/a
Wadi Group 1	305	9.6	23	0.7	8	213	6.7	70	190	6.0
Wadi Group 2	231	7.3	121	3.8	51	161	5.1	70	40	1.4
Wadi Group 3	123	3.9	25	0.8	20	86	2.7	70	61	1.9
Wadi Group 4	156	4.9	27	0.9	17	109	3.4	70	82	2.6
Wadi Group 5	428	13.5	72	2.3	13	300	9.4	70	228	7.2
Wadi Group 6	146	4.6	1	0.03	/	/	/	/	/	/
Project Area except Wadi Group 6	1243	19.2	269	8.5	18	869	27.4	70	601	18.9

The inter-annual variations of a number of observation wells have been observed since 1968. The hydrographs show a decrease of the piezometric surface from 1968 to 1973 and an increase or a stabilization of the water table since 1974 in coherence with the general climatological tendency.

The well inventory includes an estimation of the present groundwater extraction. Table 22 shows the estimated extractional rates for the different wadi groups.

The groundwater recharge rate has been calculated by an estimation of the underground flow in the lower part of the wadis according to Darcy's formula. The quantity of recharge in comparison with the surface runoff is shown in Table 23. The mean percentage of the groundwater recharge is between 20 percent and 30 percent of the surface runoff of the wadis except of Wadi Baysh where it is only nine percent of the surface runoff.

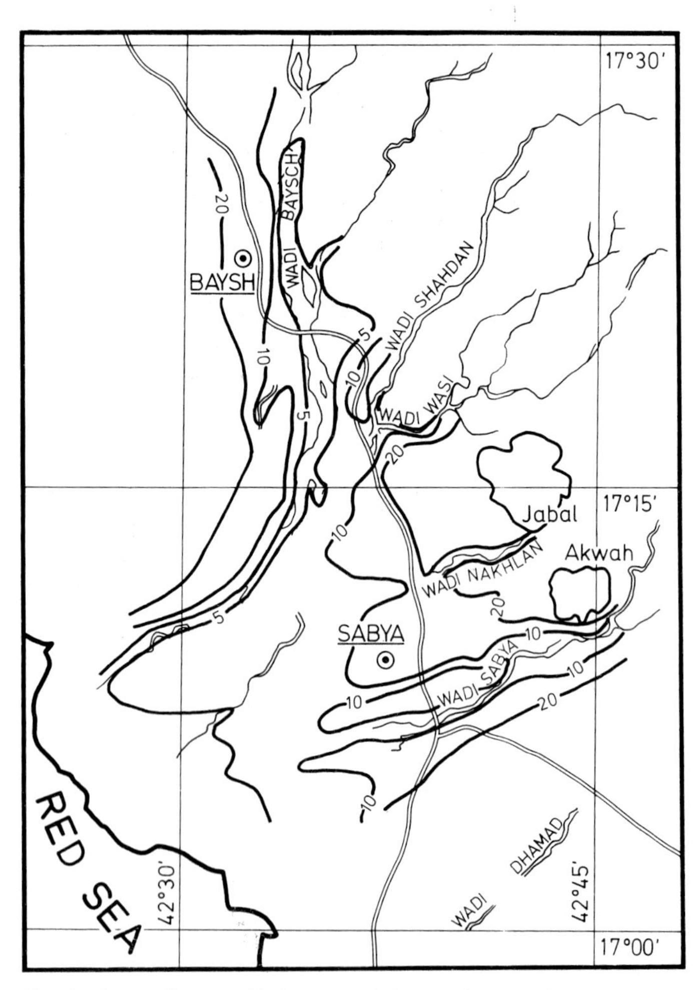

Fig. 76. Depth of groundwater table in meters below surface (northern area, Jizan region).

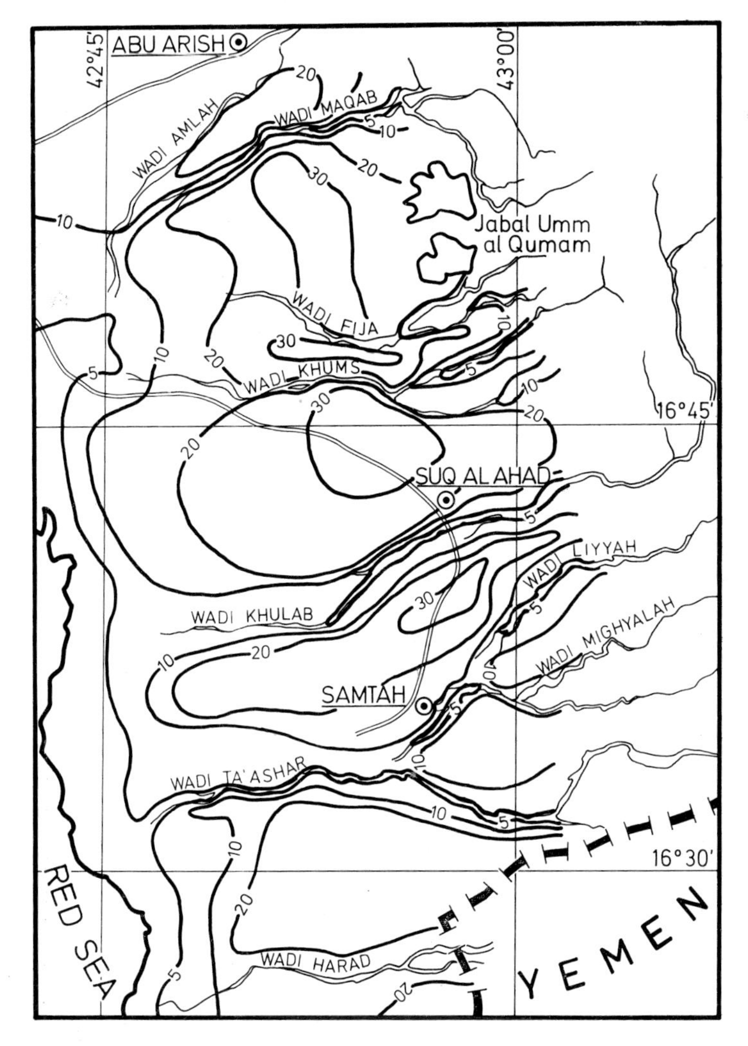

Fig. 77. Depth of groundwater table below surface (southern area, Jizan region).

The total quantity of recharge for the project area has been estimated at 1,390 l/s or 44 MCM/a. At present (1976/77), about 19 percent (8.5 MCM/a) of this quantity is used for domestic purposes and irrigation purposes. The rate of groundwater extraction is very low in the Wadi Baysh area (eight percent) whereas in the Wadi Sabya area it is about 50 percent due to the high number of pumping wells.

In the area of the wadi groups 3 to 5, the extraction rate is between 16 and 20 percent of the estimated recharge. Starting from a discharge rate of 70 percent a quantity of about 600 l/s or 19 MCM/a is available as additional groundwater extraction in the project area (see Table 24).

2.4.3.3. Groundwater quality
(E. MÜLLER)

180 samples of well waters from the Jizan Plain have been analyzed for hydrochemical studies. The analysis included the determination of the cations Na^+, K^+, Ca^{++} and Mg^{++} and the determination of the anions Cl^-, NO_3^-, SO_4^{--} and HCO_3^-.

The conductivity maps of the shallow groundwater of the project area (Figs. 78, 79) show the general situation of groundwater quality.

Groundwater of low electric conductivities is found near the wadis and in the foothill areas; the conductivity rises in a downstream direction and in the areas between the wadis.

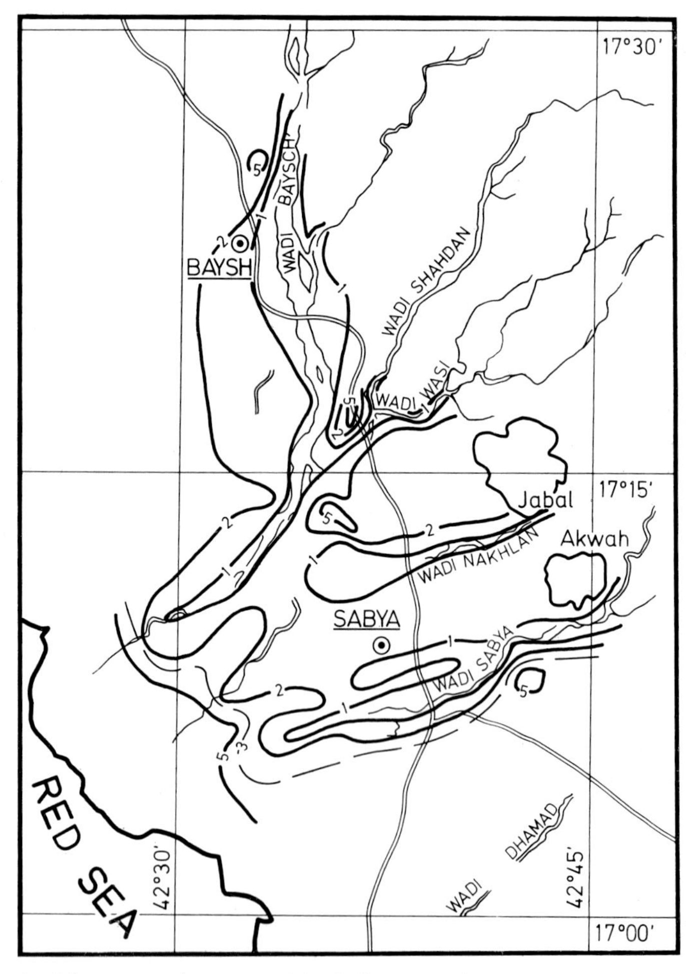

Fig. 78. Conductivity map (mmhos · cm⁻¹) of the shallow groundwater (northern area of Jizan region).

The width of the zone of low conductivity depends on the size of the catchment area and the size of the baseflow, in that order.

Generally, the water salinity increases with the length of flow.

The highest conductivities have been found in the coastal plains, where the content of diluted solids rises up to 4,800 mg/l (well No. 44).

The sea water analysis has been quoted from HALCROW and PARTNER (1972); the sample has been taken 50 meters offshore to the NE of Jizan.

The analyzed wadi water has been taken from Wadi Baysh near Masliyah, and is considered to be a representative sample. Table 25 shows the results of the two analysis.

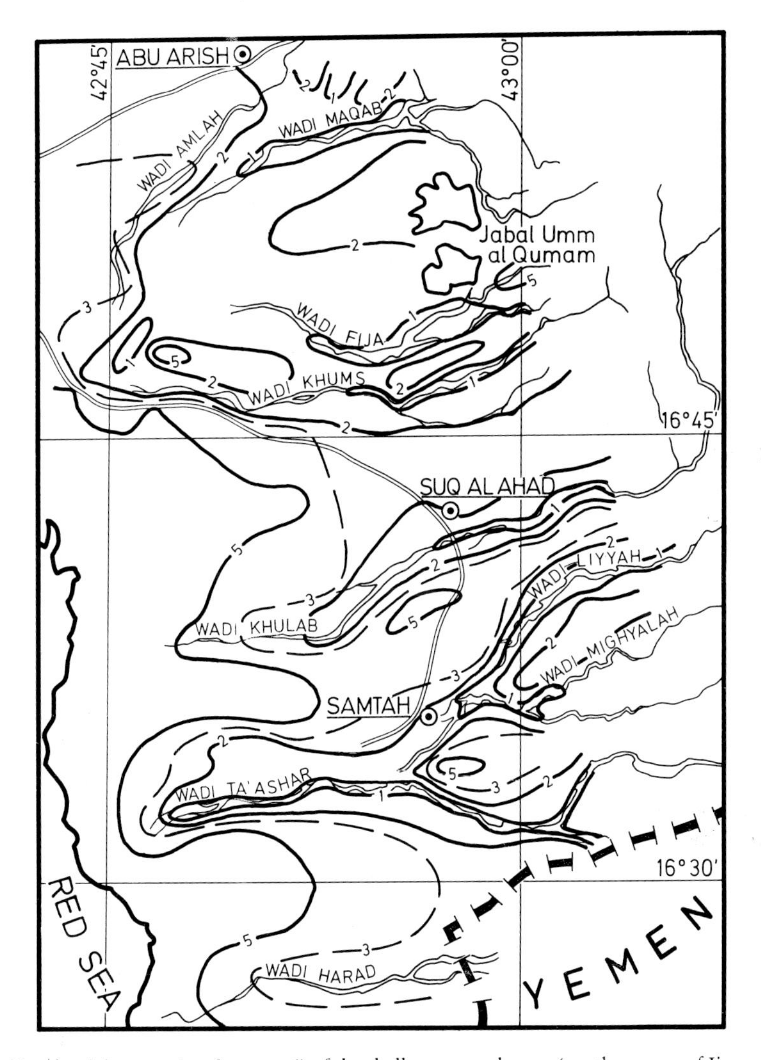

Fig. 79. Conductivity map (mmhos · cm⁻¹) of the shallow groundwater (southern area of Jizan region).

Table 25. *Composition of sea water and wadi water*

	Sea water			Wadi water		
	mg/l	meq/l	meq %	mg/l	meq/l	meq%
Na^+	13,500	580.7	76.5	34.0	1.48	27.6
K^+	664	17.0	2.2	8.2	0.21	3.9
Mg^{++}	1,886	138.3	18.3	13.0	1.07	20.0
Ca^{++}	458	22.9	3.0	52.0	2.59	48.5
Total	16,358	758.9	100.0	107.2	5.35	100.0
Cl^-	24,420	688.8	89.6	29.0	0.82	15.2
NO_3^-	420	6.8	0.9	13.0	0.21	3.9
SO_4^{--}	3,357	69.9	9.1	87.0	1.81	35.5
HCO_3^-	183	3.0	0.4	156.0	2.56	47.4
Total	2,830	758.9	100.0	285.0	5.40	100.0

Wadi water generally has low mineralization, and its conductivities vary between 0.3 and 0.7 mmhos/cm, mainly depending on the amount of baseflow.

Ca^{++} and HCO_3^- are dominant; the content of SO_4^{--} is relatively high (25 – 25 meq %); the content of Na^+ and Cl^- is generally low.

The content of Ca^{++} is equal to or slightly higher than that of HCO_3^-.

A surplus of Na^+ in comparison with Cl^- is typical.

The rocks, exposed in the catchment area of the wadis, are predominantly low-grade metamorphic schists. Generally, water originating from these rocks contains a higher amount of Na^+ than of Cl^-, as the weathering of silicate-rocks releases a higher amount of cations than of anions.

The dominant ions of the sea water are Na^+ (76.5 meq %) and Cl^- (89.6 meq %) whereas the content of HCO_3^- is only 3 meq (0.4 meq %).

Unlike the wadi water, the amount of Cl^- exceeds that of Na^+.

Generally, it is possible to distinguish three groups of shallow groundwater with respect to the dominant anion:

1. Bicarbonate water
2. Chloride water
3. Sulfate water

1. Water samples containing bicarbonate (HCO_3^-) as a dominant constituent can be found in the foothill areas near the upper courses of the wadis. The chemical composition of these waters is relatively similar to that of the wadi water. The bicarbonate and calcium ratio generally is high; the chloride ratio, therefore, is low. In the farther downstream area, calcium is partly replaced by sodium; the samples are of a typical $NaHCO_3$-type water, due to exchange reactions between the groundwater and the alluvial aquifer. The ion-exchange reactions take place in the transition zone between salt water and freshwater, due to an interaction between an ion-exchange-medium, usually Na^+-bearing clay minerals, so-called natural permutites and the Ca^{++}-ions of the freshwater.

2. Roughly 50 percent of all well water samples contain chloride as a dominant anion.

A high content of chloride is always associated with a high content of sodium.

Generally, the sodium-chloride ratio increases with an increase of the distance to the foothill area, a fact which clearly shows that the salting of freshwater is caused by a mixture of saline water and freshwater.

The water samples of the sodium-chloride-type contain a moderate to high percentage of Mg^{++} and SO_4^{--}. Generally, the content of Mg^{++} and SO_4^{--} increases in a downstream direction too.

As in the sea water, in most of the samples, the content of chloride exceeds that of Na. A large amount of Na^+-ions compared with the content of Cl^--ions has been found in a number of samples.

These waters are of a $NaSO_4^-$-type which may indicate an exchange reaction between Ca/Mg-sulfate bearing saline water and Na-bearing exchange-medium as well.

3. 25 percent of all water samples contain SO_4^{--} as a dominant anion.

In the area of the wadi groups 1 and 2, high sulfate concentrations have been found in a number of wells in the upstream area of Wadi Sabya. Here the water generally is of a relatively low conductivity.

In the middle coarse of Wadi Baysh, especially in the area of Ash Shakar and Al Goth, these water samples contain a high amount of diluted solids and a sulfate content of more than 1000 mg/l (wells No. 140, 148, 159). In the southern part of the project area, well water with a high sulfate content has been found to the south of Samtah, mainly near Wadi Haradh. The high sulfate content is always associated with a relatively high content of Ca^{++} and Mg^{++}. In all samples, the Na^+-content exceeds that of Cl^-.

Generally, the existence of Mg/Ca-SO_4-bearing water may be explained by an interaction between freshwater and sulfate rich saline water which may originate from salt-bearing strata of the Baydh Formation. Due to the restricted information about the water chemistry of the underlying formations, it is not possible, to determine definitely the source of the existing saline water.

In the project area, the groundwater is predominantly of very poor quality when compared to domestic water supply. Only 35 percent of the samples are below the WHO-standard of 50 mg/l of nitrate; 35 percent even exceed 100 mg/l. A content of 400 mg/l was found in four wells. Wells of good water quality are located close to the wadis. The high nitrate contents most probably result from a dilution of a primary nitrate content of the alluvial sediments.

As for the quality of groundwater with respect to irrigation purposes, it is found that 67 percent of the wells are classed as having low to medium saline waters. These classes of waters are suitable for the plants in the cropping pattern of the project area.

15 percent of the groundwater fall into class $C_2 - S_1$, i e., high salinity and low sodium hazard. Such water can be used to cultivate plants with a good salt tolerance, and a leaching of up to 25 percent (or 1,500 cubic meter/ha water per irrigation season) is required in order to maintain a favourable salt balance in the soil.

The remaining 37 percent of the observed wells falls into classes $C_3 - S_2$, $C_4 - S_2$, and other higher classes having predominantly high and very high salinity and low to medium sodium content. These waters can be used to cultivate very salt tolerant crops combined with a leaching of 50 percent.

There is no danger of bicarbonate hazard because the "residual sodium carbonate" (RSC) content is far below 1.25 meq/l in the samples analysed, considering groundwater and surface water together.

Sodium and chloride, on an average of 9.35 and 10.85 meq/l respectively in almost all water samples, are the dominant minerals in the shallow groundwater. In only five percent of the cases, these exceed values of 20 meq/l. The boron content in irrigation water as well as in the soil is far below the toxic limit.

Since the SAR (sodium absorption ratio) of the groundwater samples appears to be fairly uniformly low until very high salinities are reached, tube-well irrigation would seem possible, although increasingly hazardous, up to EC value of about 2.25 mmhos/cm.

2.4.4. Isotope Hydrology Tihamat Asir[1]

(E. MÜLLER, W. RAUERT, W. STICHLER, J. G. ZÖTL)

2.4.4.1. Introduction

Environmental isotopes provide an overall labelling of water throughout its cycle. Therefore, about 170 groundwater samples from the Jizan aquifer were analyzed for their ^{2}H, ^{3}H and ^{18}O content. Among these, 14 samples were analyzed for their ^{14}C content.

The fundamentals of isotopic analyses and their hydrological interpretations are presented in Vol. 1 p. 153ff. (see also H. MOSER, W. RAUERT, 1980).

2.4.4.2. Results of isotope measurements

2.4.4.2.1. Isotope data from shallow groundwater

The sampling locations are shown in Figs. 80 and 81. The samples were taken from August 1976 to November 1977.

A survey of all stable isotope results is given in the form of an $\delta^{2}H - \delta^{18}O$ diagram in Fig. 82. In addition to the groundwater data, the stable isotope data of two samples each from the Red Sea and the Maloki Dam are plotted. As can be seen, the measured points lie around or below the precipitation line $\delta^{2}H = 8\ \delta^{18}O + 10$. Fig. 83 shows the different $\delta^{2}H - \delta^{18}O$ relations for a normal precipitation line and an evaporation line.

The ^{3}H contents of groundwater samples measured are in the range of 0 to 56 TU with three exceptions reaching to 83 TU.

A few samples taken from surface water are available:

Maloki Dam (right)	8 June 1976	(16.1±1.4) TU
Maloki Dam (left)	8 June 1976	(16.1±1.4) TU
Red Sea water from south of Jizan	8 June 1976	(3.0±0.7) TU
Wadi Baysh	24 July 1976	(35.0±6.0) TU
Wadi Baysh	9 Oct. 1976	(23.0±2.0) TU

---

[1] Investigations financed by the Saudi Arabian Ministry of Agriculture and Water, Ar Ryadh.

The samples of rain water collected within the same period had ^3H contents of 7, 8 and 32 TU. Unfortunately, no isotope data are available from the station Jeddah of the IAEA precipitation network. It can, however, be estimated from the values mentioned above and from isotope data of Bahrain Island and Khartoum (Sudan) that, in 1976, the ^3H contents of precipitation generally were between about 10 and 40 TU in the area under investigation. Higher ^3H concentrations found in groundwater have to be attributed to earlier years when the ^3H contents in precipitation were accordingly high. On the other hand, very low or nil ^3H contents reveal that the groundwater had originated mainly before 1953.

Coastal area between Ad Darb and Wadi As Sirr (Fig. 84, 85).

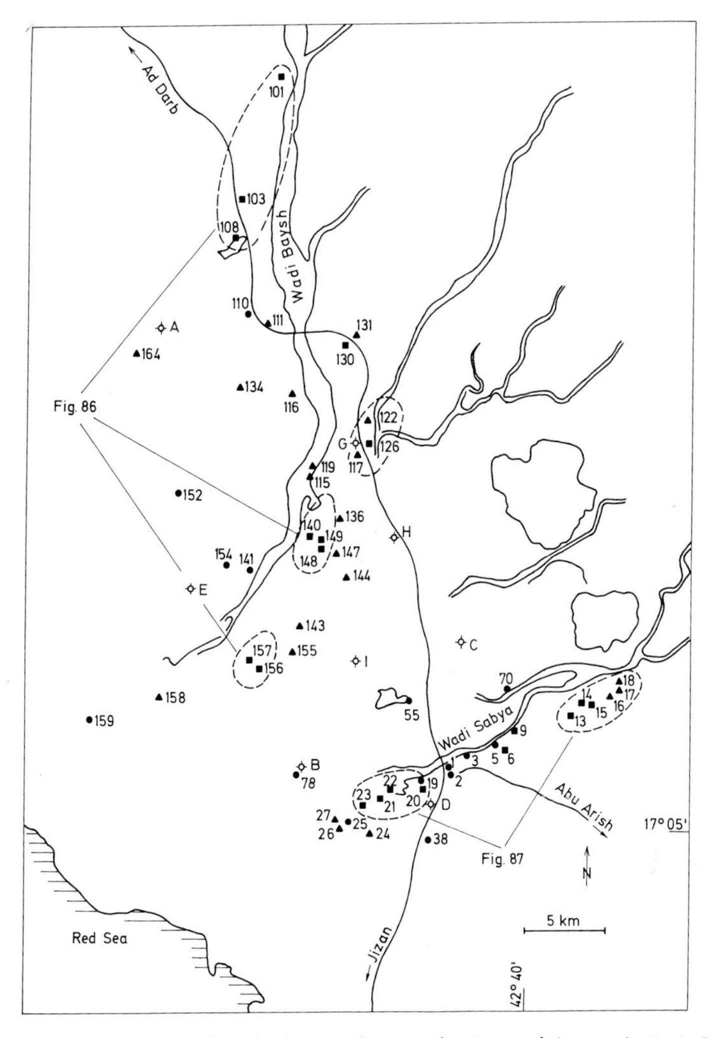

Fig. 80. Location and sample numbers for isotope data (northern area of Jizan region). *A*, *B* etc. deep boreholes.

The $^{2}H - ^{18}O$ values of the water samples (Table 26 a) lie on a straight line with a slope of 8 and a deuterium excess d of approximately +11. This line presumably corresponds to the local precipitation line. The differences of the values may be attributed to the different mean altitudes of the catchment areas. The ^{3}H contents of the samples from wells No. 800 and 803 indicate dominance of recent groundwater which has been formed by precipitation since 1953. On the other hand, the sample from well No. 802 contains groundwater formed before 1953. The remaining samples can be considered as a mixture of these two types. With

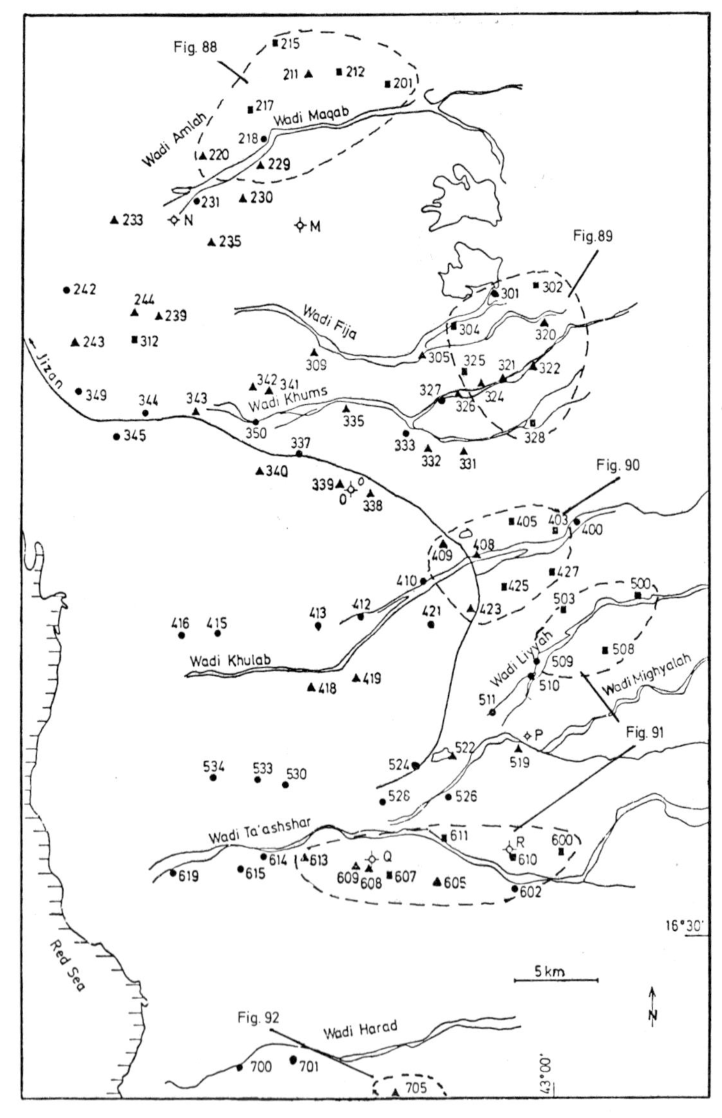

Fig. 81. Location and sample numbers for isotope data (southern area of Jizan region). Within broken lines areas of isotope data strongly influenced by evaporation (cf. Figs. 82–92 and Tables 25–31).

the exception of the sample from well No. 805, the water samples can be classified according to their electrolytical conductivities in the same manner as to their ^{3}H contents. This means that water of high salinity has a relatively long residence time.

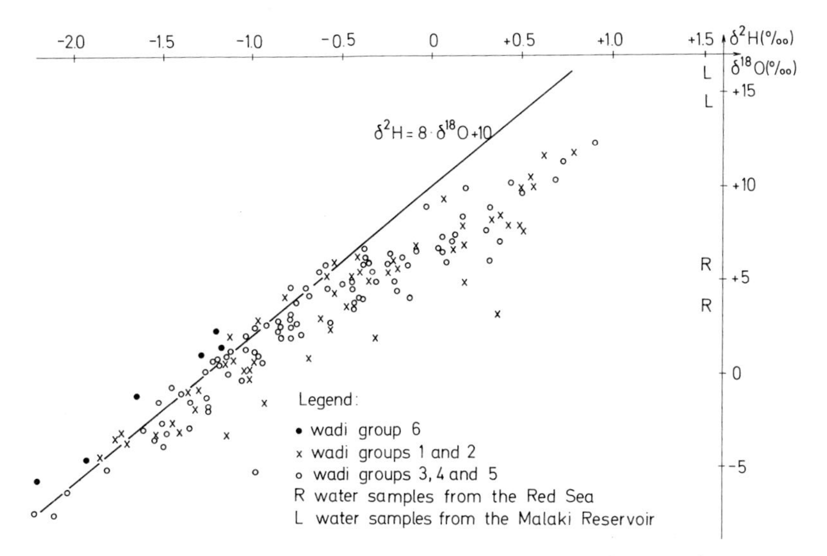

Fig. 82. δ^2H-$\delta^{18}O$ relation of all groundwater samples of Jizan region.

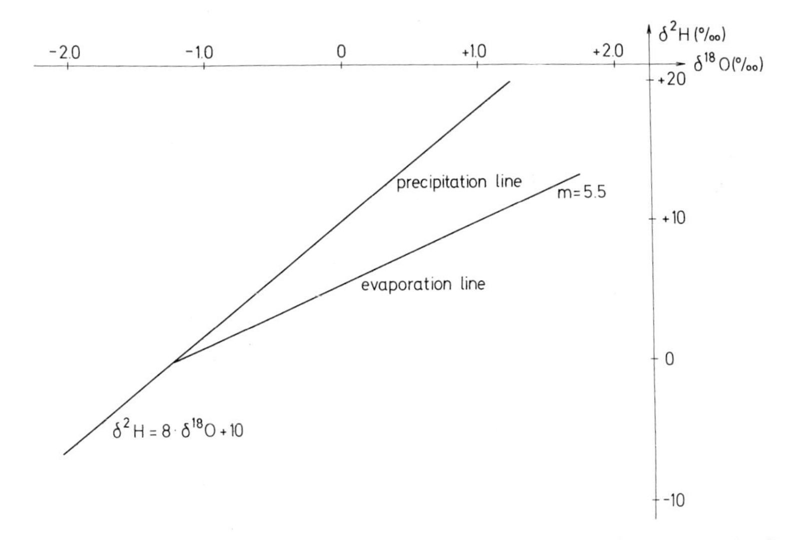

Fig. 83. Example for δ^2H-$\delta^{18}O$ relations for a normal precipitation line and an evaporation line with deuterium excess.

Wadi Baysh

The stable isotope data scatter more or less around an evaporation line with a slope of 5.5. The $\delta^2H - \delta^{18}O$ values of samples taken in the southwestern part of Wadi Baysh, near the Red Sea, liȩ on the precipitation line (e. g., wells No. 141, 152, 154, 158, 159), thus indicating little or no evaporation at all. On the other hand, the $\delta^2H - \delta^{18}O$ values of samples from the northeastern part of Wadi Baysh have obviously been subject to relatively strong evaporation (e. g., wells No. 101, 103, 116, 126, 130, 140, 164; s. Fig. 86).

For ^3H analyses only small volumes of water were available so that the precision of the results is not as high as that for the other samples. Nevertheless, it can be seen from Table 26b that the water samples can be roughly divided into

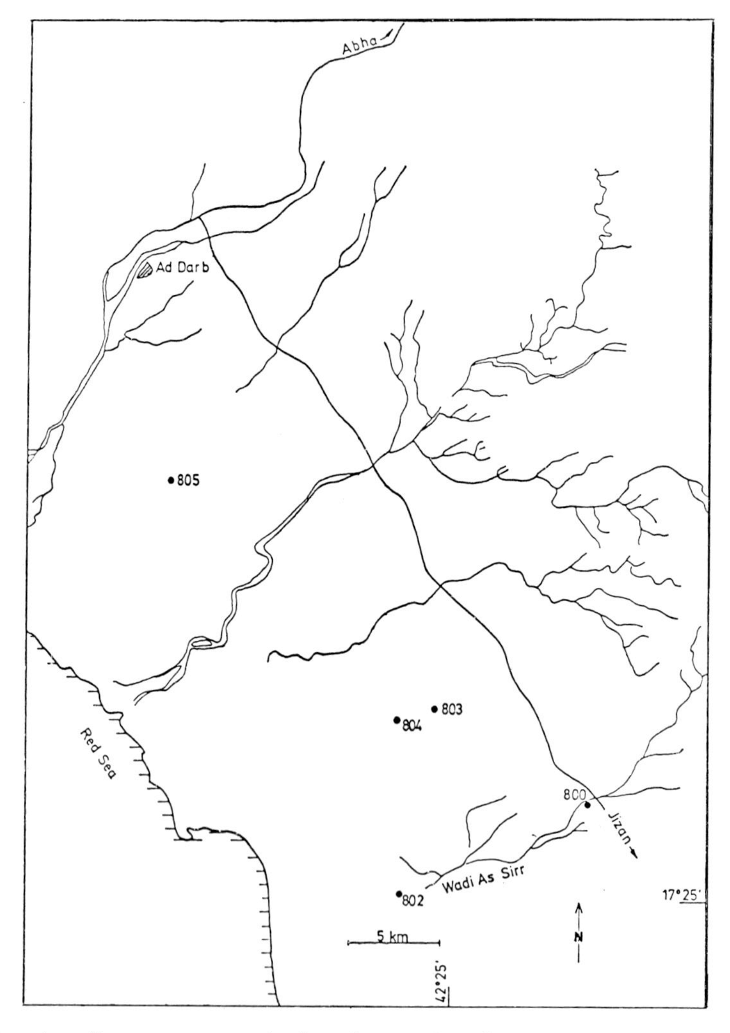

Fig. 84. Location of isotope water samples from the coastal area between Ad Darb and Wadi As Sirr.

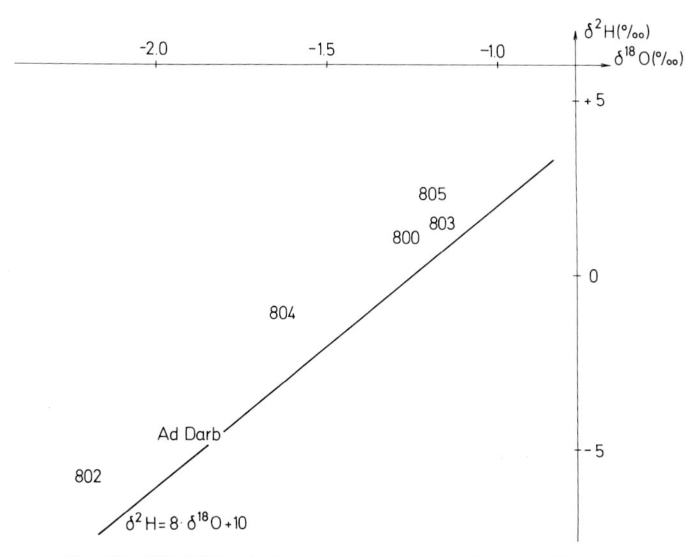

Fig. 85. δ^2H-$\delta^{18}O$ relation of water samples shown in Fig. 84.

two groups. Groundwater of high ^3H content was sampled mainly in the northern part of Wadi Baysh whereas samples of lower or negligible ^3H content were taken from the southwestern part (e. g., wells No. 144, 147, 148, 149, 155, 156, 158, 159). The samples from wells No. 152 and 154 have exceptionally high ^3H contents, a fact which could be ascribed to groundwater of a mean age differing from that of the other samples. Consequently, existence of groundwater bodies of different mean residence times has to be assumed as well as different influence of evaporation.

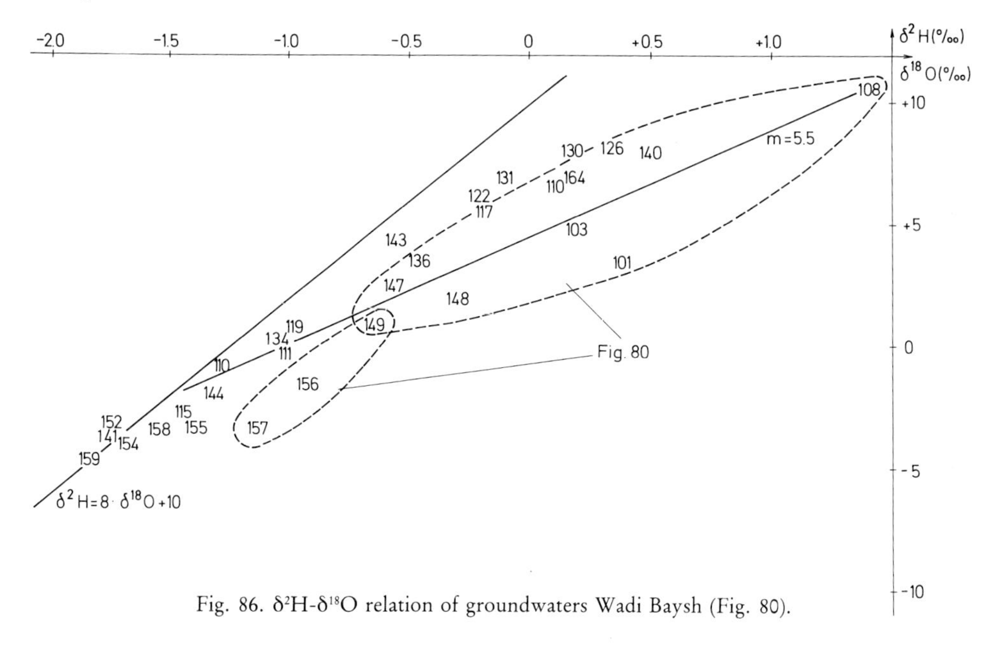

Fig. 86. δ^2H-$\delta^{18}O$ relation of groundwaters Wadi Baysh (Fig. 80).

Table 26a. *Isotope contents and conductivity of water samples from Jizan Plain north of Wadi As Sirr*

Sample No.	^3H Content (TU)	^2H Content δ^2H (‰)	^{18}O Content $\delta^{18}O$ (‰)	Electrical conductivity (μmhos/cm)
800	38.1±2.6	+1.0	−1.28	350
802	0.3±0.5	−5.8	−2.19	1,100
803	33.0±2.5	+1.5	−1.16	390
804	9.4±0.9	−1.1	−1.63	700
805	10.9±1.2	+2.3	−1.19	1,150
Ad Darb	15.8±1.5	−4.7	−1.90	n. m.

Table 26b. *Isotope contents and conductivity of water samples from the central part of the Jizan Plain: Wadi Baysh*

The ^3H contents are given together with the twofold standard deviation.

Sample No.	^3H Content (TU)	^2H Content δ^2H (‰)	^{18}O Content $\delta^{18}O$ (‰)	Electrical conductivity (μmhos/cm)
101	43±11	+3.4	+0.38	1,500
103	35±12	+4.9	+0.18	940
108	33±10	+10.7	+1.41	1,140
110	41±3	−0.8	−1.30	n. m.
111	27±13	−0.1	−1.00	980
115	25±13	−2.7	−1.44	780
116	42±3	+6.7	+0.13	n. m.
117	31±10	+5.7	−0.18	2,500
119	17±10	+0.8	−0.97	830
122	16±10	+6.0	−0.21	2,400
126	29±16	+8.3	+0.33	1,060
130	23±12	+8.0	+0.17	2,300
131	25±11	+6.9	−0.10	1,500
134	33±11	+0.2	−1.04	1,000
136	52±10	+3.7	−0.47	1,250
140	39±10	+7.9	+0.50	2,340
141	35±10	−3.6	−1.74	1,350
143	14±11	+4.6	−0.54	440
144	0±12	−1.9	−1.32	6,450
147	1±10	+2.4	−0.56	940
148	0±9	+1.9	−0.31	6,050
149	5±10	+0.9	−0.68	2,400
152	75±14	−3.1	−1.75	2,180
154	83±12	−3.7	−1.72	1,040
155	0±9	−3.2	−1.41	1,100
156	7±10	−1.6	−0.92	2,080
157	0±10	−3.4	−1.13	1,820
158	7±10	−3.4	−1.54	1,200
159	5±11	−4.5	−1.85	5,000
164	17±10	+6.8	+0.17	n. m.

n. m. = not measured.

Table 27. *Isotope contents and conductivity of water samples from the central part of the Jizan Plain: Wadi Saby a* (s. Fig. 87)

Sample No.	^{3}H Content (TU)	^{2}H Content δ^2H (‰)	^{18}O Content δ^{18}O (‰)	Electrical conductivity (μmhos/cm)
1	1.0±0.8	+2.0	−1.12	500
2	37.0±2.7	+0.6	−1.15	410
3	25.0±1.9	−1.0	−1.36	340
5	20.0±1.5	+5.9	−0.55	550
6	6.1±0.7	+7.7	+0.53	360
9	14.0±1.2	+7.9	+0.42	370
13	51.0±3.5	+10.5	+0.56	1,800
14	16.0±1.4	+9.9	+0.50	530
15	34.0±2.4	+8.4	+0.38	n. m.
16	9.2±2.2	+5.4	−0.24	n. m.
17	46.0±3.2	+4.9	−0.35	n. m.
18	9.6±1.2	+5.1	−0.44	n. m.
19	18.0±1.4	+15.0	+1.27	n. m.
20	12.0±3.1	+9.3	+0.02	n. m.
21	25.0±1.8	+11.8	+0.79	n. m.
22	38.0±2.8	+11.6	+0.62	n. m.
23	37.0±2.8	+10.0	+0.56	n. m.
24	45.0±3.3	+5.4	−0.39	n. m.
25	6.3±1.2	+6.2	−0.42	n. m.
26	0.8±1.8	+2.9	−0.62	n. m.
27	0.4±1.0	+0.2	−1.02	n. m.
38	0.6±1.1	+5.3	−0.59	n. m.
55	8.3±1.3	+2.9	−0.97	n. m.
70	0.7±0.6	+4.2	−0.82	n. m.
78	3.8±0.9	+0.8	−1.10	n. m.

Table 28. *Isotope contents and conductivity of water samples from the Jizan Plain south of Wadi Jizan: Wadi Amlah + Wadi Maqab* (s. Fig. 88)

Sample No.	^{3}H Content (TU)	^{2}H Content δ^2H (‰)	^{18}O Content δ^{18}O (‰)	Electrical conductivity (μmhos/cm)
201	18.9±1.5	+7.0	+0.38	3,000
211	0.3±0.6	+2.1	−0.71	1,150
212	27.6±2.0	+6.6	+0.03	590
215	3.1±0.7	+4.1	−0.14	1,350
217	1.6±0.5	+6.0	+0.09	1,700
218	73.6±8.8	+0.6	−1.15	600
220	0.9±0.5	+4.0	−0.40	2,500
229	1.9±0.7	+2.6	−0.78	2,200
230	4.0±0.6	+2.6	−0.78	1,550
231	1.8±0.8	+0.8	−1.10	3,500
233	3.4±0.6	+0.5	−0.92	2,650
235	3.2±0.9	−2.0	−1.23	1,200
239	19.4±1.9	−3.9	−1.48	1,200
242	0.3±0.5	+0.9	−1.11	3,000
243	12.5±1.0	−3.0	−1.34	1,900
224	34.3±2.4	−5.6	−1.80	n. m.

n. m. = not measured.

Table 29. *Isotope contents and conductivity of water samples from the Jizan Plain south of Wadi Jizan: Wadi Fija + Wadi Khums* (s. Fig. 89)

Sample No.	^3H Content (TU)	^2H Content δ^2H (‰)	^{18}O Content $\delta^{18}O$ (‰)	Electrical conductivity (μmhos/cm)
301	40.6±2.7	+7.4	+0.09	1,400
302	50.7±3.6	+10.4	+0.67	6,400
304	5.2±0.6	+4.8	−0.21	3,200
305	26.9±1.9	−1.3	−1.26	530
309	0.6±0.4	−0.2	−1.08	880
312	1.5±0.5	−5.3	−0.98	6,500
320	50.7±3.6	+4.0	−0.42	920
321	5.0±0.6	+5.9	−0.25	920
322	37.9±2.6	+8.4	+0.17	520
324	27.3±1.9	+2.7	−0.56	790
325	1.4±0.5	+6.0	+0.31	1,850
326	20.5±1.5	+1.0	−0.97	840
327	26.8±1.9	−3.0	−1.61	580
328	12.4±1.0	+6.4	+0.04	1,550
331	1.1±0.4	+2.5	−0.79	n. m.
332	27.2±1.9	+4.7	−0.24	3,700
333	38.6±2.6	+0.9	−1.14	1,500
337	0.5±0.7	−1.5	−1.53	3,000
338	1.0±0.5	+5.1	−0.31	2,750
339	0.8±0.6	+5.2	−0.34	2,900
340	0.8±0.4	−1.8	−1.23	1,800
341	0.9±0.5	−0.1	−1.12	1,700
342	0.7±0.7	−3.3	−1.50	3,250
343	0.9±0.5	−3.4	−1.53	2,750
344	0.8±0.6	−7.5	−2.22	4,300
345	0.8±0.5	−7.6	−2.11	3,900
349	10.5±0.9	−6.3	−2.06	2,600
350	5.7±0.6	−1.5	−1.34	1,400

Table 30. *Isotope contents and conductivity of water samples from the Jizan Plain south of Wadi Jizan: Wadi Khulab* (s. Fig. 90)

Sample No.	^3H Content (TU)	^2H Content δ^2H (‰)	^{18}O Content $\delta^{18}O$ (‰)	Electrical conductivity (μmhos/cm)
400	32.7±2.5	+3.7	−0.76	850
403	22.0±1.9	+12.2	+0.89	1,150
405	19.8±1.8	+8.9	+0.31	1,100
408	49.6±3.6	+5.7	−0.38	1,200
409	11.0±1.3	+4.6	−0.43	1,500
410	32.0±12.5	+2.3	−0.99	1,550
412	3.8±0.7	+2.9	−0.78	1,750
413	2.6±0.6	+2.0	−1.03	2,700
415	0 ±11.8	+2.6	−0.86	3,300
416	1.3±0.6	+0.7	−1.19	3,500
418	0.3±14.9	+1.8	−0.79	2,900
419	0 ±12.0	+4.3	−0.56	3,000
421	1.4±0.5	+6.5	−0.39	3,200
423	1.0±1.0	+6.3	−0.25	2,600
425	25.3±1.9	+7.3	+0.12	1,950
427	43.2±3.0	+9.6	+0.49	1,800

Table 31. *Isotope contents and conductivity of water samples from the Jizan Plain south of Wadi Jizan: Wadi Liyyah + Wadi Mighyalah + Wadi Ta'ashshar* (s. Fig. 91)

Sample No.	^{3}H Content (TU)	^{2}H Content δ^2H (‰)	^{18}O Content δ^{18}O (‰)	Electrical conductivity (μmhos/cm)
500	34.1±2.6	+6.5	−0.08	n. m.
503	38.6±2.7	+10.3	+0.43	1,200
508	29.2±2.1	+11.4	+0.74	2,300
509	25.4±1.8	+9.0	−0.03	790
510	55.5±3.7	+5.9	−0.58	1,150
511	3.3±0.6	+5.5	−0.60	2,250
519	29.5±2.1	+10.0	+0.19	560
522	34.9±2.4	+4.8	−0.48	850
524	3.1±1.1	+6.1	−0.36	1,900
526	16.4±1.3	+4.3	−0.67	3,500
528	11.4±1.0	+4.4	−0.69	2,100
530	1.1±0.5	+4.6	−0.79	3,900
533	1.1±0.7	+1.2	−1.00	n. m.
534	3.1±0.6	+2.4	−0.93	1,750
600	71.1±4.7	+7.7	+0.30	1,000
602	24.1±1.8	+6.0	−0.35	560
605	42.2±2.9	+6.2	−0.16	1,100
607	1.0±0.6	+4.6	−0.46	4,700
608	12.2±1.0	+3.8	−0.42	1,950
609	1.8±0.6	+1.2	−0.96	3,100
610	8.7±0.8	+7.1	+0.12	2,200
611	3.0±0.6	+5.9	−0.12	2,700
613	2.5±0.5	+3.6	−0.42	1,650
614	45.0±3.1	+2.5	−0.84	680
615	0.8±0.8	+0.7	−1.20	3,500
619	1.0±0.5	−0.7	−1.44	5,000

n. m. = not measured.

Table 32. *Isotope contents and conductivity of water stamples from the Jizan Plain south of Wadi Jizan: Wadi Haradh* (s. Fig. 92)

Sample No.	^{3}H Content (TU)	^{2}H Content δ^2H (‰)	^{18}O Content δ^{18}O (‰)	Electrical conductivity (μmhos/cm)
700	41.3±2.8	−2.7	−1.50	2,050
701	0.9±0.6	+0.1	−1.25	4,200
702	2.2±0.5	−1.1	−1.38	5,600
703	4.9±0.6	+2.5	−0.84	6,200
704	0.8±0.5	+1.9	−0.82	4,100
705	0.5±0.5	+2.6	−0.75	3,600

Table 33. *Isotope contents and conductivity of water samples from drilled wells (boreholes) in the Jizan Plain* (s. Fig. 93)

Borehole No.	Depth of filter (m)	^3H content (TU)	^{14}C content (% modern)	^{14}C age (y. B. P.)	$\delta^{13}C$ (‰)	δ^2H (‰)	$\delta^{18}O$ (‰)	Conductivity (μmhos/cm)
A	47–92	0.4±0.6	88.2±1.0	≈0	−10.6	+2.9	−0.86	2,050
B	54–114	0.0±0.6	62.9±0.7	2,420	−12.9	+2.4	−0.95	2,620
C	36–54	0.1±0.7	69.0±0.8	1,680	− 9.7	+5.5	−0.34	1,430
D	35–45	0.2±0.7	76.7±0.9	830	−11.2	+4.7	−0.62	3,250
E	55–69	0.8±0.7	73.7±1.2	1,150	−12.5	−3.8	−1.75	5,500
G	32–45	1.4±0.7	62.6±1.2	2,460	− 9.5	−1.8	−1.51	6,050
H	25–66	0.5±0.6	75.7±1.2	930	−12.8	+1.6	−0.93	1,450
I	42–124	0.5±0.7	62.5±1.2	2,470	—	+1.3	−0.80	2,350
M	37–75	0.3±0.7	77.4±0.6	750	−10.4	+0.5	−1.04	1,650
N	37–58	3.1±0.6	67.8±0.6	1,820	−10.8	−0.5	−1.31	1,400
O	48–97	0.7±0.5	80.1±0.6	480	−11.4	+1.7	−0.81	2,800
Q	39–80	1.2±0.4	77.4±0.6	750	−11.7	+4.8	−0.68	1,800
P	28–73	—	78.9±0.6	560	−13.2	—	—	2,200
R	48–60	—	82.5±1.1	240	− 9.6	—	—	2,150

Wadi Sabya (Figs. 80, 87)

The $\delta^2H - \delta^{18}O$ values suggest a classification of the water samples into groups with no, little or relatively strong evaporation. Samples with no evaporation were taken from wells around midstream, such as wells No. 1, 2 and 3 whereas samples taken upstream (wells No. 13, 14, 15) or downstream (wells No. 21, 22, 23) show relatively strong evaporation.

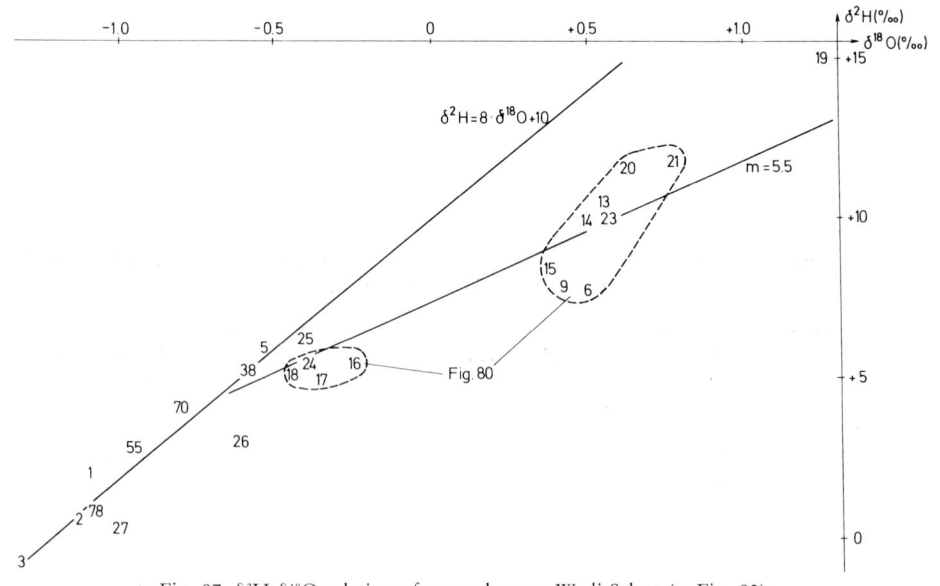

Fig. 87. δ^2H-$\delta^{18}O$ relation of groundwaters Wadi Sabya (s. Fig. 80).

The samples from wells No. 1, 26, 27, 28 and 70 are practically free of tritium and, therefore, essentially consist of water of a mean age of more than 20 years. All the other samples have ^3H contents between 4 and 51 TU, therefore, consist of groundwater formed by precipitation during the last two decades.

Wadi Amlah and Wadi Maqab (Figs. 81, 88)

Most of the $\delta^2H - \delta^{18}O$ values lie on an evaporation line which starts on the precipitation line with non-evaporated samples from the wells No. 218, 231 and 242 and ends with relatively strongly evaporated samples from the wells No. 201, 212 and 217. Another origin could be ascribed to the samples of the wells No. 235, 239, 242 and 244 due to the fact that their $\delta^2H - \delta^{18}O$ values are different from those mentioned above.

The ^3H contents of the samples scatter between 0 and 34 TU with the exception of well No. 218 (74 TU), the water of which could have a mean residence time of about 10 years.

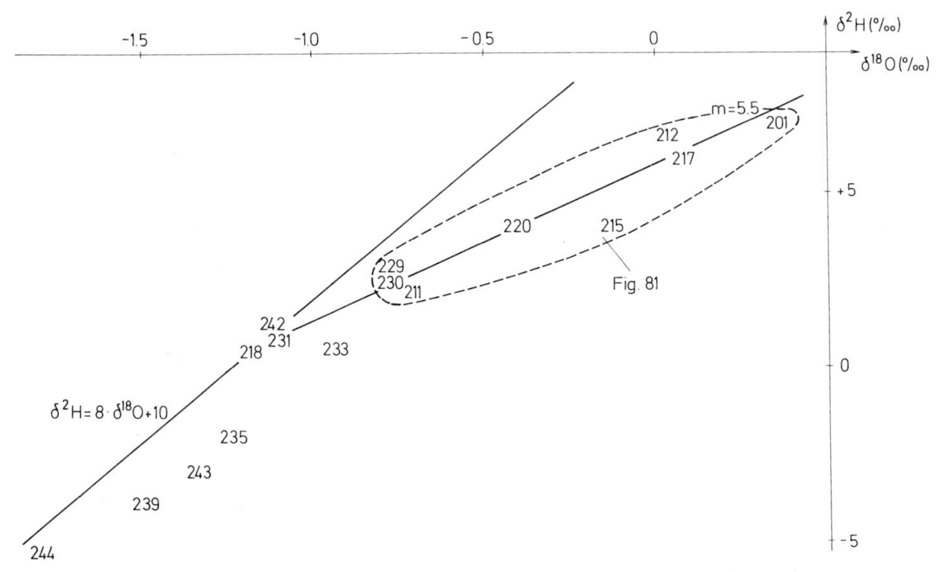

Fig. 88. δ^2H-$\delta^{18}O$ relation of groundwaters Wadi Amlah and Wadi Maqab.

Wadi Fija and Wadi Khums (Figs. 81, 89)

The $\delta^2H - \delta^{18}O$ values of samples taken from wells in the middle section of these wadis lie on the precipitation line. The $\delta^2H - \delta^{18}O$ values of the samples from the upper section of the wadis cluster around an evaporation line.

The samples taken from the wells No. 344, 345 and 349 show the lowest stable isotope content in the whole area under investigation. An unusual $\delta^2H - \delta^{18}O$-relation together with a high electrical conductivity is shown by the sample from well No. 312.

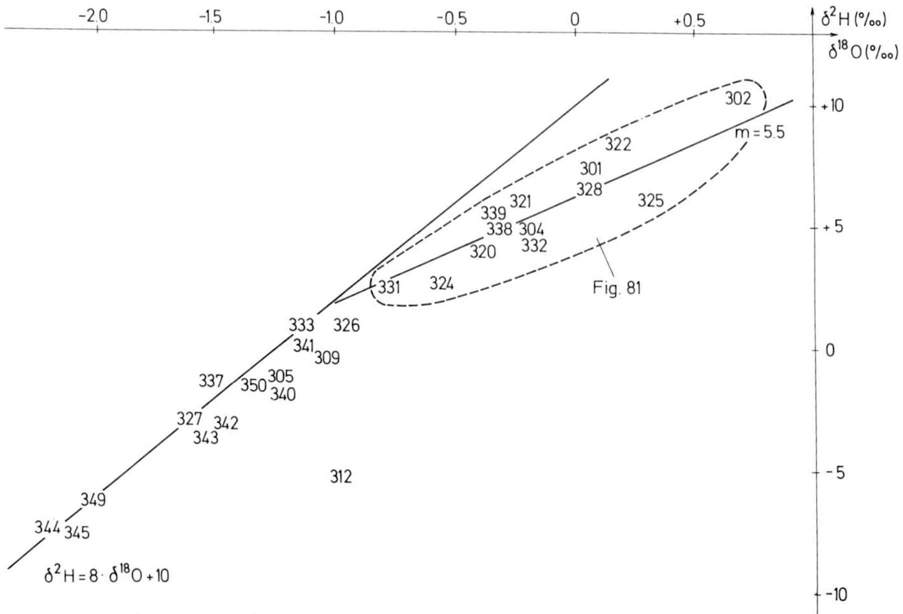

Fig. 89. δ^2H-$\delta^{18}O$ relation of groundwaters Wadi Fija and Wadi Khums.

The samples from the lower section of the wadis have no or little tritium and, therefore, contain water of a relatively long mean residence time. A wide range of ^3H contents is displayed by the samples from the upper part of the wadis. No distinct correlation can be established between ^3H content and conductivity.

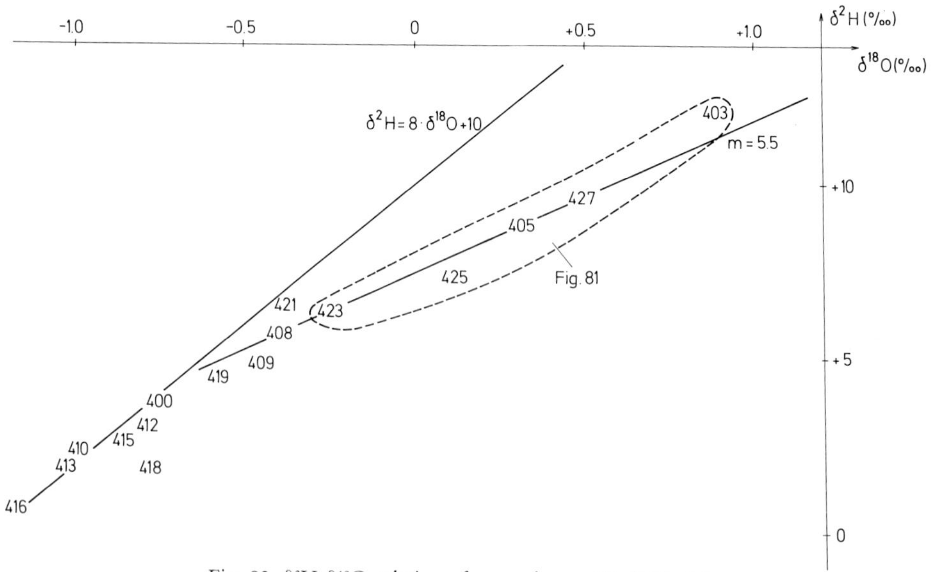

Fig. 90. δ^2H-$\delta^{18}O$ relation of groundwater Wadi Khulab.

Wadi Khulab (Figs. 81, 90)

The $\delta^2H - \delta^{18}O$ values of the samples taken from wells in the upstream section of this wadi indicate effects of evaporation. These samples have high ^3H contents along with relatively low conductivities.

The samples from the lower section of Wadi Khulab have $\delta^2H - \delta^{18}O$ values clustering around the precipitation line. Again, the sample with the lowest stable isotope content comes from the well No. 416 which is the closest one to the Red Sea. Low ^3H contents of these samples correspond to high conductivities.

Wadi Liyyah, Wadi Mighyalah and Wadi Ta'ashshar (Figs. 81, 91)

The samples taken from the upper part of these wadis have been subject to the relatively highest rate of evaporation. The samples taken from wells near the Red Sea have the lowest stable isotope contents as well as little or no tritium.

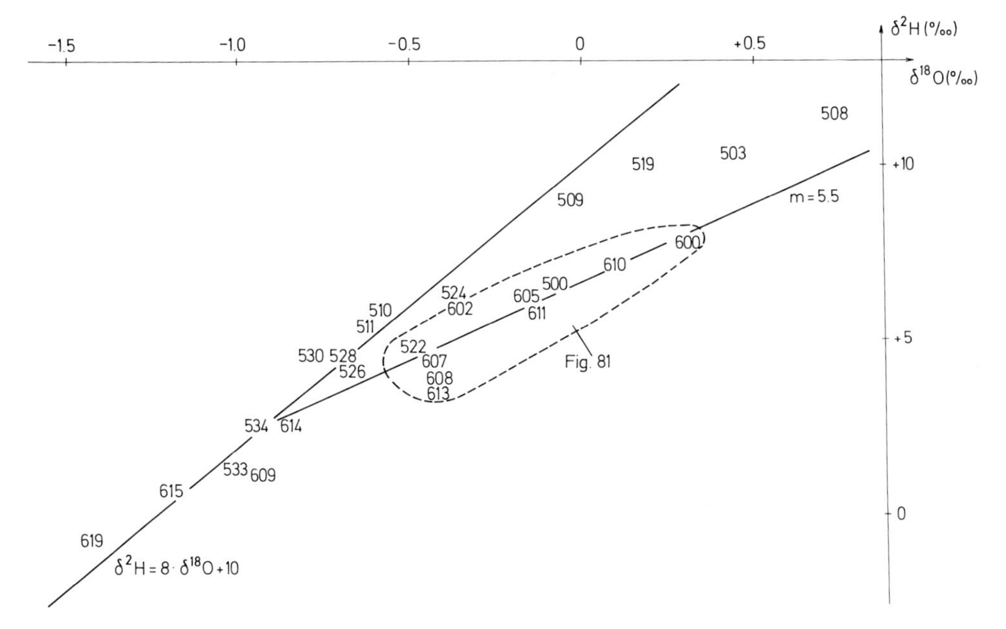

Fig. 91. δ^2H-$\delta^{18}O$ relation of groundwaters Wadi Liyyah, Wadi Mighyalah and Wadi Ta'ashshar.

Wadi Haradh (Figs. 81, 92)

The sampling points of the wells No. 702, 703 and 704 are not included in Fig. 81. The samples from wells No. 701 to 705 have ^3H contents below 5 TU and, consequently, contain large quantities of water with a mean age higher than 20 years.

The conductivities of these groundwater samples range between 3,600 and 6,200 μmhos/cm. Such conductivities are among the highest values in the whole area under investigation. The sample from well No. 700 displays a relatively high

^{3}H content (41 TU) and the lowest conductivity of all samples from this wadi. From the stable isotope data, it can be assumed that the groundwater wells No. 703, 704 and 705 have been subject to some evaporation.

2.4.4.2.2. Isotope data from deep groundwater (boreholes)

These samples were taken from the boreholes during pumping tests in September/November 1976 (wells No. 6 J 77 to 6 J 85) and in October 1977/ January 1978 (wells No. 6 J 89 to 6 J 94). The location of the boreholes is shown in Figs. 72 and 73, and Figs. 80, 81 (A–R). The isotope data are listed in Table 33. The analyzed groundwater is characterized by pH values between about 7.6 and 8.3, and total mineral content between about 700 and 1,500 mg/l.

The ^{14}C contents of the 14 samples analyzed lie between 62.5 and 88.2 percent modern. As a rough approximation, an initial ^{14}C content of about 85 percent modern is being used for calculating uncorrected ^{14}C model ages of groundwater which lie between 0 and 2,500 years before present. These ^{14}C model ages are rough estimates of the real age under the assumptions that the model, developed by MÜNNICH and VOGEL to describe the origin of the ^{14}C content in groundwater, can be applied to the Jizan aquifer, and that the initial ^{14}C content has been affected only by radioactive decay. The ^{14}C model ages represent the maximum limits of the real ages if, for example, ^{14}C was lost from the water by isotopic exchange with the aquifer matrix, then chemical water-rock interactions lead to a dilution of the ^{14}C concentration, or magmatic CO_2 was admixed. The ^{13}C contents measured on the ^{14}C samples do not contradict the MÜNNICH–VOGEL model. Admixture of bomb-produced ^{14}C would lead to ^{14}C ages being too young. This could be the case to a small extent for the samples from the wells No. 6 J 83, 6 J 90, 6 J 93 and possibly 6 J 91 as they contain very small amounts of tritium and therewith post 1953 groundwater. These low ^{3}H contents could point either to mean residence times of at least a few decades, or to an admixture of shallow and young groundwater with tritium-free old water. The admixture could possibly be a consequence of the pumping tests. Samples taken at the beginning and end of the pumping tests showed the same ^{3}H contents (e. g., wells No. 6 J 81 and 6 J 83). Samples from the other boreholes are practically tritium-free with respect to the detection limit of the analysis technique used. Therefore, these groundwaters as a whole are considered older than two decades.

When interpreting the results of the isotope measurements on water from the boreholes, one must take into account that nearly all samples contain water from different waterbearing layers or from a wide range of depths in the aquifer. Therefore, the isotope contents obtained have to be considered as mean values and cannot reflect a clear relation with depth of sampling even if there is one. For example, two of three samples with the lowest ^{14}C contents (63 percent modern) were obtained from the deepest boreholes (No. 6 J 78 and 6 J 85) but one sample came from the borehole with the shallowest depth of the lower filter edge (No. 6 J 83).

The $\delta^{2}H$ and $\delta^{18}O$ values lie in the same range as those obtained from shallow groundwater samples. The $\delta^{2}H - \delta^{18}O$ values lie on or near the precipitation line, indicating that the deep groundwater was subject to little evaporation or no evaporation at all (Fig. 93).

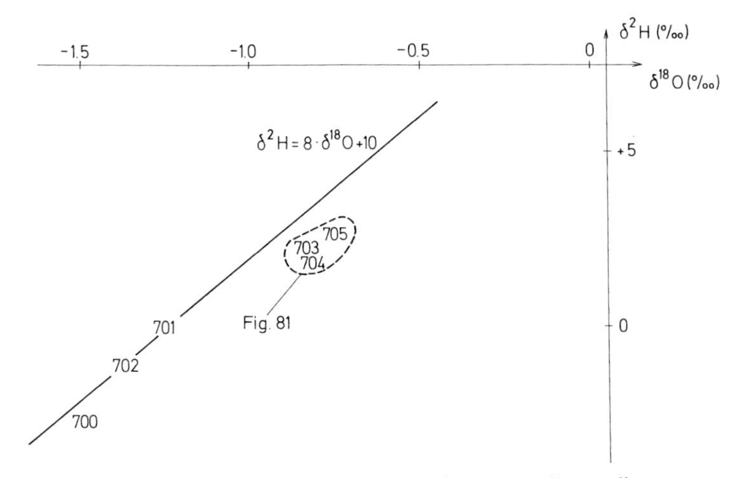

Fig. 92. δ^2H-$\delta^{18}O$ relation of groundwaters Wadi Haradh.

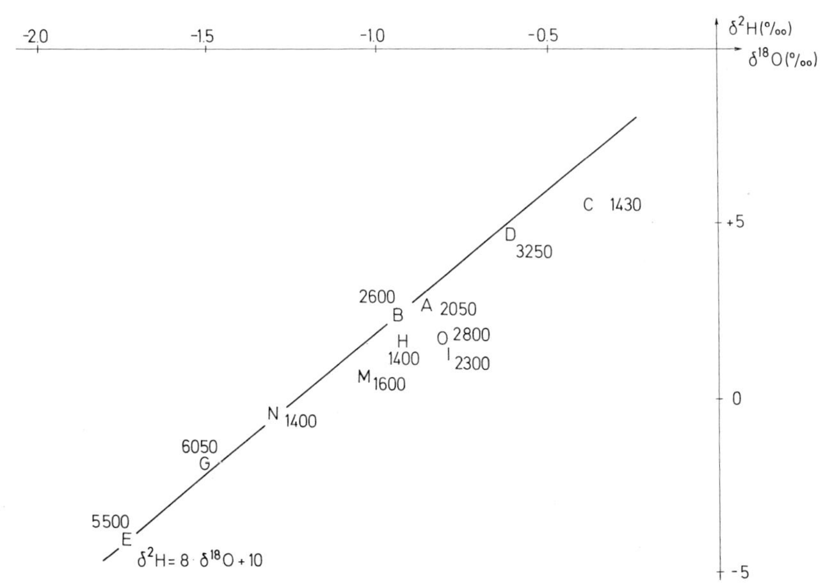

Fig. 93. δ^2H-$\delta^{18}O$ relation of groundwaters from deep boreholes. (*A-M;* depths in ft; s. Figs. 80 and 81.)

2.4.4.3. Origin of salinity

There are several theoretical possibilities to explain the high salt content in the main groundwater body of South Tihamah Plain:

a) Upturn of salt plugs and consequently dissolution processes within the aquifer.

b) Penetration of Red Sea water along faults within the basement and mixing with freshwater in the aquifer.

c) Existence of fossil sabkhah horizons, formed by pre-Quaternary or Quaternary transgressions of the Red Sea and resolution of these evaporites within the deeper groundwater body.

d) Uprise of thermal brines from the basement and intrusion into the overlaying groundwater body.

e) Increase of salt content by evaporation effects caused by flood irrigation over hundreds of years.

The following is a discussion of the above possibilities:

a) There are no geological indications of the existence of salt plugs within the Tihamah Plain except in the coastal area near Jizan. Isotopic analyses contribute to the solution of this problem only to a limited extent.

b) High contents of sulfates in comparison with chloride concentrations in some wells (e. g., wells No. 140, 148 and 149) are not in accordance with the chemical composition of Red Sea water. Therefore, the salination of groundwater is not likely to originate from the Red Sea. Isotope analyses also provide arguments against this assumptions:

– Concerning the δ^2H and $\delta^{18}O$ values, the difference between Red Sea water and groundwater is greatest only for samples from the coastal areas.

– In contrast to the values of the Red Sea water, the δ^2H – $\delta^{18}O$ values of samples from the coastal areas lie on the precipitation line.

c) The isotope data do not contradict the possibility that fossil sabkhah deposits are dissolved by groundwater. From the geological point of view, however, this possibility has to be excluded because there is no indication of any sea transgression during the Quaternary period which is evident from the lack of marine biological matter in the drilling sections.

d) Information on major sources of thermal brines (about 160 g Cl^-/kg) within the Red Sea (e. g. E. FABER, M. SCHOELL, 1978) as well as at the margins of the Red Sea depression (Ethiopia, e. g. M. SCHOELL, E. FABER, 1976, and on the eastern border of the Tihamah Plain, see chapter Tectonic Features) can be obtained from the literature above. Therefore, it appears to be a reasonable assumption that brines of similar composition ascend within the broken Precambrian basement beneath the sedimentary deposits of the Tihamah Plain. This assumption explains great differences of water chemistry in the salt water body even of neighbouring wells (e. g., wells No. 144 and 148). Also, the leakage effects which have been observed during all pumping tests can easily be explained by uprising of brines from great depths, thereby strongly supporting the existence of this phenomenon.

To explain the origin even of highest salt concentrations in the groundwater (e. g., wells No. 44, 144 and 148), a maximum admixture of about two percent of brines is sufficient. The low order of this admixture of brine with δ^2H values around +12‰ and $\delta^{18}O$ values around +1.5‰ (E. FABER, M. SCHOELL, 1978) causes a change of the isotopic composition of the groundwater which lies within the measurement accuracy. Therefore, the isotope results are consistent with the hydrogeological model concerning the origin of the salt content in the groundwater.

e) There is no general correlation between the salt concentration and the extent of evaporation estimated from the ^2H and ^{18}O contents. Consequently, the salt content of the groundwater body may have originated from evaporation only to a minor extent or in special areas.

2.4.4.4. *Summarizing discussion*

a) Two-thirds of the groundwater samples analyzed have ^{3}H contents greater than 1 TU, the maximum being 84 TU. Depending on their ^{3}H contents, these samples contain different quantities of groundwater that was formed by rain water infiltration into the aquifer after 1953.

A more detailed estimate on the mean residence times of the various groundwater bodies would require a better knowledge of the ^{3}H concentrations in the rain or soil water which recharged the groundwater during the last two decades, and the availability of time series of groundwater samples, in order to follow the change of ^{3}H concentrations with time. With only one sample from each well it is not possible to decide unequivocally on the period of rain water infiltration into the aquifer. In addition, that ^{3}H content could be the result of the mixing of two components, one being old, tritium-free groundwater and the other being young.

About one third of the groundwater samples analyzed has ^{3}H contents close to or below the detection limit of the analysis technique used. This water is considered older than two decades.

The following generalizations are observed regarding the local distribution of the ^{3}H contents. No tritium or very little tritium (below TU) imply long residence times in most of the groundwater samples from the downstream sections of the wadis (which are situated closer to the Red Sea). Most samples of groundwater in the upstream show ^{3}H-contents (above 10 TU) corresponding to shorter residence times (see Tables 26–33).

b) The samples taken during pumping tests from 14 boreholes represent mixed groundwater from various depths between 25 and 124 meters. The ^{14}C contents lie between 62.9 and 88.2 percent modern which correspond formally to ^{14}C model ages or maximum limits of real ages, respectively, between 0 and 2,500 years before present. At the same time, these ages represent an upper limit for the mean residence times of shallow groundwater in the area under investigation. Small traces of tritium were detected in a few samples of the deep groundwater, thus indicating very small quantities of young precipitation water.

The δ^2H and δ^{18}O values of the deep groundwater samples lie within the range of the values for shallow groundwater. This suggests that the shallow and the deep groundwater have the same recharge areas. Most of the samples from deep groundwater were not subject to evaporation.

c) In groundwater of the Jizan Plain, north of Wadi As Sirr displacement, a high conductivity or high mineral content frequently corresponds to a low ^{3}H content and vice versa; but this relationship does not exist in every part of the area under investigation. One of the possible explanations for the salination of groundwater is the increase of salt content by evaporation during irrigation. The salination, however, cannot exclusively be attributed to evaporation because no direct correlation exists between deuterium excess d (as a measure of evaporation) and conductivity.

Hydrological considerations favour the thesis of brines ascending within the broken basement and introducing salt into the groundwater in the sedimentary deposits of the Tihamah Plain. The results of the isotope analyses are consistent with this thesis.

d) The deuterium excess $d = \delta^2H - 8\ \delta^{18}O$ indicates that the water was subject to evaporation ($d < +9$). In Figs. 81 and 82, three different ranges of d are indicated by different signatures. It can be seen that groundwater from the upper part of the wadis generally shows marked evaporation effects. On the other hand, groundwater from the downstream part of the wadis or from near the coast has generally d values between +9 and +11 which indicats that the water was not subject to substantial evaporation.

The evaporation in the upper part of the wadis (areas roughly bordered by broken lines on Figs. 80, 81) occurs obviously at the surface as a consequence of slow infiltration of the wadi discharge, and is being increased by irrigation measures. In these areas also the depth of groundwater table below surface is lower than downstream (compare Figs. 76 and 77). This is caused by the irrigation measures and by a step in the water daming layers between east of the large terraces and the coastal area. The flexure-like dipping of the young layers can be seen in the profile of Fig. 62.

It can be assumed that the groundwater in the downstream parts of the wadis is being recharged when surface discharge reaches the downstream parts of the wadis without essential evaporation. From the relatively low δ^2H and $\delta^{18}O$ values, however, it seems possible that the groundwater in the downstream sections of the wadis is also being recharged by a subsurface channel flow coming from mountainous catchment areas (see Figs. 72 and 73). The very low or missing ^3H contents would be consistent with this possibility.

Evaporation of groundwater becomes only effective in the coastal zone (sabkhahs) where the groundwater surface crops out, but no water samples were collected from this area for isotope analyses.

2.4.5. Farasan Islands

(A. Dabbagh, H. Hötzl, H. Schnier)

2.4.5.1. General considerations and geological structure

In the southern part of the Red Sea, where the sea reaches its maximum width of 360 kilometers, remarkably flat shoals lie off both coasts. The Farasan Bank, situated on the Arabian side, extends from 16° to 20° N Lat. and attains a width of up to 120 kilometers. Here, the depth of the sea is almost always less than 100 meters. Often the water is only a few meters deep. The Farasan Islands are located within this bank and some 40 kilometers from the Arabian coast, between 16° and 17° N Lat., in corresponding position with Jizan. On the African side, the Dahlac Islands are to be found in a similar position of almost mirror-image symmetry.

The two main islands are Farasan Al Kabir and Sajid (Fig. 94). The length of both islands runs noticeably parallel to the Red Sea from NW to SE. The larger of the two, Farasan Al Kabir, is over 60 kilometers long and has a width of five to eight kilometers. Sajid is 35 kilometers long and some 10 kilometers wide. Other larger islands are Ad Dissan, Zufaf, Qummah and Dumsuk. All are surrounded by a number of small islands.

The two main islands are permanently inhabited; the few inhabitants' livelihood comes mainly from fishing and cultivation of small oases. There is little left to recall the period of economic prosperity at the turn of the century, when the islands were important for pearling.

Geologically, the Farasan Islands consist of an originally more or less uniform reef flat. Tectonic processes led to shattering and uplift, whereby salt diapirism played a significant role. Generally, the islands are flat and protrude only 10 to 20 meters out of the sea. Only in isolated locations are altitudes of up to 75 meters to be found.

The first geological studies by W. A. McFayden (1930) and L. R. Fox (1932) already mention the young Plio-to-Pleistocene age of this reef. In structure and fauna, it differs very little from the recent reefs surrounding the islands. The

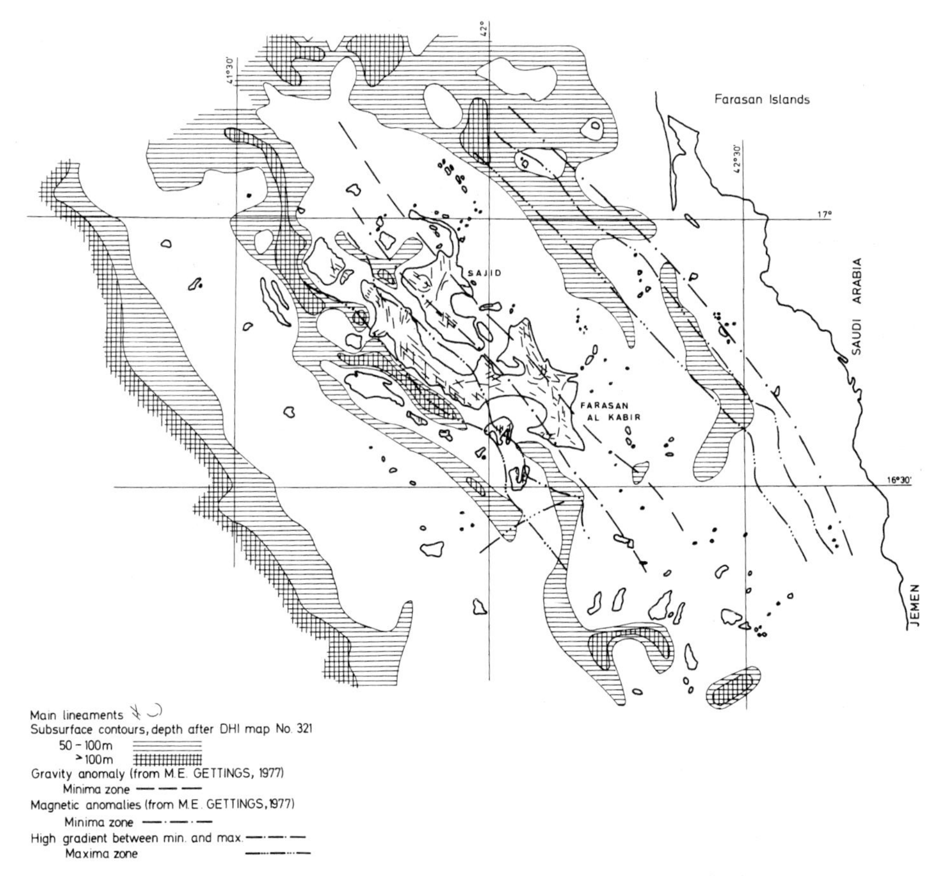

Fig. 94. Bathymetric and geophysical features of Farasan Islands.

elevated reef flat is some 15 meters thick and can be divided into different growth periods.

The reef limestones are underlain by bright white marly limestones. Isolated, firmer fossil-rich layers are interstratified. Farther down, these are replaced by yellow-green sequences with isolated clay interlayers. Thicknesses of up to 50 meters are exposed. W. A. McFAYDEN (1930) also reports "fish-diatom marls" which are exposed in a few places. They are finely laminated and white to light green in color. They take their name from the numerous fish and diatom remnants. Although the outcroppings are not unequivocal, this series with a thickness exceeding 20 meters is nonetheless interpreted as a lenticular inclusion in the marly limestone.

At the core of individual domed eminences as, for example, north of the main locality on Farasan Al Kabir, gypsum and anhydrite rock are outcropping.

This total series, as determined from the outcroppings, is confirmed by bores made to find oil (Petroleum Development Western Arabia, Ltd.). P. SKIPWITH (1973) unfortunately gives only very condensed profile descriptions. The series show great variations in thickness which in part are surely a result of halokinesis. Thicknesses for the topmost reef flat and marly limestone are given as 42 to 136 meters, and for the clay, gypsum and anhydrite series, with some salt in places, as 20 to 180 meters. Below that, drillings were made into but not through a rock-salt layer.

2.4.5.2. Photogeological evaluation and tectonic analysis

Although the Farasan Islands are distributed very irregularly in a nearly quadratic area of 40 by 40 kilometers, the form and position of the main island parallel to the Red-Sea graben system indicate a tectonic development and disposition. The Plio-Pleistocene age of the sediments implies very young tectonic movements. Here, there is an apparent contradiction to the general opinion that with the opening of the central graben in the Pliocene, the tectonic activity in the bordering areas of the old rift system weakened considerably. This is also expressed in the tectonic picture of the neighboring southern Tihamah (cf. chapter 2.4.).

A brief visit to the main island, Farasan Al Kabir, showed that the islands, in addition to their general delimitation, include numerous faults and fault bundles, accompanied in places with graben and horst structures. In the brief time available, it was not possible to make a systematic evaluation. An aerial photographic evaluation seemed to be appropriate to the nearly barren and ideally exposed island. Available for the stereographic evaluation were black-and-white photographs with a scale of 1 : 60,000 which had been taken in 1971 from a height of approximately 5,000 meters and which covered the entire area (Plate IV, see insertion at back cover).

The primary purpose of the photogeological study was to cover the tectonic lineaments. Additionally, lithological mapping was attempted where possible. This, however, could only be supported through terrain comparison done by inspection of the main island of Farasan Al Kabir by two of the authors before the interpretation of aerial photos. A subsequent ground check and correction of the map were not possible in the time available.

Lithology

The various shades of gray and the roughness of the surface were used to differentiate the lithological structure of different areas. The shades of gray can be divided into three large categories:

- Very light to white shades:

These are found mainly in the immediate vicinity of the coast and indicate sand beaches, sandbars and shallow areas with sand sedimentation. In the central area of the island the light spots are always very small; they are usually related to relatively recent sediment accumulations in the form of drifted or alluvial calcareous sands and silts. These occur in the area of the salt diapirs in the sinkholes there, and in places where the reef limestone flat is sectioned into wide crevices and graben structures by sliding and cracking. When the flanks of a graben or fracture are somewhat flat but high, the white marly limestones under the reef limestones can be seen as very light stripes.

For cartographic purposes, except for isolated, wide graben structures, only the young sand accumulations on the coast can be taken into consideration. They consist almost exclusively of bioclastic light calcareous sands.

- Middle to light shades of gray

These are mainly found adjoining the coastal strips, in places where there are more extensive flat to gently upward-sloping surfaces. They are often separated from the darker rock areas by a distinct erosion step. Terrain study shows this to be a marine-cut terrace, revealing the high-water mark of a sea transgression which occurred after the island had arisen from the sea (cf. chapter 2.4.5.3.). Fine gray shading parallel to the coast or contour lines on this transgression surface indicates younger erosion steps.

They either document individual regression phases, or are due to more recent marine invasions. This transgression surface is generally developed within the Plio-Pleistocene reef flat. The surface in this part shows an induration by a relative weak duricrust, presented by a relatively light patina.

There are also small areas of medium shades of gray in the vicinity of the outcropping salt diapirs and the block zone. Medium gray also appears in the transitional area between shallow and deep water, but interpretation is difficult in the subaquatic area, as algae and seaweed produce medium to dark shades of gray in relatively shallow water.

- Dark gray shades

Dark gray is mainly found in the more uplifted parts of the island. It usually indicates relief formation by shattering, and solution cavities. The intensive, long-term weathering of the limestones has led to a characteristic duricrust formation.

The upper main crust shows dark gray to dark brown color. This causes finally the dark grays in the aerial photographs.

The deeper part of the sea and the terrestrial vegetation appear to be almost black. The latter is limited to small fields in isolated depressions, occasional palm groves and a few small mangrove swamps on the coast.

The roughness structure of the outcropping rock surfaces as seen in the aerial photographs offered an additional possibility for differentiation, especially of the solid rocks. Here, the Plio-Pleistocene reef flat was notably uniform and shows only few and indistinctly structured gray shades on the surface otherwise there were also areas with a remarkably fine structure of lighter and darker shades of gray. Their sharp, nearly oval borders identify them as the subrosion depressions developing at the core of salt diapirs. We were able to visit one of these structures north of the main locality on Farasan Al Kabir. The irregular relief is formed by numerous dolines and solution cavities with intermediate crumbled reef limestones, gypsum and anhydrite, as well as marls.

Up at the edge of this salt diapir the overlying sediments above the evaporites are dragged upward. The varying sequence of marly and clayey rocks on the one hand and gypsum and anhydrite layers on the other sometimes creates ring-shaped structures on the outcropping bassets. They are mapped together with the Plio-Pleistocene reef flat, and only these set off bassets appear as ring-shaped structures.

In addition to the subrosion depression north of the main village of Farasan Al Kabir, there are more such diapirs in the subsoil (see Plate IV, insertion at back cover):

- southeast of Farasan Al Kabir; the eastern part of this depression has, however, sunk into the sea;
- the dome-shaped elevation NNW of the Huseini Oasis on the Segir Peninsula (northern part of Farasan Al Kabir); the beginning subrosion is apparent here in a small central depression;
- between the Segir Peninsula and Ad Dissan Island (NW of Farasan Al Kabir); leaching has interrupted the originally continuous crest of the island and caused it to sink below sea level; only the southern and northern outer rings remain;
- the dome with the central depression in the NW part of Sajid Island; and
- the oval bay in the SE part of Sajid Island.

Fault structures

Photogeological evaluation of the lineaments quickly shows that several causes are responsible for the development of the various fault structures. Though the tectonic blocking of the Farasan Bank was responsible for the delimitation and arrangement of the islands, the present-day fracturing of the island is due to the ascending salt and so in part to the associated gravitational sliding processes.

This is strikingly evident in the northern part of the Segir Peninsula on Farasan Al Kabir (see Fig. 94 and Plate IV [insertion at back cover] for the details). The blocking and sliding of the Plio-Pleistocene reef flat in the central area caused by salt ascent followed four diamond-shaped fault systems, which overlaid one another laterally to a greater or lesser extent. The slightly warped, outcropping fault surfaces are apparently lystric surfaces, whereby the movement is often accompanied by external rotation with corresponding antithetic tipping of the blocks. At the same time, there are large crevices and small special graben systems (Fig. 95). The diamond-shaped pattern is not necessarily typical. In the dome directly opposite on the northern end of Sajid Island there is a triple fault system the individual systems of which are at some 120° to each other.

This sort of sliding at lystric surface intersections can also appear on the steep island flanks without direct influence of salt diapirs. These secondary fault structures make it very difficult to determine the primary tectonic directions. A preferred direction, to be found on almost all of the islands, runs at about 140° parallel to the Red Sea system. As the main islands are also arranged parallel to this direction, it is necessarily also emphasized in the secondary fault structures caused by slippage.

The N–S direction also has its particular significance. The two larger islands in the south, Qummah and Dumsuk, are mainly blocked in this fault direction.

These particular faults are, in accordance with the slippage processes, tensional downthrowing faults. There are no indications of larger lateral movements. In places, slight lateral coastal shifts can imitate horizontal movement. This, however, is only due to variations in the slippage processes of the individual subblocks.

Fig. 95. Farasan Islands, special graben in the Plio-Pleistocene reef platform of the northwestern part (Segir) of Farasan Kabir. (Photo: H. HÖTZL, 1977.)

Both the subsurface contour map of the Red Sea and available measurements of the gravitational and magnetic fields for this area were used to analyze and interpret the main tectonic directions.

The marine map of the German Hydrographic Institute was used for bathymetric purposes. The course of the most important isobaths clearly shows the SE–NW direction of the sub-basins and shallows around the islands, as well as the parallel steep slope down to the axial deep trough of the Red Sea. Beside of this main direction of the graben, only the N–S direction appears more distinct in isolated shallows and sub-basins, as north of Ad Dissan Island and south of Farasan Al Kabir.

The distribution of both the gravitational and magnetic fields emphasizes the significance of the graben trend for the tectonics of the entire Farasan Bank. The magnetic field shows fine stripes parallel to the coast, and thus also to the axis of the central graben. The minima for magnetic anomalies and gravity distribution are mainly in the domed structures and subrosion depressions of the islands, while the maxima are found in the zones of subsidence between the islands. These measurements thus serve to confirm the ascent of salt in the form of diapirs and elongated diapirs along tectonic lines parallel to the main graben.

This shows that the tectonic pattern influences the occurrence and direction of the islands. The mobile salt uses the existing weak areas, with halokinesis causing further shattering of the originally continuous Plio-Pleistocene reef flat.

Gravitational slippage due to halokinesis or simply to the steep flanks of the individual island lamellas are further movement components contributing to the development of the existing structures.

2.4.5.3. Development of the island group in the Quaternary

L. R. Cox (1932) and A. G. Brighton (1932), examined in detail the mollusc and echinoderm fauna collected by W. A. McFayden (1930) from the marls, marly limestones and reef limestones of the exposed layers on Farasan. Most of the species identified (some 90% of the molluscs and 100% of the echinoderms) are known recent forms in the Indopacific area. Yet only these species have changed very little in recent geological history; many of the species described are also known in the Pliocene of the Red Sea area and the adjacent Indian Ocean.

The fauna studied thus are not very suitable for establishing a time limit for the Pliocene-Pleistocene. It seems remarkable that two mollusc species, *Pecten vasseli* and *Ostrea tridacnaeformis,* were not found on Farasan (L. R. Fox, 1932). Both are very significant for the Pliocene in the Red Sea. *Chamys isthmica* and *Chamis lessepsii,* which are known from the Pleistocene shore sediments in the Sudan, were, however, also found here. L. R. Fox and A. G. Brighton thus suggest a classification in the "Pleistocene, or in any case not much older than Pleistocene".

Indications for the age classification are also provided by lithological comparison with the drill logs from Durvara 2 on the Sudan coast and Mansiyah 1 on the Arabian coast north of Jizan (cf. M. Gillmann, 1968). In both logs, the carbonate development at the top of a height of up to 180 meters, and the subsequent upper continental series of 700 meters are classified as Quaternary. There are, however, no distinct time marks.

On Farasan, the entire marl-to-limestone sedimentation following the gypsum and clay series would be comparable with the above-mentioned young series. It was noted in former chapters that Pliocene age is not out of the question for part of the clastic continental series. If this is extended to Farasan, the portions of marl superjacent to the clay and gypsum series also belong to the Pliocene. Some of the white marly limestones and above all the overlying reef limestones would, however, belong exclusively to the Pleistocene.

This chronological classification shows that the tectonics in the vicinity of the Farasan Bank have remained active through very recent history. Certain shifts in very young beach terraces suggest that the tectonic salt ascent is still taking place at present.

There are currently no further data for a more precise dating of the reef limestones. Some of the samples we took from the upper part of the main reef limestones were negative with ^{14}C dating (cf. Table 18). Considering the last glacial low water mark, this would indicate a minimum age of more than 100,000 years. The beach terraces cut in places into the reef flat, along with the intensive and mature duricrust at the top, suggest that the reef flat is more likely Middle to Early Pleistocene.

After this reef flat had been lifted and broken to form the islands, a number of marine invasions occurred. As shown in Chapter 2.4.5.2., they caused out transgression surfaces and cliff lines which show the earlier beach lines.

W. A. McFayden (1930) mentioned elevated beach terraces. He cites two terrace levels of 1.30 and 2.70 meters near Ra's Hassis in the NE of Farasan Al Kabir, and another, higher beach terrace some six to eight meters above present sea level. As regards the arrangement by height, it should be noted that these fossil beach formations can develop independently of eustatic variations in sea level as a result of tectonic lifting and salt ascent. The height of marine terraces formed by eustatic variations will, however, be altered. This is shown by heights that fail to match one another on one and the same island.

An even more precise survey of the entire island area would be required for an exact evaluation and establishment of chronological parallels. This was not possible in the short time available to us on the island. Only the NW part of Farasan Al Kabir (Segir Peninsula) could be studied more closely.

Starting from the present-day average sea level, one first encounters a barrier beach up to 1 meter high; it is made up mainly of sandy to fine-gravelly shell and coral detritus. This is followed by a gently sloping somewhat older deposit of similar nature, which is cemented in places as beach rock. This leads to a wave-cut notch at 1.50 meters above sea level eroded into a cliff in the old reef body.

The upper edge of the cliff and the planation beyond it are some 3.5 meters above the present-day sea level. Noteworthy on this planation are accumulations of large, thick pelecypod and gastropod shells. These form either loose piles of shells, or wider, flat shell banks.

Although the shells no longer form a continuous strip, this seems to be an older high tide barrier beach. The individual piles may, of course, be man-made. Dating of these shells with ^{14}C produced an age of 4,700 years (Table 18 sample

No. IRM 7616). It is possible that these shells were propelled by storm tides over the then flatter cliff during the Flandrian transgression, which was responsible for the formation of the deeper notches.

Immediately beyond the cliff there is a gentle uphill slope 20 to 40 meters wide. The coral colonies of the reef flat forming the cliff cleared out by weathering in places on this surface. This is followed by another somewhat overprinted cliff step some 1.5 meters high. Its upper edge is thus some 8 meters above the present sea level. It is remarkable that the flat just in front of this step is dissected by individual deep canyons. These canyons do not, however, extend back into the next-highest cliff step. The reef flat adjacent to the second cliff step shows the typical duricrust-weathering profile with a hard and still light gray upper crust and the yellowish, unconsolidated undercrust. Apparently this crust prevented further canyon-like erosion, while the undercrust permitted marine erosion which back-cut the old cliff line.

Proceeding into the interior of the island, this second cliff step is followed by an up-sloping transgression surface one to two kilometers wide. This shows isolated edges which may readily be interpreted as less clearly developed older cliff steps. The rather uniform duricrust coating, however, prevents an absolutely definite interpretation.

This transgression surface then ends at a pronounced hill step some 25 meters above sea level. Its course, which is readily evident from the aerial photograph, indicates a considerably older shore line. This is made even more apparent by the differences in the duricrust in front and in back of the hill step. The transgression surface still shows a compact surface structure light gray in color, while the highest planation shows small-sized deterioration of the hard crust. It is also mostly of the dark brown color seen in mature calrete.

The chronological course may be summarized as follows. After the uplift and shattering of the reef flat formed in the Middle or Early Pleistocene, there was intensive duricrust formation (old crust). During a later transgression, part of the reef flat with the old crust eroded and an extensive transgression surface was formed. After the sea had retreated, duricrust formation resumed (new crust). Another increase in sea level later eroded parts of this young crust and formed the cliff at eight meters. During the subsequent regression, canyons were incised in front of the eight meters cliff line, which were based on a considerably lower sea level than prevails today. With the last increase in sea level the 3.5 meters cliff was formed, and the canyons in the lower part filled with fine material. Since the last increase in the sea level, which ^{14}C dating shows to be identical to the Flandrian Transgression, the sea has receded only slightly.

2.4.6. Hydrochemical Studies of Thermal- and Groundwaters in the Hinterland of Jizan and on the Farasan Al Kabir Island

(W. KOLLMANN)

In the period between February 20th and February 28th, 1977, a total of nine water samples of most diverse hydrogeological origin were taken in the Jizan hinterland and on Farasan Island (Fig. 96). On the mainland, the study involved thermal springs and coverage of potential mixing processes with shallow groundwater in the Quaternary sediments. Owing to the deliberations of the Institute for Applied Geology, Jeddah (1975), the possible influence of a salt-water intrusion on the ground-water lens in the Quaternary deposits on Farasan Island was to be investigated by preliminary sampling (1977/8 and 9).

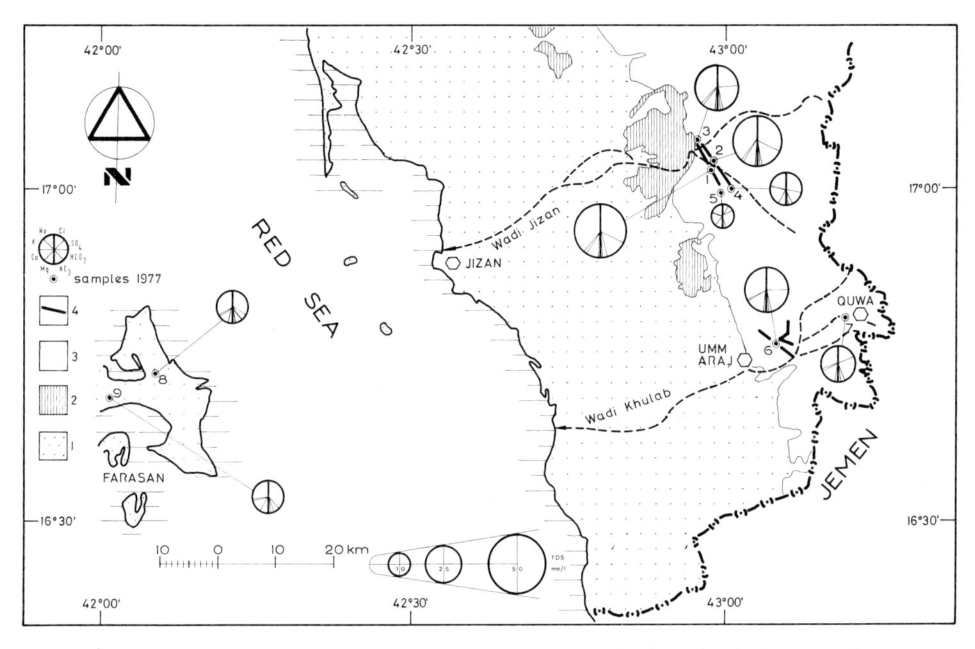

Fig. 96. Hydrochemical data of thermal- and groundwater in the hinterland of Jizan and Farasan Kabir (*1* Quaternary sediments, *2* Quaternary basalt, *3* basement, *4* fault).

The thermal waters samples 1977/1–3 discharge in the area of the Maloki Dam. Their outlets are obviously connected to the more or less parallel arrangement of the tectonic fault bundle striking from NW to SE (Fig. 96). It was not possible to determine to what extent the young basalt eruptions cause the genesis of the thermal springs. Indications of exclusively tectonic effects without volcanism include low values for free carbon dioxide with a resultant deficit of carbonic acid, as well as the existence of thermal waters to the south-southeast (samples 1977/6 and 7) which are not related to the basalt flows. For purposes of comparison a sample (1977/4) was also taken in the area of the tectonic fault zone of the superficial and cold valley groundwater in the Quaternary wadi filling. Sample 1977/5 came from a joint aquifer which developed from a well in shattered Pre-Cambrian granite-gneiss and its weathering detritus.

2.4.6.1. *General hydrochemical description*

Absolute and relative composition

It may be generally observed that the thermal waters (1977/1–3, 6, 7) with 1,300 to 1,500 ppm are more highly mineralized than the shallow, cold groundwaters. The latter waters from loose rock or joint-aquifer (1977/4, 8, 9 and 5, respectively) with total mineralization of 700 to 1,200 ppm are clearly lower.

The relative composition gives further indication for the differentiation of these two types of water. The thermal waters are quite uniformly of the Na- or Ca-Cl-SO_4-type, while the mineralization of the cold groundwaters varies. Depending on the soluble substances available in petrographically different aquifers, distinction may be made between Na-Ca-Cl-SO_4-, Na-Ca-Mg-HCO_3-, Na-Cl- and Na-Mg-Cl-types.

2.4.6.2. *Trace elements*

What is true of the concentrations of the main components of the thermal waters applies to the concentrations of trace elements as well (Table 34). Only strontium and zinc, as well as silicic acid, which is listed for comparison, appear to be less specific. Their solubility in this milieu appears to be dictated less by the temperature than by the concentration.

Table 34. *Trace elements in the thermal springs and cold groundwaters in the Jizan area and on Farasan Al Kabir Island (all values in ppm)*

Sample No. 1977	Li^+	Sr^{2+}	Zn^{2+}	Mn^{2+}	F^-	SiO_2	Water temperature °C
Thermal waters:							
1	0.74	1.04	0.05	0.08	4.60	34.0	50.8
2	0.72	1.07	0.07	0.11	4.90	21.8	54.2
3	0.32	1.42	0.02	0.03	4.90	7.5	56.6
6	0.38	3.28	0.01	0.05	4.40	7.3	74.4
7	0.30	0.59	0.07	0.01	4.20	11.8	55.2
Fresh waters:							
4	0.014	0.75	0.05	0.01	0.18	3.9	28.1
5	0.010	0.18	0.02	0.01	0.48	28.2	—
8	0.03	2.26	0.07	0.001	0.79	6.4	—
9	0.03	3.28	0.07	0.001	0.97	4.3	—

2.4.6.3. *Ion exchange*

With the exception of analysis 1977/9 from Farasan Island, there are no significant signs of ion-exchange processes between water and the rock it flows through. The well in Huseini shows a notable chloride excess as compared to the alkalis. Magnesium also predominates among the alkaline earths. Both of these facts may be viewed as evidence of a sea-water intrusion followed by ion-exchange processes. This mechanism is complicated by mixing with fresh water from the island's groundwater lens. To model this, it may be assumed that upon penetration of sea water with an original ion relationship of $Na^+ + K^+ = Cl^-$ and a high Mg concentration, the contact with the Quaternary, in part limestone aquifer of the island produces the following reaction. Upon exchange of alkalis from the probably small quantities of infiltrating salt water for alkaline earths (particularly calcium) from the rock, the equivalent sum $Na^+ + K^+ < Cl^-$ changes. Quantification of this descriptively suggested process is possible on the one hand with positive values for the "positive" base-exchange index

$$\text{pos. } I_{BE} = \frac{(Cl^- - (Na^+ + K^+))}{Cl^-}$$

with +0.3, but on the other hand, with +1.1 as well, via negation of the "negative" base-exchange index

$$\text{neg. } I_{BE} = \frac{(Cl^- - (Na^+ + K^+))}{(SO_4^{--} + HCO_3^- + NO_3^-)}$$

For the sample from the thin fresh-water lens on Farasan Al Kabir Island this mechanism is confirmed by the alkaline-earth chloride changes (1.1), by the alkali-chloride ratio (0.7), and the hydrogen carbonate-saliniferous ratio (0.1). The stable isotopes in sample 1977/8 were present in a remarkable concentration, indicating direct involvement of sea water.

2.4.6.4. *Mixed waters*

The results of the hydrochemical and hydrological-isotopic analyses showed that the total mineralization of the thermal waters cannot be explained by mixing processes. This then suggests a rather isolated ascent of tempered deep waters. A mixing of thermal waters among themselves, or with cold superficial groundwaters is definitely not the case in the area of the Maloki Dam (1977/1–5). The temperature differences at the outlets can probably be interpreted as a function of the rate of ascent. Table 35 gives the according values for the observed thermal springs. For a more reliable statement the number of five springs and their different geological situation is too restrictive.

Table 35. *Relation between the outlet temperature of thermal-spring water (°C) and outputs (l/s)*

Sample 1977	Outlet temperature	Output
1	50.8	0.1
2	54.2	0.3
3	56.6	0.22
6	74.6	0.5
7	55.2	0.23

An exception to the model case of isolated paths of ascent is found in the thermal-spring region of Áyn Al Hara (1977/6, cf. Fig. 97). Here, there is a diffuse distribution of the rising thermal water in a small space. Since the tectonic

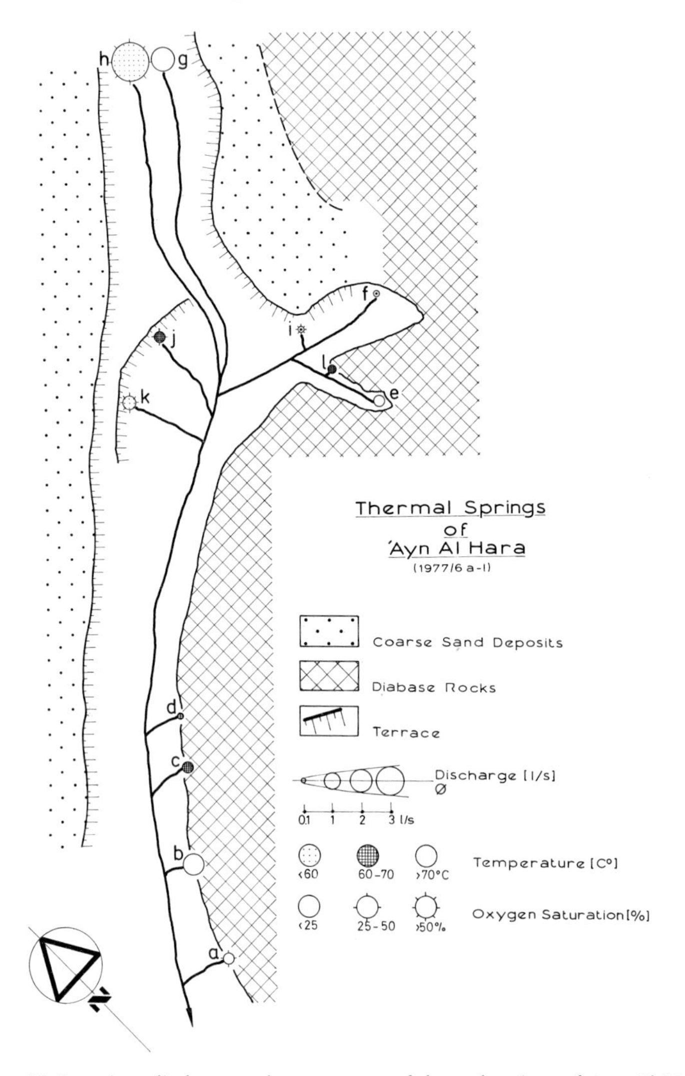

Fig. 97. Location, discharge and temperature of thermal springs of Ayn Al Hara.

fault system crosses the terrain surface in the wadi itself – and of course the wadi erodes with preference along a tectonically given lineament – local mixing of cold superficial groundwater from the Quaternary aquifer is possible. Since full analyses are not available, an attempt will be made to use the measured outlet temperatures, output and dissolved oxygen to find indications of the involvement of shallow-lying groundwaters (Table 36).

A multiple regression can be used to work out the relationship between output, outlet temperature and oxygen saturation for the waters mixed with thermal water:

$$x = a + by + cz$$

$$x = 8.6 + 0.13\,y - 0.03\,z$$

where x = output in l/s
y = temperature in °C, and
z = O_2 saturation in %

Table 36. *Discharge, temperature and O_2 concentrations of the individual outlets of Áyn Al Hara 1977/6 a-l (barometric pressure 749 mmHg = 9.986 10^4 Pa)*

	Q l/s	Temp. °C	O_2 corr. ppm	O_2 Saturation %
a)	0.5	71.1	1.1	31
b)	2.5	75.1	0.1	3
c)	0.5	69.6	0.5	14
d)	0.05	65.4	0.4	10
e)	0.5	74.6	0.1	3
f)	0.05	54.6	not determined	–
g)	3	74.0	0.7	22
h)	7	32.6	9.4	137
i)	0.05	59.2	2.3	50
j)	0.5	62.6	1.1	26
k)	0.8	54.3	3.3	65
l)	0.3	60.1	1.0	22

Table 37. *Mixture representation of deep thermal water (A) and shallow groundwater (B) in the outlets in the Áyn Al Hara thermal-spring region*

Sample 1977 6	A % under conditions of Temp.	O_2	B % Temp.	O_2
a)	91%	89%	9%	11%
b)	100%	100%	0%	0%
c)	87%	96%	13%	4%
d)	77%	97%	23%	03%
e)	99%	100%	1%	0%
f)	52%	n. d.	48%	n. d.
g)	97%	94%	3%	6%
h)	0%	0%	100%	100%
i)	63%	76%	37%	24%
j)	71%	89%	29%	11%
k)	51%	66%	49%	34%
l)	65%	90%	35%	10%

n. d. = not determined

This shows that with temperature increase of 1° C and accompanying decrease in O_2 saturation, the amount of thermal water mixed with fresh groundwater increases by 0.13 l/s. Thus, with greater output, a cooling in the more superficial regions is scarcely possible owing to the faster rate of ascent. Table 37 shows a quantification of the two components of the mixture, under the assumption that the hottest outlet (1977/6 b) transports only deep-lying thermal water, and the coolest, oxygen-richest outlet (1977/6 h) represents the groundwater of the wadi valley fill.

In water samples 1977/6 d, j and l, the differences in the respective mixture compositions are larger. A slight cooling probably took place in the immediate vicinity of the outlet, as the water temperatures were measured on February 23rd, 1977, in the early morning hours with a low ambient air temperature of some 20° C. The small output also produces a cooling effect. For these reasons, the mixture compositions obtained by calculation of the oxygen content are more reliable.

2.5. Hydrogeological Studies in the Upper Wadi Bishah

(W. KOLLMANN)

2.5.1. Area Around Bishah

2.5.1.1. Characteristic geological features

The country surrounding the locality of Bishah has been thoroughly studied by different companies in the course of extensive hydrological and geophysical research sponsored by the Ministry of Agriculture and Water. It is thus possible to interpret the random water samples, taken by the author within the Quaternary research program between 13 and 15 March, 1977, from this section of the wadi, in the light of previous information on the aquifer and underground situation, and to compare them with the hydrochemical findings from 1967 and 1968.

These previous studies were generally used as the basis for the following characteristic features and boundary conditions or the ground-water resources which are plentiful in places (ITALCONSULT – Final Report 1969).

Precambrian basement complex

The Wadi Bishah valley forms a connecting element between the regional units of the Al Hijaz plateau in the SW and the Najd pediplain in the NE (S. S. AL-SAYARI, J. G. ZÖTL, 1978). The wadi course, which itself appears to be marked by tectonic faults, is accompanied in the southern hinterland by a mountainous area with differences in elevation of 200–300 meters between the valley floors and summit area. The northern section is characterized by broad alluvial plains in a low hill country with isolated outliers protruding from the cover of debris.

The basement complex is made up mainly of granite-gneiss and granodiorites, along with metamorphics such as sericite schist and green schist ("Baysh greenstone"). Owing to intensive superficial weathering which has caused cleavage with open joints, the granitoid rocks are generally to be evaluated as having good permeability and absorptive capacity for water in the disaggregation zone. In contrast, the north-south striking and steeply pitched schists are narrowly jointed with open joints only a few millimeters wide. The absorptive capacity for water is accordingly somewhat less than that of the weathered acidic rocks. The basement complex shows three main tectonic directions:

1. Most significant is the NE–SW structure, which the wadi course also generally follows.
2. There is a smaller, but more compact, WNW–ESE system.
3. A N–S direction predominates mainly in the southern section of the Bishah area.

Loose rock deposits

The valley floor is hundreds of meters to some kilometers wide; in it, three types of valley filling may be distinguished on the basis of their genesis and morphology. The geophysical soundings and numerous borings made in the course of the ITALCONSULT research make possible a description of the three generations of sedimentation in the underground as well, and of hydrologically significant data. The following simplified model may be assumed (Fig. 98):

1. an older silt-sand-gravel (flood-plain) valley filling was dissected and
2. filled to the rim with sand-gravel (terraced alluvium) deposits. There was then another, younger phase of trough and terrace formation, which, perhaps due to climatic-morphological changes, was in turn followed by
3. aggradation and filling of the channel with recent, generally roughly clastic material (channel alluvium).

These loose sediments from cycle one overlap the basement complex directly and form a body 10–40 meters thick filled with ground water; this body is made up of sands and gravels with pebbles, and the diameters of the components are given as 20–40 cm. This leads to the climatic-morphological conclusion that flow speeds greater than at present must generally be assumed for this thickness of the coarsely granular deposits and their transport. In the roof of this roughly clastic series, cycle one is concluded with a sandy-silty covering and protective layer 10–20 meters thick with generally good filtration properties for the lower-lying ground-water deposit. This cycle has been more or less completely conserved, and has only been removed by erosion of the wadi channel itself, which is 5–10 meters deep. In those cases, cycle one and cycle three (channel alluvium) form a single, undivided ground-water body. On the underside of the silty covering layer there is usually a tufa horizon.

A usually thin terrace-sediment layer from cycle two (terrace alluvium) was conserved mainly at the edges of the older sediments described. This layer is some 3–5 meters over the recent wadi bed, is not very thick (8 meters) and tapers off to the sides. The structure is characterized by slanting layers of sand and pebbles with diameters of 10–15 cm. There is also a weakly carbonatically cemented layer at the base of the fine-clastic and older silt-sand filter layer mentioned previously (cycle one).

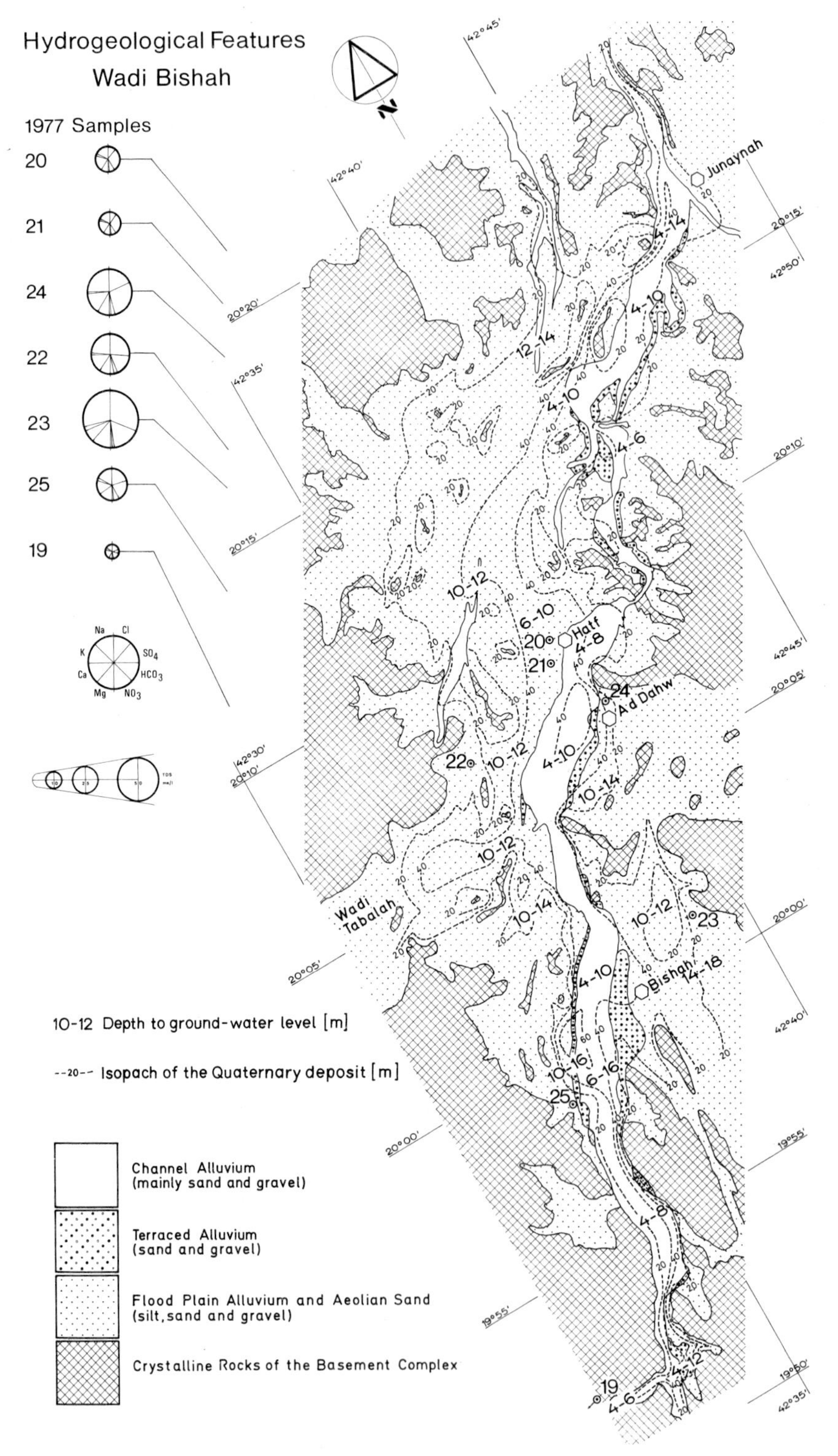

Fig. 98. Simplified hydrogeological and hydrochemical map of the basin of Bishah (after ITALCONSULT, 1969) with locations of samples (vertical numbers).

The channel alluvium (three) derived from the young channel filling is rather coarsely clastic (coarse sand, gravel and rocks) at the base and probably contains mainly reassorted terrace gravel (two). This third sequence, with a highly variable thickness of 1–15 meters, is in places directly superimposed upon the deepest roughly clastic older sediments from cycle one and with them forms a more or less uniform groundwater aquifer. A division into levels occurs when the recent gravel filling (three) is separated hydraulically from the deeper, coarse series (one) by a sandy-silty intermediate layer of non-eroded material from the upper part of cycle one.

The youngest sediments deposited in the wadi are finely granular (sand and silt), are repeatedly reassorted by the annual floods and are also transported by wind.

2.5.1.2. Characteristic hydrogeological features

The studies made by ITALCONSULT in 1969 showed that only the loose sediments and the underlying disaggregation zone of the basement complex, which is two–five meters thick and geophysically differentiable, are to be viewed as potential groundwater aquifers. The sandy-silty upper layers of the flood-plain deposits (one) can sometimes increase the pressure level; this was often seen when these layers were bored through and the ground water in the lower horizon rose three–seven meters in the bore holes. The isopach presentation for the Quaternary deposits (Fig. 98), taken from the ITALCONSULT report and substantially simplified, shows that numerous, sometimes parallel, subsurface channels must be incised into the basement, the bottom of which is to be expected at 40 meters, or in places in excess of 60 meters depth. It should be noted that these data are mainly based on geophysical measurements (seismic and geoelectrical), along with consideration of 60 additional bores down to a maximal depth of 60 meters, covering the disaggregation zone of the crystalline basement, which is also to be given a similar hydrogeological evaluation.

Depth of the groundwater table

The area numbers shown in Fig. 98, with dimension in meters and validity for the survey period of May–June 1967, were chosen to display the distance from ground level to the surface of the ground-water table because these numbers allow two forms of information to be given:

a) Data on the approximate subsurface depths of the surface of the groundwater table;

b) Magnitude of possible fluctuation.

The presentation is based on data from a total of 831 wells studied in the Bishah area; only the depths of the quiescent water tables, measured at a time when pumping was not taking place, are given.

It is apparent that the surface of the groundwater table in the area of the central channel, which branches to the northeast over Junaynah, is relatively shallow and is generally to be found at a depth of less than 10 meters. There are a few sample measurements from the vicinity of the parallel but shallower channel to the west and the mouth of the Wadi Tabalah, which generally suggest a thicker overlay (10–14 meters). At the edges of the flood plain (1) and near the crystalline basement, the ground-water table was generally found at depths greater than 10 meters; the greatest depths were found east of Bishah (14–18 meters).

Slope of the groundwater table – surface

The 270 geodetic groundwater bores made by ITALCONSULT in 1967 for the flood-plain zone (one) showed a groundwater table slope of 1–2%. The average slope increased to 3–4% in the upper course near the mouth of the Wadi Tarj. The slope for Wadi Tabalah is 0.5%, which is considerably lower. Generally, the flow direction of the ground water follows the course of the main wadi, and deviates only in areas where larger amounts of water are removed. Owing to the slight slope of the groundwater table surface of the tributary wadis and their slight transmissibility, no significant lateral supply to the main groundwater flow can occur in Wadi Bishah.

Groundwater table variations

Monthly sample measurements from May 1967 to October 1968 and from 53 bores produced highly variable data. In general, there was an increase in ground water from November 1967 to about June 1968; the ground water in most cases remained constant until completion of the measurement series, but in some places either continued to rise, or dropped abruptly. A significant observation is that the central wadi bed, which covers about the same area as the channel alluvium (three), showed the greatest variation in groundwater maximum and minimum, amounting to five–six meters during the observation period. The range of variation decreases with increasing distance to the outside, and near the basement outcropping shows only slight (one meter) or insignificant differences (Fig. 98).

The ITALCONSULT studies showed that along river courses with coarsely clastic, permeable bottoms, the variations in groundwater table were directly related to the floods which are a significant factor in groundwater renewal, in that they enrich groundwater reserves through streamwater loss; this may raise questions of hygiene. These influent factors should be quantified and localized because of the need not only to establish a protection area, but also to set up a water balance by installing a number of effluent measuring stations and collecting data over a longer period of time.

Owing to the occasional intercalated sandy-silty layer (one) and the resultant pressure-level tension, the effect of the infiltrating floods on the groundwater level can be determined distant from the recharge area, but almost simultaneously owing to the rapid increase.

Previous measurements of the groundwater level suggest that the groundwater surface, e. g. in the area of the Rawshan Oases, would have sunk by 0.5–2 meters; in view of the larger seasonal variations and increased usage since then, this is by no means significant, and more likely meaningless. On the other hand, it should be noted that the 1967/68 season was considerably moister and showed heavier flooding.

Aquifer parameters

Hydrological and hydrochemical studies showed that the unconfined aquifer in the channel alluvium is in close hydraulic contact with the confined aquifer of the lower flood-plain deposit, so that changes in the regeneration or withdrawal from both groundwater aquifers are reflected in both almost simultaneously. Ten long-term pumping tests and the evaluation of the results from twenty-two sounding tubes in their subsidence area provide perspective on the order of magnitude and variability of the permeability factor k_f. Values for k_f from approximately 2×10^{-3} to 2×10^{-2} m/s were found for the channel-alluvium area. In contrast, the aquifer of the flood plain deposit shows generally lower k_f values of approximately 1×10^{-4} to 1×10^{-3} m/s.

Should these given k_f values, however, be extrapolated – perhaps for the entire breadth of the valley – then it must be borne in mind that the results will look better than they actually are, as it may be assumed that the bores were located on the basis of favorable geophysical soundings.

Considering the heterogenity of the underground with regard to thickness and permeability of the aquifer, a groundwater flow F_{gw} (RICHTER and LILLICH, 1975) of about 150–300 l/s may be estimated for a cross profile at the level of Al Hifa. A groundwater flow of 30–60 l/s from Wadi Tabalah enters this underground.

Enrichment and regeneration of the aquifers

Flood-water seepage has been shown to contribute significantly to the regeneration of both confined and unconfined aquifers. It is to be regretted that simultaneous hydrometric measurements at short intervals along the river courses were not made to localize and quantify the seepage processes. Experience in other arid areas has shown that such information can provide important data for the establishment of a water balance (Cyprus Mission, 1975. A Reconnaissance Survey of Northern Oman Water Resources and Development Prospects). In the three-year observation period (1966–1968), the flow difference between two measuring stations at the Hifa and Sada gorges showed groundwater alimentation due to surface water seepage to total at least (because inflow from the intermediate catchment areas was disregarded) some 80 million cubic meters with, however, considerable variability from year to year (0.4 million m^3/y in 1966 and 62 million m^3/y in 1968).

There are possibilities of increasing the quantity of seepage by means of artificial recharge measures, because the water-uptake capacity of the porous surface layers and potential groundwater aquifers is generally large. There is, however, the empirical fact that, depending on the size and duration of the floods, the infiltration rate decreases with increasing run off. The creation of artificial regeneration seems possible on the basis of the combined hydrological studies by ITALCONSULT (1969); this would involve additional seepage of water flowing downstream from the Sada gorge, which makes up some 60% of the inflow at the Hifa water gauge.

Knowing what we do about groundwater replenishment by seepage of surface waters, before any plans are made for further artificial regeneration or greater usage of groundwater, we should consider means to ensure appropriate protection from pollution for the portions of the river involved in the seepage.

2.5.1.3. Survey of groundwater utilization

Based on hand-dug wells with depths of generally 8–15 meters and bottoms usually only 2–3 meters below the upper edge of the aquifer in the Wadi Bishah between the Hifa and Sada gorges, ITALCONSULT calculated or extrapolated a daily yield totaling approximately 200,000 m^3/day during the observation period of May–June 1967, when there were only a few drilled wells. Of the 831 wells studied, 88.5% were equipped with motor pumps and were used for irrigation. The remaining wells served domestic purposes and were operated with a handwinch. A yearly withdrawal of some 100 million m^3 was estimated for all the 1,575 wells found in the farther environs of Bishah in 1967.

2.5.1.4. Hydrochemical studies

The interpretation of extensive hydrological-isotope and hydrochemical studies permits the distinction of two typical kinds of groundwater with the following characteristics:

Component 1: Groundwaters in the area of the central deep channel (channel alluvium and terraced alluvium in Fig. 99).

Component 2: Peripheral groundwaters in zones with thin Quaternary deposits (Flood plain alluvium in Fig. 99).

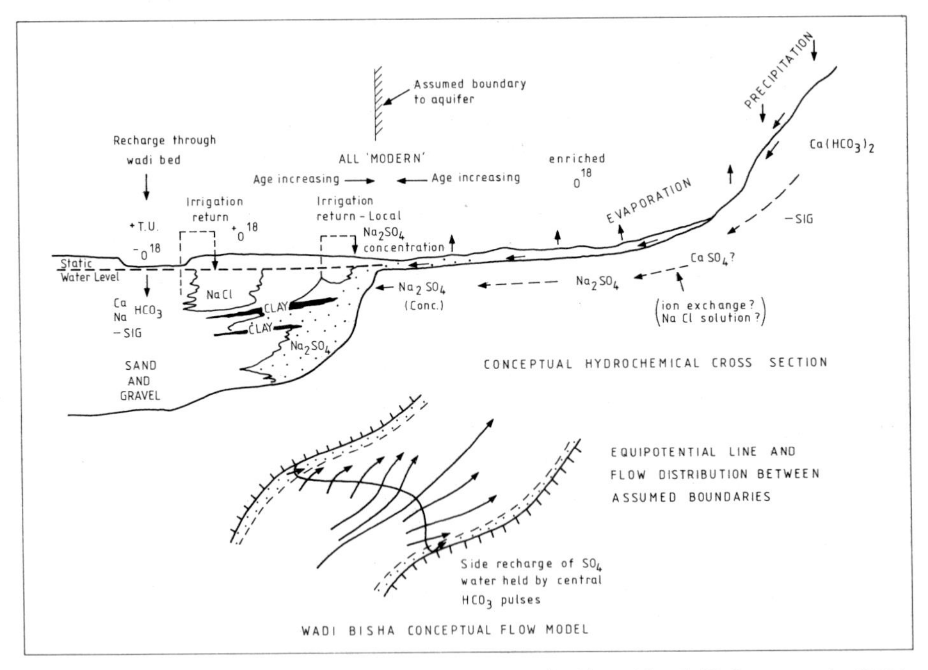

Fig. 99. Hydrological model of the valley cross-section Wadi Bisha. (After J. W. LLOYD et al., 1980.)

Main groundwater current in the central deep channel

The water from the area of the central deep channel (component one) may be described as follows:

– The coolest groundwater temperatures (24–26° C) in the Bishah area were measured in May and June 1967. The reason for this could be infiltration of floods in November with a water temperature of only some 15° C. The seasonal constancy of the groundwater temperatures, seen throughout the entire length of the deep channel, the lack of anomalies and the slight deviation of only 20° C would seem to indicate that there is a uniform, connected and significant groundwater current in this area.

– This is confirmed by the distribution of the electric conductivity measurements; the smallest values 2,000 μmohs/cm were also found uniformly and with little variation along the course of the deep channel, including the area of the parallel, filled deep furrow and the mouths of the larger tributary wadis (Wadi Tabalah).

– A slight chloride dominance occurred among the anions, shown in the Cl^-/SO_4^{2-} ratio, which was generally less than 2, and usually less than 1; this is distinct from peripheral waters and can be interpreted as a greater and faster water drive with diminished chance of solution.

– As may be expected with significant groundwater replenishment by floodwater seepage, the oxygen-18 isotope contents were determined to be less (–3 to –6 $\delta^{18}O$) during the test period of September 1967 to May 1968 in the zone along the deep channel.

Isolated unconnected groundwater occurrences in peripheral stretches of the wadis

The groundwaters in peripheral areas with generally thin Quaternary sedimentation (Component two) may be described as follows:

– Although the surface of the groundwater table was at a lower depth, warmer groundwater temperatures (26–29° C) were registered.

– Considerably higher values for electric conductivity up to 5,000 μmohs/cm and high variability over short distances lead to the conclusion that more or less isolated, stationary groundwater bodies are predestined to allow more salts to get into solution owing to the prolonged retention period. The relatively slight permeability of the mainly fine-clastic sediments involved in the structure of the gravelly-sandy-silty flood-plain alluvium can be used as an argument for a longer retention period.

– The Cl^-/SO_4^{2-} ratios are from 2 to 3, with chloride predominant. This indicates almost complete saturation of the solution components with great local difference, and confirms the existence of groundwater types forming two different groups. One exception is a narrow margin verging directly on the basement outcropping, with relatively more sulfate than chloride – expressing incomplete chloride saturation – and leading to Cl^-/SO_4^{2-} ratios smaller than 1.

– The concentrations of the oxygen^{-18} isotope are generally less negative (−2.5 to −0.4 $\delta^{18}O$) and their state of enrichment suggests that the peripheral, isolated and generally stagnant groundwaters are mainly fed by slowly seeping precipitation highly subject to evaporation.

2.5.1.5. Construction of a simplified hydrochemical aquifer model

Presentation of the situation with a wadi cross profile

Using the example of the valley section in Wadi Bishah, J. W. LLOYD et al. (1980) show an interpretation of hydrochemical and hydrological isotopic coefficients adjusted to a model of a valley cross section. Emphasis is placed on the significance of the supply of flood-water seepage for the main groundwater flow in the central wadi. The thin aquifers peripheral to the flood-plain alluvium, with ions in concentrated solution in the water, show mixing processes in the area of contact with the channel filling. This two-component model is complicated by superimposition of reinfiltrated irrigation-seepage waters with additional soluble substances and very enriched with stable isotopes.

A tritium profile perpendicular to the axis of the wadi shows that the small ^3H values on both flanks could also indicate a lesser flow speed, due to their greater age – in comparison to the bank-filtered accompanying groundwater which may be classified as modern. A shallower groundwater gradient shown previously together with smaller permeability coefficients k_f for the silty flood-plain deposits on the wadi flanks is in any case responsible for the lesser filter speeds.

Determination of the mixture components in the samples taken in 1977

Seven water samples were taken in the upper catchment area of Wadi Bishah, east of Al Alayyah and in the area between Qal' at Bishah and Hatf on 13–14 March 1977, in the course of Quaternary geological studies on terrace bodies. As shown in Fig. 98 (Nos. 19–25), the points of withdrawal were chosen in the above mentioned contact area, where mixing of the two groundwater types must occur. The purpose of this small sample series was the determination of the components involved and their relative quantification. Points of reference in this system were the least mineralized sample 1977/19 from the central bed of the upper course of the wadi, representing the main groundwater flow (A), and the sample 1977/23 taken at the greatest distance from the wadi, representing the flood-plain groundwater near the basement (B). According to their general hydrochemical characteristics, case A may be classified as Na-Ca or Mg-Cl-HCO_3 waters, and case B as having a Na-Ca-Cl-SO_4 composition, which also occurs in mixed waters.

The substances in solution in these waters occur in concentrations which vary within wide limits; sodium and chloride with 2–54 and 2–57 meq · kg^{-1} respectively, are predominant. As may also be seen in the mixture diagram (Fig. 100), lithium, sulfate, calcium and magnesium determinations clearly indicate differing concentrations of the primary water types involved in the mixture. Although the composition is very similar with regard to potassium, strontium, manganese, zinc, bicarbonate and fluoride concentrations, a slight rise in the curve none the less shows that these values depend on the particular water involved. This is confirmed by moderate correlation coefficients.

Ferrous iron and ferric iron were not found in any case, and silicic acid only in small quantities in solution (3.2–6.8 ppm SiO_2). The lithium concentrations in water types A and B were very different, allowing a clear componential analysis

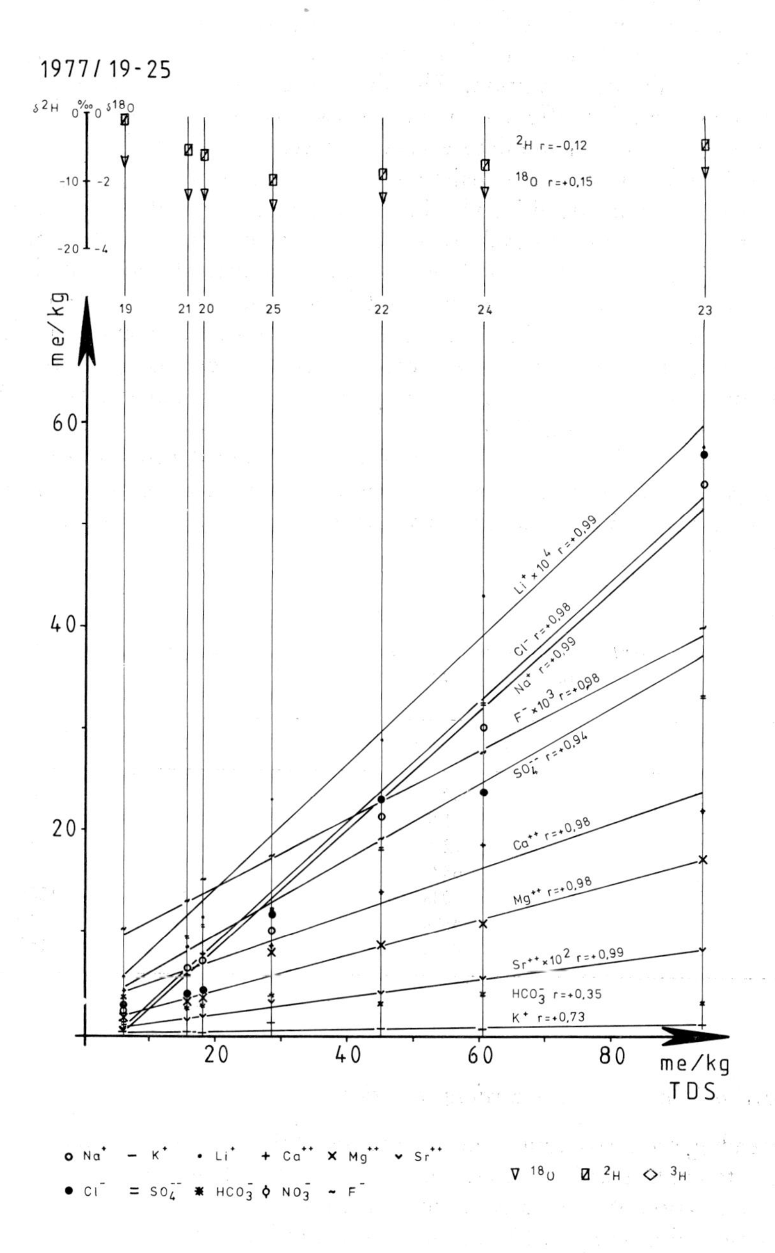

Fig. 100. Chemical composition of the groundwater in the basin of Bishah (1977/19–25) and the possible mixing components between the water from the channel-alluvium represented by 1977/19 and waters from the flood-plain deposits (Type 1977/23).

of the mixed types. They form a nearly linear relationship with a highly significant correlation of coefficient r = +0.99 (Fig. 100). The strontium and fluoride contents in all the groundwaters studied also show very good agreement with an ideal mixture condition. The differences in concentration of stable isotopes do not permit a differentiation of the two geologically and geochemically different aquifers and their mixture types, especially in that the groundwater outlet 1977/19 followed a very shallow course (1 m subsurface) and was greatly subject to evaporation. Additionally, in this section of the wadi it is not a matter of mixture components created by dilution and replenishment of the same primary waters, but of two catchment areas of similar altitude but different lithological structure (channel alluvium and flood-plain deposits). Groundwaters used for irrigation seep back into the ground and are re-used, creating an effect of secondary to tertiary evaporation and causing pronounced changes in the stable isotope contents and proportions in comparison to the original situation (S. S. AL-SAYARI and J. G. ZÖTL, 1978).

Table 38 shows the percent content in the samples of the two water types of different geological origins. As a basis, the parameter with optimal agreement with a linear model was chosen; in this case it was the sodium distribution with r = +0.994.

Table 38. *Percent content of mixture components A for channel alluvium and B for flood-plain deposits in the groundwaters in the Bishah area*

1977	A (low conc.)	B (high conc.)
19	100%	0%
20	91%	9%
21	92%	8%
22	63%	37%
23	0%	100%
24	46%	54%
25	85%	15%

Main hydrochemical types according to ratios

Depending on the solubility and weathering of the rocks in the catchment area coming into contact with the water, both the individual and total mineralization and the calculation of ratios permit the distinction of two groups. The groundwaters represented in samples 1977/19–21 from gravel aquifers with good permeability in the central wadi filling show small total amounts of substances in solution of 410–1,220 ppm with 540–1,520 μmohs/cm at 20° C. The groundwaters from the peripheral silty-calcareous sand and gravel bodies are more highly mineralized with 1,820–5,750 ppm, and 2,450–6,880 μmohs/cm (samples 1977/22–25). The ratios shown in Table 39 also show the groups formed by the dominance of one or the other of the two components.

Table 39. *Regional distribution of hydrochemical ratios*

	1977/19–21	1977/22–25
Ca/Sr	460–910	210–350
(Na + K)/(Ca + Mg)	0.6–0.7	0.7–1.4
(Na + K)/Cl	0.9–1.9	0.9–1.3
Cl/NO_3	19–38	36–135
Cl/F	226–258	639–1435
HCO_3/Cl	0.8–1.1	1.1–0.3
Ca/Cl	1.1–1.9	0.3–0.8
Na/HCO_3	0.8–2.3	2.6–17.7

As Table 39 also shows, the difference found with a number of specific quotations is in part due to evaporation – enrichment processes dominated by chloride, and in part to ion-exchange processes.

As the bicarbonate content is in general relatively small (4.1 mmol/kg), Na_2SO_4 waters are to be expected as exchange types, the comparison between sulfate and noncarbonate hardness (Table 40) is valid for larger values of the former according to B. HÖLTING (1980) as an indication of ion-exchange processes of the type:

$$\frac{Ca}{Mg} SO_4 + Na_2A \rightleftarrows \frac{Ca}{Mg} A + Na_2SO_4,$$

where A represents the cation exchanger. The results of the ratio: $(Na^+ - Cl^-)/SO_4^{2-}$ are between −0.3 and +0.3 and thus satisfy the criterion 1 for an indistinct classification as Na_2SO_4 waters.

Table 40. *Sulfate and noncarbonate hardness (indication for Na_2SO_4 ion exchange)*

1977	Sulfate hardness °dH	Noncarbonate hardness °dH
19	2.5	3.2
20	30.1	22.3
21	26.5	17.6
22	51.2	55.8
23	92.4	101.3
24	90.8	71.0
25	34.2	36.4

The references given for samples 1977/20, 21 and 24 were confirmed by alkali dominance in the ratio (Na + K)/Cl in the range of 1.3 to 1.9, with primarily strong sodium excess as opposed to chloride, which, with the slight carbonate content, appears mainly to bind sulfate.

Additionally, the results for the positive base-exchange index for 1977/20, 21 and 24 ($+ I_{Ba}$: −0.7, −0.9 and −0.3) and the results of the negative base-exchange index for the same samples ($- I_{Ba}$: −0.2, −0.3 and −0.2) are all negative, indicating an exchange of alkaline earths dissolved in water for alkalis from the sediment. The positive values from 0 to +0.1 calculated for the remaining samples 1977/19, 22, 23 and 25 do not offer clear indications of any ion-exchange processes.

In summary, the ion-exchange processes described for samples 1977/20, 21 and 14 can be interpreted to mean that waters containing calcium sulfate (gypsum water), in the course of their underground passage through very fine-grained sediments, have taken up sodium from this exchange material and deposited calcium in its crystal lattice.

2.5.1.6. Technical evaluation of the water samples

Suitability of the waters for water supply

For the practical evaluation of these groundwaters with regard to their technical properties and their use for water-supply purposes, in addition to iron and manganese determinations, free carbonic acid and hydrogen-ion concentration were determined on the spot. As mentioned previously, iron contents greater than 0,01 ppm were not found. Manganese concentrations varied between 0.001 and 0.02 ppm and are thus below the limit of 0.05 ppm given by W. RICHTER and W. LILLICH (1975) for potentially dangerous substances in groundwater.

The lime – carbonic-acid balances generally show these waters to be deficient in carbonic acid; they thus tend to deposit lime, which may lead to rapid incrustation of pipes. The samples taken in the area of the main groundwater flow of the central wadi (1977/19–21) show, however, with their moderate deficits, that these economically interesting waters are nearly in calcium – carbonic-acid equilibrium, so that an advantageous thin rust layer would be formed. The values analyzed and calculated according to J. TILLMANS (in G. MATTHESS, 1973) are shown in Table 41. A comparison of the measured values and equilibrium values according to R. STROHECKER and W. F. LANGELIER also shows positive values for the calcium carbonate saturation index, indicating a more or less pronounced calcium oversaturation.

Table 41. *Balances for carbonic acid (in ppm) and hydrogen-ion concentrations in waters in Wadi Bishah*

1977	Titrated free CO_2	Related CO_2	Excess CO_2	Measured pH	Equilibrium pH	Saturation index
19	1.3	4.7	− 3.4	8.1	7.7	+0.4
20	12.3	18.4	− 6.1	7.55	7.1	+0.45
21	12.3	14.0	− 1.7	7.45	7.2	+0.25
22	12.3	31.4	−19.1	7.5	6.9	+0.6
23	24.6	35.4	−10.8	7.3	6.8	+0.5
24	33.0	67.7	−34.7	7.25	6.6	+0.65
25	25.5	33.3	− 7.8	7.3	7.0	+0.3

An evaluation of the practically and technically relevant water hardness must indicate that, with the exception of two central wadi groundwaters (1977/19 and 21), all of the values for total hardness according to KLUT-OLSZEWSKI (1945) are to be classified as "very hard". In addition to high total hardness (31–110° dH), some of these gypsum waters show extremely high noncarbonate hardness of 22–101° dH, meaning that when used as steam-boiler feed water they may form the dreaded and especially hard gypsum boiler scale which, when overheated, can cause the boiler to explode (K. HÖLL, 1970).

Suitability of the waters for agricultural purposes

Owing to the dominance of sodium chloride among the dissolved salts and a total content in solution of less than 6 g/l, the sampled groundwaters are basically suitable for watering livestock (G. MATTHESS, 1973). For use in optimized poultry breeding, however, the waters with more than 2,860 mg/l (1977/22–24), from the periphery of the wadis, would be disadvantageous.

The hydrogeological prerequisites for irrigation projects in Wadi Bishah are to be classified as favourable, insofar as the groundwater table is usually deep and permeable covering layers permit the removal of dissolved salts from the soil. There are also sufficient receiving streams to provide the necessary drainage when the groundwater periodically runs high. The most important groundwater flow in the central channel, unlike the more or less stationary groundwater occurrences in the flood plains, is capable of combating salt enrichment by evaporation and repercolation. The rather shallow position of the groundwater table along the central channel (4–10 meters below ground, cf. Fig. 98), which at sample point 1977/19 was even less than one meter, can, however, lead in places to concentration of stable isotopes through secondary evaporation (Fig. 100).

The problem of displacement of exchangeable alkaline earths in the soil by irrigation water high in sodium, leading to lowered permeability of the soil and resultant difficulty in tilling or even infertility, can be quantified with the sodium-absorption relation

$$SAR = \frac{Na^+}{\sqrt{(a^{2+} + Mg^{2+})/2}}$$

Values of two to three for the main groundwater flow of the wadi seem to give less ground for concern than values between 6 and 12 in the peripheral area. The suitability of the central wadi groundwaters (1977/19–21) for irrigation is confirmed with a classification according to L. V. WILCOX (1948) of "excellent" to "good to usable" (Fig. 101). According to this criterion, the peripheral groundwaters are clearly unsuitable with one exception (1977/25 "doubtful to unsuitable").

Indications of man-made pollution

Since the samples in the lower Wadi Bishah were taken after several hours' pumping, the indications offered by the pollution-specific ions are valid not only

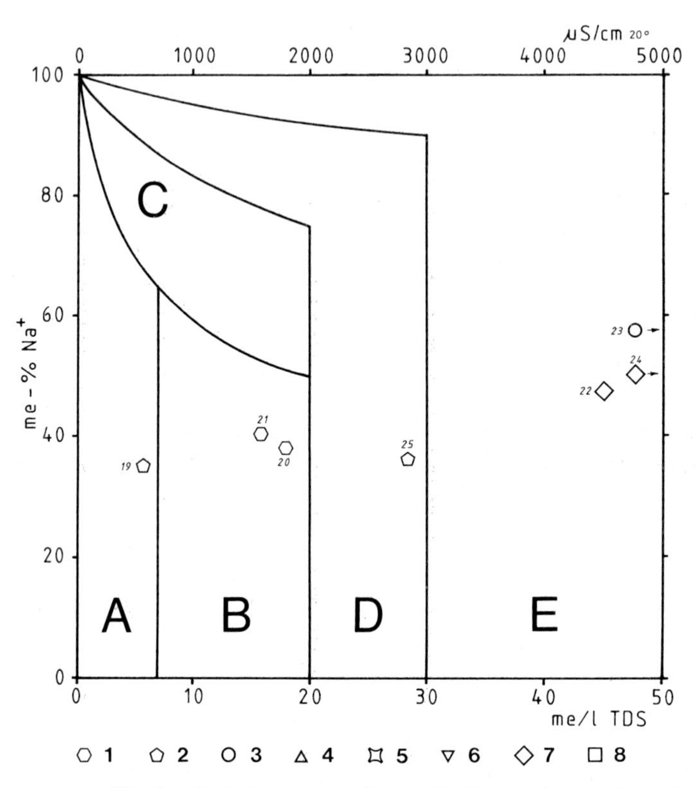

Fig. 101. The water qualification for irrigation use after L. V. WILCOX (1948): *A* excellent – good, *B* good – practicable, *C* practicable – doubtful, *D* doubtful – not applicable, *E* not applicable.

for the immediate vicinity of the well, but also for pollution occurring in the area of the cone of depression. Elevated concentrations of potassium (7–42 ppm), nitrite (2–10 ppm) and occasionally nitrate (7–26 ppm), and phosphate (5–8 ppm) in samples 1977/20, 23 and 24 indicate moderate human influence, particularly leaching of fertilized fields.

2.5.2. Hydrochemical Measurements of Groundwaters in the Uppermost Wadi Bishah near Khamis Mushayt

Between 10 and 11 March 1977 a total of nine groundwater samples was taken from the upper course of Wadi Bishah (Abha – Khamis Mushayt – Bani Thawr area) for a hydrochemical longitudinal profile (Fig. 102). Care was taken to sample waters from a variety of geological-lithological units. Samples Nos. 10 and 11 come from basalt catchments (Miocene – Oligocene). The other mildly mineralized waters in samples Nos. 12, 13, 17 and 18 are typical for joint aquifers in magmatites (gabbro, diorite, granite). In the more highly concentrated groundwater samples Nos. 14–16, seepage through a thicker Quaternary sand and loess covering is responsible for a hydrochemical modification of the water type from metamorphic rocks (green schist, chlorite-sericite slate, granite-gneiss).

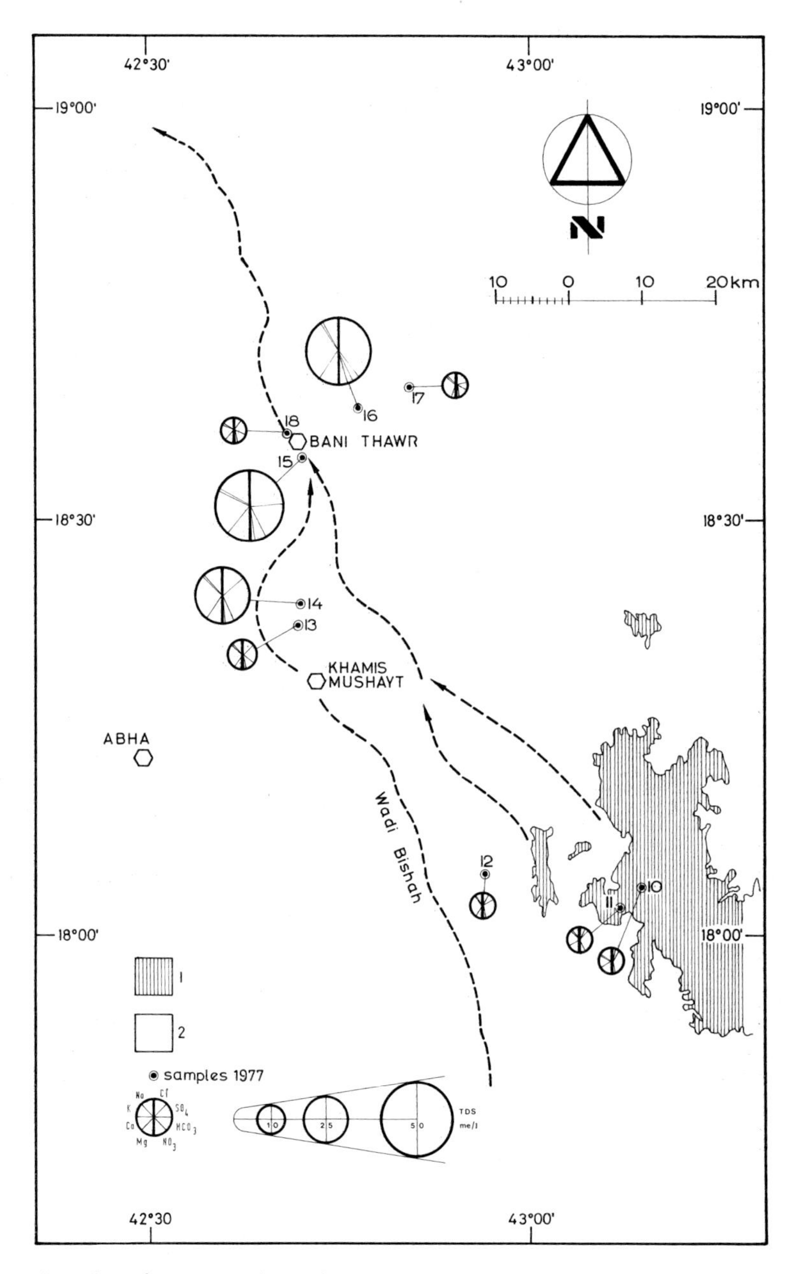

Fig. 102. Location of water samples in the uppermost Wadi Bisha near Khamis Mushayt (*1* basalt; *2* crystalline bedrock and wadi channel).

2.5.2.1. *General hydrochemical classification*

Absolute and relative composition

This grouping into three water types according to the geological structure of the catchments cannot be verified by the total mineralization alone, owing to the slight solubility of basaltic and magmatic rocks. But a differentiation is indeed possible between those waters coming almost exclusively from joint aquifers with total dissolved contents of 479–769 ppm, and those from metamorphites with essentially loose-rock covering and elevated mineral contents of 1,773–2,980 ppm.

These groups formed on the basis of the geological-lithological composition of the catchments may, however, be clearly recognized in the relative composition. The two waters from the basalt flows southeast of Khamis Mushayt (Nos. 10 and 11) with their higher magnesium content are typical for a Mg-Ca-Na-HCO_3 water.

The samples from magmatic catchments (Nos. 12, 13, 17 and 18) are not of such uniform composition and represent Ca-Na-Cl-HCO_3-SO_4, Ca-Na-Mg-SO_4-Cl-HCO_3 or Ca-HCO_3 water types. The third group of more highly mineralized crystalline waters covered by sandy-silty deposits (Nos. 14–16) also lacks uniformity in the relative ion distribution. The three waters samples represent three individual types: Ca-Na-SO_4-Cl, Ca-Na-Mg-Cl-SO_4 and Ca-Cl.

Trace-element contents and their relationship to the hydrogeological water types

Samples from basalt catchments show low concentrations of trace elements (Table 42). Silicic acid, however, showed the highest values around 30 ppm. This is thus in agreement with the frequency maximum given by G. MATTHESS (1973) of 20–40 ppm SiO_2 seen in waters in gabbro, basalt and comparable crystalline rocks. Sample No. 12, taken orographically and geologically from the transitional zone from the basalt flows to or upon Precambrian granite-gneisses, documents with concentrations similar to basalt waters the effect of seepage through the lava flow.

Table 42. *Trace-element and silicic-acid contents (ppm) in the three water types determined hydrogeologically (A = basalts, B = magmatites, C = metamorphites and loess)*

1977	Type	Li^+	Sr^{2+}	Zn^{2+}	Mn^{2+}	F^-	SiO_2
10	A	0.004	0.10	0.01	0.00	0.20	28.0
11	A	0.004	0.02	0.002	0.02	0.15	34.2
12	B	0.004	0.18	0.02	0.01	0.28	28.0
13	B	0.01	0.42	0.01	0.01	0.36	7.5
14	C	0.02	1.02	0.01	0.01	0.54	6.0
15	C	0.04	1.77	0.01	0.001	0.93	3.9
16	C	0.05	3.28	0.03	0.01	0.82	5.4
17	B	0.006	0.40	0.002	0.01	0.46	12.0
18	B	0.004	0.23	0.006	0.001	0.12	6.4

These dissolved contents in magmatic rocks are a contrast to the strontium and, in places fluoride compounds in the volcanic rocks that are only very weakly water soluble. While there is here a concentration distinctly above the detection limit, the silicic-acid concentrations, as seen in other acid rock complexes (G. MATTHESS, 1973), are relatively slight. In both hydrogeological water types (A and B), there are generally similarly slight concentrations of lithium, zinc and manganese ions.

Seepage through the sandy-silty Quaternary deposits on the crystalline slates causes an increase in almost all the dissolved contents. The trace-element concentrations in type-C water (Table 42) are considerably elevated, as are the main components. An exception is to be found in the silicic acid which, in comparison to the afore mentioned type B (magmatite), shows similarly slight, or even lesser values. The variation in the composition of waters significantly influenced by loose Quaternary rocks permits the hydrogeologically decisive conclusion that there are rather independent loose-rock aquifers. The extent to which these waters circulate in complete isolation in a porous groundwater aquifer, or are affected by a slight inflow of joint waters from the basement, can be determined by making up a potential mixture model.

2.5.2.2. *Studies to detect mixing processes*

Samples Nos. 10 to 18 were evaluated with a concentration diagram (Fig. 103) to demonstrate mixing processes. As there is a linear relationship of the concentrations of the individual ions to the total mineralization, a simple two-component model may be used. There is a quite strict relation in all the samples with the exception of sample No. 15, the most highly mineralized Quaternary groundwater.

In this case it seems permissible to assume the existence of an isolated lenticular groundwater occurrence within the Quaternary accumulation. In the remaining crystalline and loose-rock aquifers, a limited exchange zone could determine the nature of the waters in samples Nos. 13 and 14. A slight alimentation of the Quaternary groundwater aquifer by joint waters from crystalline catchments seems here to be the case (Table 43).

Secondary evaporation in the course of repercolation of irrigation water in the lower course (area north of Khamis Mushayt) is responsible for a distribution of stable isotopes that does not follow the linear hydrochemical relationship.

Table 43. *Percentage amount of joint waters from basalt or magmatite catchments (low concentration) in groundwaters from Quaternary-covered metamorphites (high concentration), with the exception of the independent groundwater type represented by sample 1977/15*

1977	low conc.	high conc.
10	100%	0%
11	99%	1%
12	89%	11%
13	82%	18%
14	42%	58%
16	0%	100%
17	89%	11%
18	96%	4%

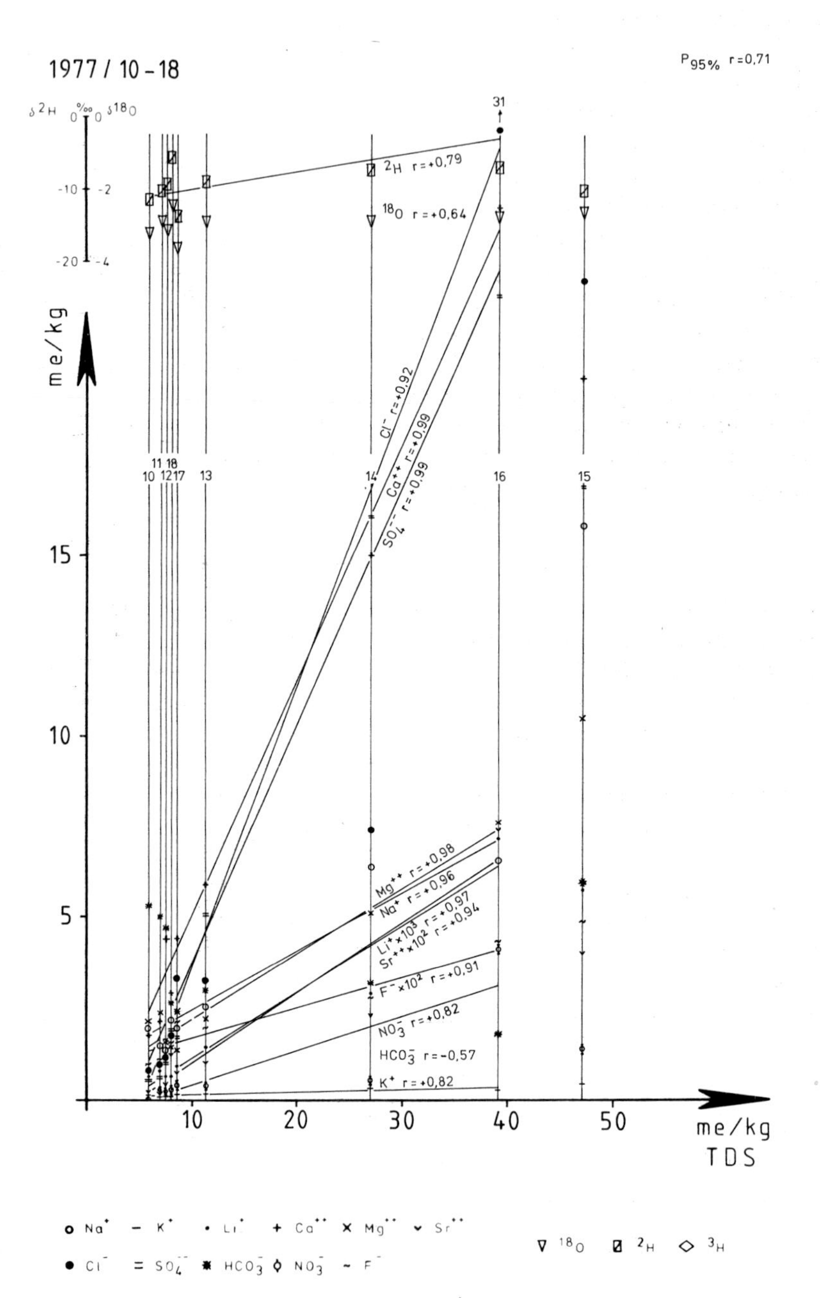

Fig. 103. Chemical composition of the groundwater in the upper course of Wadi Bishah in the vicinity of Khamis Mushayt (1977/10–18). A mixture process may be assumed between the waters from basalt flows (represented by sample, No. 10) and those from joint aquifers made of magmatites and metamorphites (No./16). Sample, No. 15 presumably comes from an isolated groundwater lens.

2.5.2.3. Ion exchange between water and rock

In the absence of permutites with exchange capability, ion-exchange processes are not to be expected a priori in waters from basaltic and magmatic rocks.

Sample No. 10 from the basalt flows southeast of Khamis Mushayt is an exception (Fig. 103). The higher sodium content is not in stoichiometric equilibrium with equivalent amounts of chloride or sulfate, but appears to be bound to bicarbonate as well. Negative values of the base-exchange indexes ($+ I_{BE} = -2.4$; $- I_{BE} = -0.2$) are the basis of the assumption that alkaline earths in water – perhaps during seepage through silty or loess-like deposits – have been exchanged for alkalies from these sediments. Generally, negative values for the base-exchange indexes for groundwaters from crystalline rocks ar not necessarily due to ion-exchange processes, so that some other explanation could be given as the cause of the primary water. When silicates weather, more alkali ions than chloride ions are liberated and can thus make any inequilibirum plausible.

Important secondary changes in the hydrochemical nature of the initial solution have very probably taken place in sample No. 16. This groundwater from the Quaternary covering of the metamorphic basement is remarkable for having almost exclusively chloride as an anion, and slight alkali content. The calcium-chloride water type allows on the basis of high positive values in the range of +0.8 to +3.0 for the exchange indexes an interpretation in the direction of an exchange of alkalies in water for alkaline earths from the rock. This can be explained hydrogeologically by the leaching of wadi-sabkhah sediments (S. S. AL SAYARI, J. G. ZÖTL, 1978) and subsequent passage of the highly mineralized groundwaters through a carbonaceous medium.

3. Hydrogeology

(H. HÖTZL, J. G. ZÖTL)

The groundwater occurrences in Saudi Arabia can be divided from their geological situation into a number of types. There are the enormous underground aquifers in the cuesta landscape of the Shelf, and the reservoirs in the thick sediment bodies in the northern part of the country. The coastal landscapes along the Red Sea have hinterland wadis with sediment fillings; although their capacity varies with precipitation and size of the catchment area, they usually offer unconfined groundwater, the largest amount of which is to be found in the vicinity of Jizan.

Finally there is the highland. It was raised up in the recent geological past and shows an extreme scarcity of water in the extensive volcanic plateaus. In the crystalline, however, there are minute oases in the uppermost roots of the wadis, and water is taken from wadi basins at medium altitudes for large settlements (Al Madinah, At Taif, Bisha), whereby the groundwaters under basalt flows are also significant.

A number of individual studies will be required for water management suited to the circumstances to ensure a modest but sufficient water supply for the country in the future. Within our Quaternary research project directed hydrogeological studies were performed in selected areas. One of the aims was to get additional information on the Quaternary, especially on climatic changes. On the other hand hydrogeological studies were just intensified to show one practical aspect of such basic scientific research.

This summarizing chapter on hydrogeology will give a short description of the general situation in the Kingdom of Saudi Arabia. It covers first the Arabian Shelf in the east, then the northwestern parts of the country, and the Arabian Shield with the Red Sea coast. Within these three large areas, the discussion will follow the stratigraphic order as far as possible (Plate V, see insertion at back cover). The simplified presentation of the rock distribution given in Plate V is based on the Geologic Map of the Arabian Peninsula (scale 1 : 2,000,000, 1963), the 1 : 500,000 partial maps and, where available, maps with larger scales. The signatures were generalized.

3.1. Central and Southern Arabian Shelf

3.1.1. Hydrogeology of Quaternary Deposits

3.1.1.1. Eolian sands (Qe)

Eolian sand deposits cover the greatest parts of the land, with the Great Nafud, smaller Nafud areas between the ranges of cuesta landscape in the west, the wide strip of the Ad Dahna trending N–S in a curve of more than 1,000 kilometers, the Jafurah, and finally, the vast surface of Ar Rub' Al Khali (Plate V, see insertion at back cover).

There are various types of dunes to be found, but in general these are purely desert areas. The hydrogeological conditions, however, are by no means identical. It may be assumed that the grain-size distribution is about the same (mainly 0.2–1.0 mm; see Vol. 1, fig. 91, p. 270) with only limited variation in permeability. The frequency and intensity of the episodic precipitation, however, differ according to the geographic positions of the deserts.

The eastern and central **Rub' Al Khali** is dominated by high dunes up to some hundreds of meters in height; in the southwest there are mainly seif dunes. The groundwater seeping underground from east and southwest down slight gradients (0.5–1‰) to the Arabian Gulf is a brine lying slightly under the surface in the upper unconfined aquifers. Fresh water exists only in confined aquifers at greater depths. When borings open up artesian wells of deep groundwater, however, vegetation will sprout even in the central hyperarid desert (Fig. 104).

The shallow groundwater is replenished only by quickly absorbed short runoffs from the surrounding mountains; seepage of the very infrequent episodic precipitation is inconsequential.

The conditions in the central and northern desert areas differ from the Rub' Al Khali. Not only are the dune forms different, but the smaller thickness of the eolian sand and the more frequent precipitation in comparison to the Rub' Al Khali are significant. In years with above-average showers or thunderstorms we saw a downy covering of fine grasses tinging the dunes in the northern **Ad Dahna.** T. DINCER (1978) and T. DINCER et al. (1974) show with isotope measurements of the moisture in sand samples taken in 1972 and 1973 from the Ad Dahna that while most of the precipitation evaporates, a small amount seeps down into the earth. Systematic samples from sand profiles from different depths showed increasing moisture just a few decimeters below the surface, with the isotope content of the retained water in good agreement with that of the precipitation in Ar Riyadh. In the meanwhile, seepage of rainwater also in the sand deserts of the Sahara has been confirmed (L. SONNTAG et al., 1978). Even though measurements have as yet only been made in a few localities and have only shown seepage quantities of 1–2 millimeters per year reaching the bedrock under the dunes, this nonetheless means, considering the extent of these sand deserts, an enormous groundwater replenishment.

Table 44. *Generalized hydrogeologic section of the Arabian Shelf*
Prepared by G. Otkun after R. W. Powers and L. F. Ramirez (1968)

Age				Formation	Generalized lithologic description	Thickness (Type or reference section near outcrop)	Aquifer characteristics
Cenozoic	Quaternary and Tertiary			Surficial deposits and basalt	Gravel, Sand, Silt and Basalt		Produce variable quality and quantity of water depending upon recharge by rainfall. Basalt yields little water in western Saudi Arabia.
Cenozoic	Tertiary	Miocene and Pliocene		Al Kharj	Limestone, lacustrine Limestone, Gypsum	28 m.	Generally called Neogene aquifer. Irregular occurences of water. Artesian and non-artesian conditions. Prolific aquifer in the areas of Al Hasa, Wadi Miyah and some others in Eastern Province.
				Hofuf	Sandy marl and sandy Limestone	95 m.	
				Dam	Marl, Shale, subordinate Sandstone	91 m.	
				Hadrukh	Calcareous, silty Sandstone	84 m.	
		Eocene	Lutetian	Dammam	Limestone, Dolomite	33 m.	Produces moderate amount of water with artesian and non artesian conditions.
			Ypresian	Rus	Marl, chalky Limenstone	56 m.	Productivity unknown.
		Paleocene	Thanetian / Montian(?)	**Umm er Radhuma**	Limestone, dolomitic Limestone	243 m.	One of the **most prolific aquifer** of the Kingdom with high transmissibility varying between 500,000 and 3 million gpd/ft.
Mesozoic	Cretaceous		Maastrichtian / Campanian	Aruma	Limestone	142 m.	Yields little water of low quality.
			Turonian(?) / Cenomanian	**Wasia** (Sakaka sandstone northwest Arabian)	Sandstone, subordinate Shale	42 m.	Low productive or even dry near outcrop, very **high productive artesian and non-artesian conditions in Eastern Province.** Hydraulically interconnected with Biyadh near outcrops.
			Aptian / Barremian	Biyadh	Sandstone, subordinate Shale	425 m.	**Moderately productive sandstone aquifer,** hydraulically interconnected with Basia near outcrop.
			Hauterivian	Buwaib	Biogenic Calcarenite and calcarenite Limestone	180 m.	Productivity low or unknown.
			Volanginian	Yamama	Biogenic Calcarenite	46 m.	Productivity low or unknown.
			Berriasian	Sulaly	Chalky aphapitic Limestone	170 m.	Productivity low or unknown.
	Jurasic		Tilhonian	Hith	Anhydrite	90 m.	Yields always mineralized water.
				Arab	Calcarenite, calcarenite & aphanitic Limestone	124 m.	Yields little amount of water, mostly mineralized. Irregular occurence of water.
			Kimmeridigian	Jubaila	Aphanitic Limestone	± 118 m.	Similar to Arab Formation above.
				Hanifa	Aphanitic Limestone	113 m.	Unknown productivity.
			Oxfordian / Callovian	Tuwaiq Mountain	Aphanitic Limestone	203 m.	Productivity unknown.
			Callovian(?) / Bathonian / Bajocian	Dhruma	Aphanitic Limestone Sandstone south of 22° N, and north of 26° N	375 m.	Produces moderate amount of water north of 26° N and south of 22° N where it is generally represented by sandstone. South of 22° hydraulically connected with Minjur.
			Toarcian	Marrat	Shale and aphanitic Limestone	103 m.	Yields little water fair to poor quality.
	Triassic		Upper	**Minjur**	Sandstone, Shale	315 m.	Generally **highly productive sandstone aquifer** with flowing and non-flowing artesian conditions.
			Middle	Jilh	Aphanitic Limestone Sandstone and Shale	± 326 m.	Mostly hydraulically interconnected with Minjur, produces low quality water.
			Lower	Sudair	Red & Green Shale	116 m.	Aquiclude.
Paleozoic	Permian		Upper	Khuff	Limestone and Shale Sandstone south of 21° N	171 m.	Moderately productive limestone aquifer, mostly mineralized water.
			Lower / Undated	**Wajid**	Sandstone	950 m. Calculated	**Highly productive sandstone aquifer** with flowing and non-flowing artesian conditions.
	Devonian		Lower	Jouf	Limestone, Shale & Sandstone	299 m.	Productive generally in Al Jouf area.
	Ordovician and Silurian			**Tabuk**	Sandstone and Shale	1072 m.	**Productive sandstone aquifer** with flowing and non-flowing artesian conditions.
	Cambrian			**Saq** (Umm Sahm, Ram, Quwiera, Siq)	Sandstone	± 600 m.	**One of the most productive Sandstone aquifers of Saudi Arabia,** with flowing and non-flowing artesian conditions.
Precambrian Basement Complex							

Fig. 104. Outflow of poor mineralized groundwater from a borehole in the Central Rub Al Khali. Vegetation comes up by wind transported pollen over large distances. (Photo: J. G. ZÖTL, 1972.)

3.1.1.2. Sabkhah deposits (Qs)

Sabkhah deposits are silt, clay and/or muddy sand, saturated with brine and encrusted with salt. In the regional descriptions in Vol. 1 we covered coastal and inland sabkhahs and, most recently, wadi sabkhah sediments (after H. HÖTZL et al., 1978). Coastal sabkhahs are due to infiltration of seawater as a result of flooding; inland sabkhahs are the result of evaporation of unconfined groundwater lying near the surface. Wadi sabkhahs are limited local inclusions of sediments with a sabkhah-like composition in lenses or layers of fluvial wadi sediments. They are naturally of small size. All sabkhah sediments are completely lacking in vegetation owing to their high salt content, and do not present any possibility of obtaining low mineralised ground water. If they still carry water, they also present nearly insurmountable problems for roadbuilding. When they dry out form the surface of the desert to meter-thick salt flats, they are as hard as cement where larger transport planes may be landed upon them.

3.1.1.3. Quaternary surficial deposits (Qu), excluding Qe and Qs

Recent wadi deposits

Both the entire cuesta landscape of central Arabia and the mountains in the north along with the edge of the less steep eastern slope of the Shield are dissected by a large number of usually dry wadis.

Gravel and sand on the surface of the wadi bed are generally of more or less local origin, but material from distant areas may be transported by heavy floods. Occasional stagnation may lead to silt deposits, and over flat stretches evaporation may lead to the formation of wadi-sabkhah lenses. In general, the wadi filling is, depending on its thickness, the result of layers carrying more or less groundwater.

Vol. 1 briefly covered wadis in the central part of the Tuwayq Mountains (pp. 216–222, fig. 72). The former main waterworks of the capital city of Ar Riyadh was located here in the wadis Hanifah, Al Luhy and Nisah before water was yielded froom deeper aquifer, and before began to be delivered from the large desalination plants on the Arabian Gulf. We thus know that the thickness of the fluvial accumulation reaches approximately 60–80 meters in Wadi Hanifah and some 300 meters in Wadi Nisah. As the latter follows a graben, it may be concluded that a tectonic subsidence occurred here in the Quaternary. The wells in Wadi Hanifah are some 60 meters deep and the bores in Wadi Nisah reach the sandstones of the Biyadh Formation; the water shows remarkably slight mineralization.

Very little information is available on the absolute amounts of water available and actually tapped. An order of magnitude is suggested by a report published by R. WOLFART in 1965 that a bore in Wadi Nisah produced 800 cubic meters/day during four days' test pumping, with a drawdown of only 5 cm. Closer examination of the situation shows that replenishment is still going on today.

R. WOLFART (1965, p. 175) assumes for the catchment area for the groundwater in Wadi Nisah (approximately 1,500–1,800 square kilometers) a seepage rate of 5% of the total annual precipitation. That would produce a regeneration of 7–8 million cubic meters of groundwater. Taking safety factors into account, he estimates a possible yearly yield of 5 million cubic meters.

Fig. 105 shows the residual runoff and related groundwater replenishment in Wadi Hanifah the day after a cloudburst with a thunderstorm and high water on March 28, 1974. The wadi is normally dry. There are hundreds of such wadis, often with small oases on terrace remnants impervious to floods. We shall see them again on the western edge of the Shield in similar form. They do not appear on our hydrogeological map, and even the large geological map (1 : 2,000,000) has too small a scale to show these wadis[1].

Pleistocene deposits

As the small and medium-sized wadis do not appear on the map owing to its small scale, the Quaternary surficial deposits designated as Qu are summarized as follows: (1) the unmistakable estuary cone and lower-course sediments of the Ar Rimah river system, which previously crossed the entire Shelf area from west to east (alluvial fan at Al Batin in the west); (2) the wadis from the central hinterland

---

[1] Wadi Hanifah and Wadi Nisah are found on the geological map of the Northern Tuwayq Quadrangle (R. A. BRAMKAMP and L. F. RAMIREZ, 1958), 1 : 500,000. For detailed geological sketches see Vol. 1, p. 203, fig. 63.

Fig. 105. Residual runoff and related groundwater replenishment in Wadi Hanifah one day after a cloudburst and flood on March 28, 1974. (Photo: J. G. ZÖTL, 1974.)

(Wadi Birk, Wadi Nisah and others), joining at the basin of Al Kharj to form the Wadi As Sha'ba which opens with a former delta fan some 300 kilometers wide on the coast from the west to the south of the Qatar Peninsula; (3) the fluvial sediments of Wadi Ad Dawasir, which still occur but are mainly covered by the dunes of Ar Rub Al Khali. Ample substantiation was given in Vol. 1 for their age classification in a so-called "Late Pliocene / Early Pleistocene Phase". There are on the one hand limestone and quartz pebbles in channels which currently carry water, while on the other, there are gravels and limestone pebbles of various sizes on terraces and elevated coastal areas whose position cannot be explained by the current situation. The hydrogeological circumstances of these plains have scarcely been studied, and there are hardly any wells or settlements. West of the Tuwaiq mountains there are vast Qu plaines that belong to the southern Nafud.

3.1.2. Tertiary Sediments in Eastern and Northern Saudi Arabia

3.1.2.1. Middle and Upper Tertiary deposits of larger distribution (Table 44)

The Young Tertiary sediment sequence (Tms), including the Al Kharj, Hofuf, Dam and Hadrukh Formations consists of alternating layers of marly sandstone, sandy marl, and sandy-to-pure limestone (cf. Geologic Map 1 : 2,000,000, 1963). These sediment sequences, whose thickness reache 250 m and more at the Gulf Coast and thin out about 200 km farther inland, cover an area from the Iraqi border to the Rub Al Khali amounting to more than 200,000 square kilometers. Though the permeability of this sequence is not sufficent some intercalations with higher hydraulic conductivity are of regional importance as aquifers, especially in the coastal zone. But the well known prolific springs in the Neogene of the Gulf Coast have its origin in the connections with underlying main aquifers. The situation there is explained in the chapter 3.1.4.

To the west (As Summan Plateau, cf. Vol. 1, p. 166ff.) the permeable carbonate layers thins out and clastic sequences with some evaporitic intercalations prevail. The Young Tertiary on its western margin (As Sulb Plateau), directly superposes the karstic Paleocene limestones. Karstification of the calcarous Tsm with a network of caves and shafts there connects hydraulically the Tsm with the underlying limestones. Insofar as rain falls, in the area with the vertical caves most of it flows into the ground without evaporation loss and replenishes confined groundwater (Fig. 106)[1].

The Middle Tertiary (Eocene) the **Dammam Formation** (limestones) and **Rus Formation** are generally outcropping outside the country (Iraq or Emirates; see geological map). Therefore the areas of Eocene deposits are not separated in Plate V at the end of the book. But the Dammam formation, with Alat and Khobar carbonate member belongs to the most important aquifer in the Gulf Coast area. High yield is obviously caused by upward leakage from the underlying Umm Er Radhuma Formation. The chalky and evaportic Rus Formation contains small local aquifers with high mineralized water.

3.1.2.2. The Paleocene Umm Er Radhuma Formation

The **Umm Er Radhuma Formation** (Map Plate V, TU) is exposed even in the floor in the area of the Summan Plateau; it also forms the outermost curve of the cuesta landscape of central Saudi Arabia extending to the east. It may be **the most important freshwater aquifer in Saudi Arabia** from every point of view. The curve has an average width of 50–100 kilometers and extends from the Iraqi border to the Rub' Al Khali. In spite of layers of dolomite and dolomitic limestone, there is more rather pure limestone and G. ÖTKUN says of its waterbearing properties "especially its high transmissibility has never been met in other aquifers" (1969, p. 12)[2].

The thickness of the Umm Er Radhuma Formation is considerable. An exploratory well drilled in Wadi Miyah crossed 490 meters of limestone and dolomite; in Haradh the thickness exceeds 300 meters.

Large part of the surface of the outcropping Umm Er Radhuma Formation are covered by dunes of the Ad Dahna desert belt. This lends special significance to the studies performed here on the seepage of rain water through dunes (T. DINCER

---

[1] The collection of wells around Jubat Yabrin and Al Khunn (23°15′ N, 49°00′–49°15′ E) and related inland sabkhahs (Sabkhat Al Buth) may also extend their catchment area into the central Summan Plateau (cf. Geologic Map 1 : 2,000,000).

[2] G. ÖTKUN (1969, p. 13) also gives numerical measured values of the transmissibility in bores in the Umm Er Radhuma Formation: Haradh 478,240 gpd/ft, Dhahran 529,600, Shedgum 26,600. The result of a test of 50 day's duration in Wadi Miyah exceeded all these data (3 million gpd/ft).

et al., 1973), as regards replenishment of water in the Umm Er Radhuma Formation in view of the size of the area. Replenishment of 2 millimeters per year over 100,000 square kilometers amounts, approximately, to 200 million cubic meters of water.

3.1.3. Aquifers in Cretaceous Formations (K)

Of the Cretaceous formations (K Plate V, see also Table 44), the upper layer (**Aruma Formation**) is still connected hydrologically to the Umm Er Radhuma Formation, as long as it is made up of (thin) limestones and dolomites. According to P. BEAUMONT (1977, p. 47), the main aquifers are the **Wasia and Biyadh Formations.** Both formations are made up mainly of sandstone interbedded with thin layers of shale and dolomite. The two aquifers often cannot be distinguished hydrologically, and especially near the outcrop Wasia is hydraulically interconnected with Biyadh (G. ÖTKUN, 1969, p. 10). Data on the thickness

Fig. 106. Vertical shafts in the Summan Plateau. (Photo: J. G. ZÖTL, 1974.)

show increase from west to east, i. e. from the outcrop with its downward dip; Wasia with a thickness of 42 meters at the outcrop reaches 600 meters under the Rub Al Khali and is overcovered by 800 meters. This gives a gradient of 2.5–3‰ for the dip of the layers to the top east and southeast.

The sandstones show generally good but very irregular transmissibility; A. I. NAIMI (1965) and G. ÖTKUN (1969) mention values of 20,700 to 69,000 gpd/ft, and 1,443,000 gpd/ft (?) for a bore near Abqaiq.

A SOGREAH study in 1967 (Riyadh Water Supply Report) estimated an infiltration rate between 2–6%.

As far as the quality of the water is concerned, in both cases the water has ion-concentration sums of 300–900 ppm near the outcrop of the aquifer and is to be designated as very good. The slight solution concentrations are typical for sandstone aquifers. The concentrations increase toward the east and with greater depth: in the Wasia Formation to more than 200,000 ppm (brine!) near the Kuwait border. The water in the Biyadh Formation reaches a salinity of more than 6,000 ppm in the Eastern Province (G. ÖTKUN, 1969, p. 10ff).

No detailed studies are available for the Lower Cretaceous formations, including the **Buwaib, Yamama and Sulaiy formations** (mainly calcarenite and calcarenite or chalky limestones). More detailed information on the lithologic sequence give stratigraphic profiles in Table 44 and 45.

3.1.4. Jurassic Formations and Aquifers (J)

3.1.4.1. The Hith and Arab Formations

The outcrops of the **Hith Anhydrite,** with the best examples of a gypsum karst to be found on the Arabian Peninsula, are of hydrological interest. The best known is the overhanging face of the Dahl Hith near Ar Riyad; the downslanting cave is of considerable size and reaches down to the groundwater table.

Another impressive example of the gypsum karst are the small lakes near Aflaj, which developed in the area of the outcrop of the central-Arabian Hith Anhydrite. Here there is a row running SSW–NNE of five small open waters, the northernmost and largest of which has a length of some 400 meters and a breadth of 150–200 meters; the second and third lakes are also some hundreds of meters long. The water is very clear and tastes very bitter. The high sulfate content prevents any significant plant growth, and only goats drink occasionally at shallow places along the shore[1].

The anhydrite outcrop of Aflaj (22°10′ N, 46°45′ E) is in the middle of a funnel-like Quaternary surface that crosses the cuesta of the Cretaceous formations some 100 kilometers to the southeast (east of Layla). It extends into the Tertiary of the southwestern offshoots of the As Summan Plateau (Wadi ’Al Jadwal). Whether this is a pre-Quaternary surface runoff – a similar but narrower breakthrough through the Cretaceous formations is found some 100 kilometers

[1] Here there are interesting ruins of two kanat systems differing in age and today covered by dunes, as well as a younger, abandoned attempt at a channel system. At the time of our visits in 1971 and 1973, this phenomenon was becoming a tourist attraction.

to the south (Wadi Al Maqran) – or the result of extensive solution tectonics, is a matter demanding closer attention.

The **Arab formation** yields only a small amount of water. In the eastern province it is famous for its oilbearing: there the lower layers are among the most important.

3.1.4.2. Tuwayq Mountain and Lower Jurassic Formations

The morphologically most impressive escarpments are the cuesta scarps of the Jabal Tuwayq showing in places a series of steps from west to east of **Tuwayq Mountain Limestone,** and the **Hanifah** and as the youngest the **Jubailah Formations,** with a shallow backslope to the east (Plate V, see also Table 44). The transmissivity in these aquifers changes in dependance from the development of secondary porosity. Though the limestones provides water to many wells in the surrounding of Ar Riyadh and Al Kharj the total storage in the limestones is rather small.

Preceding the Tuwayq escarpment to the west there are the outcrops of **Dhruma** and **Marrat Formations.** At the time of the survey, very little was known about the hydrological characateristics of these layers. G. ÖTKUN (1969) writes of the **Dhruma Formation,** "this is a formation which shows great variation in lateral facies. It is generally represented by sandstone north of 26° N and south of 22° N. Between the two regions tight limestone and shales prevail. South of 22° N parallel it is sometimes indistinguishable from underlying Minjur" (p. 9). Ground water storage is good where sandstone facies are developed but is less favourable in the central carbonatic to shaly area.

The bores made in 1968 in the outcrop area of the 375-meters thick Dhruma Formation somewhat south of 26° N produced appreciable amounts of water with quality similar to that of the water from the underlying Minjur Formation.

It may very well be that the lower layers of the Dhruma Formation are hydraulically connected to the underlying Minjur Sandstone (Lower Jurassic and Upper Triassic).

3.1.5. Triassic Formations

The **Minjur Sandstone** (Plate V, thickness at the outcrop 315 m, see Table 44) is an important aquifer. The quartzitic sandstone is of continental origin. Characteristic structures of bedding proves eolian nature for some of the sandstones, while small intercalations of shale and limestone are of lacustrine origin.

As the Minjur Sandstone outcrops west of the capital of Ar Riyadh, a number of bores exist. These show that the Minjur sandstone does not underlie the entire area of the Shelf, but is more or less limited to the central-Arabian area. A bore showed maximum thickness at Ar Riyadh (400 m). North and south of Ar Riyadh it gradually becomes thinner, then pinches out. The situation is similar toward the east, where deep bores approximately 350 kilometers southeast of Ar Riyadh did not encounter any Minjur.

The water bearing properties of Minjur depend on its tectonic behavior and lithology. Thick layers of coarse sandstones show good productivity.

Wells in the outcrop area have partly an unconfined water table; toward the east, the water is confined. In bores, it may rise above ground level like an artesian well because of the lower surface elevation. According to G. ÖTKUN (1969), the quality of the water varies in the wells near Ar Riyadh with mineralization of 1,000 to 5,000 ppm.

A long-term pump test in a bore near Ar Riyadh showed a transmissibility of 96,800 gpd/ft. For others bores, the values vary between 12,000 and 44,000 gpd/ft.

An average carbon-14 age of 25,000 years for water samples from various wells is not especially reliable, as there are neither carbon-13 measurements nor data in percent modern.

The Triassic **Jilh Formation** is mostly hydraulically interconnected with the Minjur, but produces water of low quality. The **Sudair Shale** is of very low permeability.

3.1.6. Paleozoic Aquifers

While the **Khuff Formation** (limestone, shale and sandstone) is a moderately productive aquifer with mineralized water, the **Wajid Sandstone** is one of the highly productive sandstone aquifers. It occurs only in the southern part of the Shelf and its thickness is given as nearly 1, 000 meters; to the north the sequence thins out, a well near As Sulayyil (20° 29′ N, 45° 32′ E) showed a thickness of only 160 meters. G. ÖTKUN (1969) explains this with two fault systems west of As Sulayyil. The layers must dip rather steeply to the east. East of Khamasin (some 80 km west of As Sulayyil) the water overflows with only a slight artesian pressure of 0.5 atmosphere, while bores around As Sulayyil showed excess pressures of 9.1 atmospheres (G. ÖTKUN, 1969, p. 7). As this water is of good quality (450–990 ppm total dissolved solids), it is to be hoped that the aquifer will not be subjected to overpumping and permanently demaged (there were already some 50 bores in 1969).

3.1.7. Large Groundwater Developments on the Shelf of the Arabian Peninsula

The largest oases in Saudi Arabia are found around the largest water outlets in the country. They were discussed in detail in Vol. 1. The following additional information mainly includes data which have been published after 1977 (W. J. SHAMPINE et al., 1979; H. HÖTZL et al., 1979; B. VOSS, 1979).

3.1.7.1. Al Hasa Oasis

The largest oasis is Al Hasa (some 200 km^2) with its local capital of Hofuf[1]. D. UHLIG (1971) gives the total production of the 162 wells as 12.4 cbm/sec. To this are added another 336 bores with a total production of 1.7 cbm/sec. As not all the wells can be included in the development of an irrigation system (some of them are privately owned), the maximal output for the system is around 8.9 cbm/sec

[1] Cf. Vol. 1, p. 61, fig. 12.

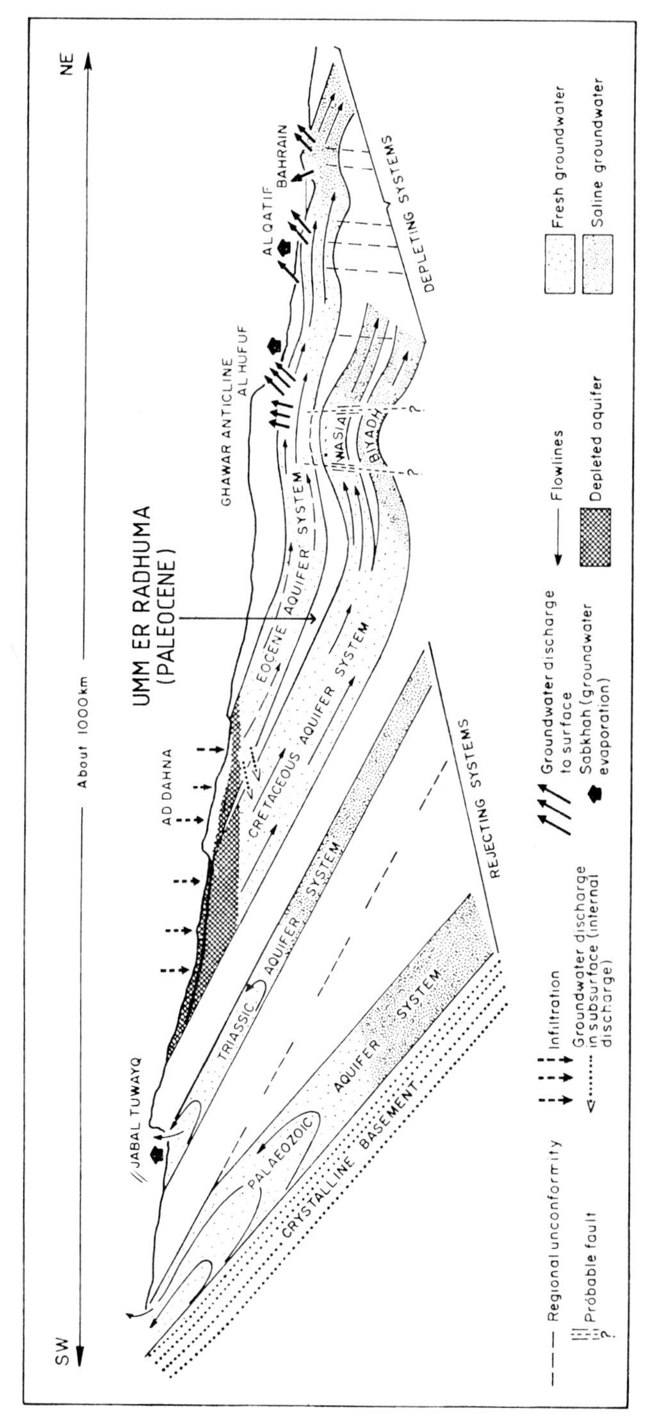

Fig. 107. Cross-section through the central Shelf. (Modified after D. BURDON, 1973.)

(as of June 1976). The output of the wells varies greatly and some of them produce more than 1 cbm/sec (Ayn Haql, 1.3 cbm/sec). The water temperature also varies; in 1971 it was between 35° and 40° C. Between 1973 and 1975 C. Job (Vol. 1, p. 119ff) made a detailed study of the chemical composition of the wells using all the material available. Even then he found that depending on the location of the wells, the waters of different aquifers are mixed to various extents. Later oxygen-18 studies (W. J. Shampine, T. Dincer, M. Noory, 1979, p. 458) showed that water from the Umm Er Radhuma Formation is moving upward in

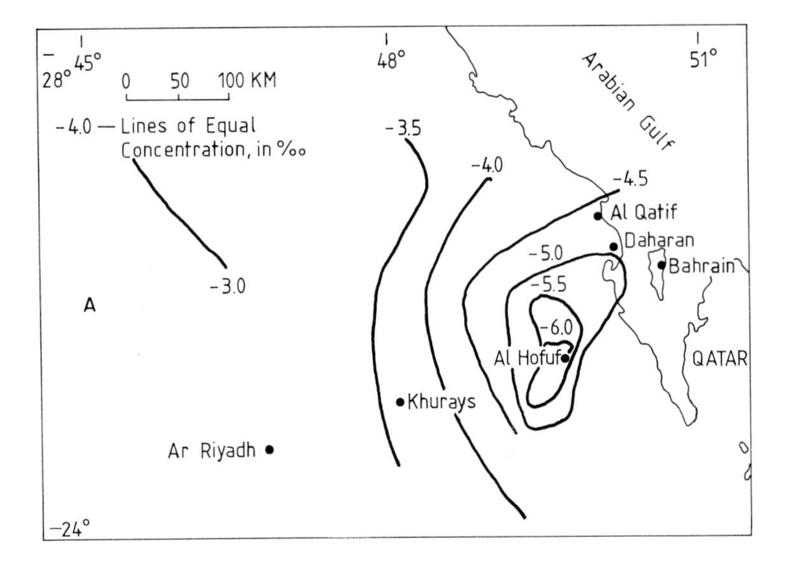

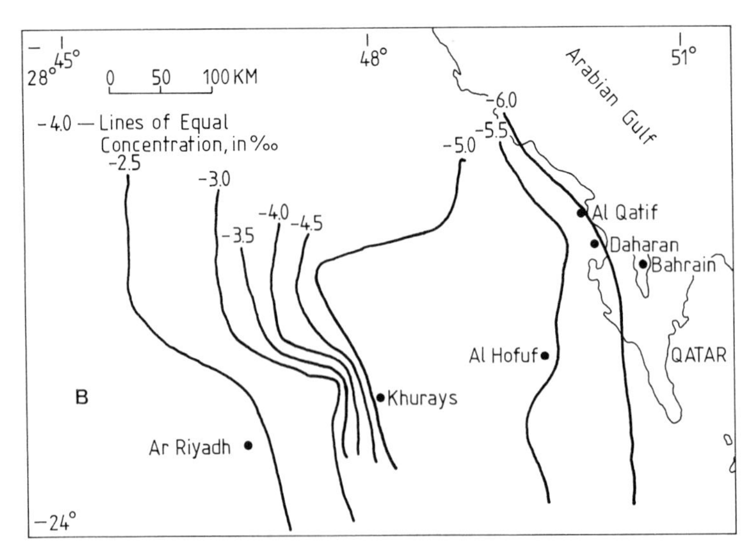

Fig. 108. *A* ^{18}O isopleth pattern of aquifers overlying Wasia Formation. *B* ^{18}O isopleths of Wasia aquifer. Wasia is completely isolated from the overlying formations. (After D. Burdon, 1973.)

this area, spreading into the Neogene aquifers. On the other hand, the ^{18}O isopleth pattern for the Wasia aquifer shows features completely different from those of overlying aquifers, i. e. the Wasia aquifer is completely isolated from the overlying aquifers (Figs. 107 and 108). Boreholes in recent years also reached the Wasia Formation (Table 45).

Piezometric maps corresponding to the BRGM measurements were published by B. Voss (1979, pp. 48–50). They show that the piezometric head of the Umm Er Radhuma aquifer in the Al Hasa area is between 165 and 115 meters a. s. l., with a SW–NE gradient southwest of Hofuf of 0.65‰, and northeast of Hofuf of 1.0‰. What the piezometric level of the Umm Er Radhuma aquifer only suggests, appears as a distinct step in the Neogene aquifer in the Al Hasa area.

Table 45. *Litho-stratigraphic profile of the ARAMCO borehole WW-22 3 km west of Hofuf.* Modified after B. Voss (1979, p. 24)

Depth (m)	Period			Formation	Lithology	Hydrology
	Tertiary			Neogene	limestone dolomite marl --discordance--	local aquifers
	Tertiary	Eocene	Lutetian Ypresian	Dammam Rus	limestone dolomite	local aquifer
500	Tertiary	Paleocene	Montian	Umm Er Radhuma	limestone and dolomite --discordance?--	regional important aquifer
	Cretaceous	Upper Cretaceous	Campanian -Maastrichtian	Upper Aruma	limestone dolomite	
1.000	Cretaceous	Upper Cretaceous	Campanian -Maastrichtian	Lower Aruma	marl and marly lime- stone --discordance--	aquiclude
	Cretaceous	Upper Cretaceous	Cenomanian Turonian?	Wasia	marly limestone	
1.500	Cretaceous	Lower Cretaceous	Albian	Wasia	sandstone --discordance--	regional important aquifer
	Cretaceous	Lower Cretaceous	Aptian			water temperature 86° C

There, from 30 kilometers west to the city of Hofuf in the east the piezometric level drops from 160 meters to 140 meters a. s. l.; to the east there is then a drop within 4 kilometers from 130 meters to 110 meters a. s. l., followed by an levelling-off from 110 meters to 85 meters a. s. l. over a distance of 30 kilometers. This rather clearly defined change in the piezometric gradient from 0.66‰ to 5.0‰ and then back to 0.8‰ suggests tectonic steps on the eastern flank of the Ghawar Anticline, as shown only schematically in Fig. 107. The static water level in a borehole into the Wasia Formation is 40 meters higher than that of the Umm Er Radhuma; this is a further indication of the absence of a natural hydrological connection between this aquifer and the Umm Er Radhuma Formation (cf. Table 45).

The comparative data on the number of wells given by Voss certainly depends only on the reliability of surveys taken at different times. That the well outlets are typical karst wells was discussed in detail in Vol. 1 (p. 73, fig. 19).

Proven variations in output of the large Al Hasa wells were also mentioned in Vol. 1. B. Voss (1979, p. 115) believes that the diurnal variations are due to opening and shutting of the gates of the dammed wells according to the irrigation rhythm. This is an indication of a hydraulic connection between the aquifers of the system which have their outflow through the springs.

The less permeable layers of Neogene insertions can cause local pressure differences and delayed reactions of water outlets between the northern and eastern parts of the area. The same explanation is valid for the temporal and quantitative differences in output; they are due to the altered irrigation times both in the individual years and during the course of the year.

We already presented our cautious evaluation of **direct age determination** by carbon-14 measurements. There are special reservations for water, especially as many aquifers are made of carbonate rocks. Carbon-13 and Carbon-14 measurements were made for six wells. It is unfortunate that of them, only 'Ayn Khudud was tested twice (see Vol. 1, p. 61, fig. 12). Neither for ^{13}C nor for ^{14}C are the values especially revealing (1964: 14% modern, or 20,000 years; 1972: 16.3%±0.36% modern, or 14,400±180 years). It is difficult to agree with the conclusion of B. Voss (1979, p. 84) that is it possible to differentiate the waters from the eastern and northern parts of the oasis with Carbon-14 studies. The same is true for the temperatures, which are determined by the various mixture components from the aquifers; B. Voss himself assumes a vertical circulation of at least 600 meters. We are more inclined to agree with the evaluation of isotope measurements given by W. J. Shampine, T. Dincer and M. Noory (1979), that the "^{18}O data indicate that water from the Umm Er Radhuma-aquifer is moving upward into overlying aquifers", and thus emerges in the Al Hasa wells in mixed water. We also agree with the conclusion of these authors that the lower-lying Wasia aquifer has no connection with the Al Hasa well outlets.

When water is injected in the course of oil production, aquifers with poorer-quality water should be used whenever possible, and such important aquifers as the Umm Er Radhuma and Wasia aquifers should be spared (cf. Plate V, see Fig. 107 and 108).

3.1.7.2. Al Qatif Oases

The waters of Al Qatif were studied and described in great detail by C. Job (Vol. 1, p. 93 ff).

The Al Qatif Oasis covers an area some 10 kilometers long and 2–3 kilometers wide along the coast of Tarut Bay on the Arabian Gulf (at 26°35′ N and 60°00′ E).

The differences in water temperature from hand-dug wells (called 'Ayn, 10–35 m deep) and drilled wells (75–250 m deep, usually with only a few meters' casing) show that both kinds of wells carry mixed water from the Neogene (Hofuf, Dam and Hadrukh formations) and water from the limestones of the Eocene Dammam Formation. A connection with the Paleocene Umm Er Radhuma Formation is also provided by the natural path of the joints through overlying layers. The Umm Er Radhuma aquifer is also important, and is fully exploited by the drilled wells. The waters have temperatures of 36–38° C. This aquifer is also quite thick here and lies at a depth of 230–600 meters. As the ground level in northern Al Hasa is at about 120 meters a. s. l., this would give a theoretical gradient of 2.5‰, but the gradient is certainly also determined by tectonic step faults going down to the Arabian Gulf. Both hand-dug and drilled wells show an artesian outflow.

C. Job based his detailed chemical studies in part on extensive data from the ITALCONSULT studies, and in part on his own findings; the latter led to the most important conclusions.

In 1973/74, 32 'ayns had a total discharge of about 36,000 cbm/day, and 219 boreholes, 168,500 cbm/day. This is a total of about 205,000 cbm/day for a cultivated area of 18.5 square kilometers, and 2.3 cbm/sec. C. Job estimates a utilized water quantity of 800,000 cbm/day (about 9.26 cbm/sec) for the entire cultivated area of 75 square kilometers.

The water in the wells and boreholes in the Al Qatif Oasis are relatively high in minerals; this means that there are two important problems. One is whether there is repercolation of highly mineralized waste water from the older drainage channels, which are usually not lined with concrete, into the rising well waters; the other is whether there is intrusion of sea water into the aquifer being used for irrigation. C. Job's extensive work showed conclusively that sea-water intrusion and seepage of drainage waters have no significant influence on the salt content of the karst springs and wells. The water in the Umm Er Radhuma aquifer is of older storage (^{14}C content of two samples: 3.5 ± 2.4 and $<1.2\%$ modern). These values are considerably lower than in Al Hasa and allow the cautious assumption of a longer residence and flow time in agreement with the greater distance to the outcrop.

Tests in fully cased wells showed a piezometric water level in the Umm Er Radhuma Formation of 13–15.6 meters a. s. l., and even the higher-lying Khobar aquifer showed a piezometric level of 8–11.5 meters a. s. l. This suggests that freshwater springs also occur along faults in the Arabian Gulf. These paths developed all the more easily as the erosion basis for the underground waters in the paleoclimatic Ice Ages was considerably lower, due to the decrease in sea level.

The submarine springs in the present-day Gulf have not yet been located precisely. In similar hydrogeologic position are the well-known **artesian springs** on the **islands of Bahrain.** They are located on the flat northern side of the main island and on the island of Sitra, while the oil boreholes are concentrated in the center of the main island, in conformity with the tectonics, i. e. the presence of a dome-shaped anticline. At its core (Jabal Ad Dukhan, 135 m a. s. l.) we find limestones from the Dammam Formation in the middle of a basin in rocks from the Rus Formation, and then, around the Rus Formation, again limestones, dolomites and marls from the Dammam Formation and sediments from the Dam Formation, out of which the largest wells – surrounded in places by Quaternary deposits – flow. The total discharge was estimated at 1.5 cbm/sec in the 20's (A. HEIM, 1928). Around 1925, 19 wells had been drilled successfully; their depths are given as 50–100 meters. There is also an important and distinct indication of submarine water outlets on the southern tip of the small island of Al Muharraq in the northeast. The highest piezometric level of 7 meters a. s. l. is given for this island.

Bahrain is outside of the Kingdom of Saudi Arabia. The sea is only 10–20 meters deep between the mainland and the islands. There is no doubt that the water outlets on Bahrain belong to the aquifer system in Al Hasa and Al Qatif; the pressure conditions in these aquifers indicate the presence of submarine freshwater springs in the Arabian Gulf.

3.1.7.3. "King Faisal Settlement" project near Haradh

ARAMCO (Arabian-American Oil Company) oil wells some 160 kilometers southwest of Al Hasa encountered the very thick (about 300 meters) water-bearing sequence of the Umm Er Radhama Formation mentioned in previous chapters as being the most important in the Tertiary series in eastern Saudi Arabia (Table 46).

This circumstance led to the plan for the building of the "King Faisal Settlement" project in Wadi As Sah'ba, a dry valley running east to west between Al Kharj and Haradh Station (at about 24°10′ N and 49°05′ E) on the Dammam – Ar Riyadh railroad.

Tectonically, this is within the area of the Ghawar anticline, whose N–S saddle axis here dips to the south. This dip is probably accompanied by innumerable tension joints, giving increased transmissibility as a result of solution processes.

Table 46 shows a stratigraphic breakdown of the layers.

Two ARAMCO bores were followed by three test wells and after test pumping and studies on productivity, optimal drilling depth and size of the cones of influence, 47 more holes were drilled, so that a total of 52 deep wells are available.

The casing pipe of the production hole is 26″ at the outlet and 20″ where it reaches the Umm Er Radhuma aquifer. The final depth of the boreholes is 458–516 meters, depending on the elevation of the terrain.

The trial pumpings described in detail by G. MARTIN and L. KRAPP (1977) show extremely good transmissibility and a result of a long-term pumping test

Table 46. *Stratigraphic profile in Wadi As Sah'ba in the Haradh area* (after G. MARTIN and L. KRAPP, 1977)

Period	Epoch	Formation	Thickness (m)	Lithology	Remarks
Quaternary			25–35	silt, sand, gravel, anhydrite and gypsum	discordance
Tertiary	Mio-Pliocene	Hofuf Dam Hadrukh	75–100	limestone, sandstone, red and gray sandy mudstones	main elevation phase of the Ghawar anticline
	Eocene	Dammam Rus	40–75	mudstone, limestone, dolomite and anhydrite	discordance
	Paleocene-Eocene	Umm Er Radhuma	300	dolomite limestone, dolomite and thin layers of clay shale	main aquifer
Upper Cretaceous		Aruma	200	clayey shale limestone, dolomite	discordance
		Wasia	450	sandstone alternating sequence of carbonate – clay shale and sandstone with clayey shale inclusions	

with a constant output of 161 l/sec from one well (1977, p. 53). The cones of influence had diameters of 70–85 meters.

In the area of the project, the total mineralization of the Umm Er Radhuma water is between 1,320 and 909 ppm. As the sodium adsorption rate is only about SAR 3, the water is very suitable for irrigation purposes. C. JOB (Vol. 1, pp. 130–132) has described the water chemistry in detail. The aim of this project is to produce feed crops and to maintain large sheep and camel herds. The 400 hectares of irrigated surface are on the edge of the Rub' Al Khali, and this is the largest project in the Kingdom that is supplied exclusively with pumped groundwater.

3.2. Northern Saudi Arabia

Areas of the "Shelf" of the northern Arabian Peninsula no longer bear the characteristics of cuesta landscape. The sedimentary rocks deposited in the north of the Shield will thus be handled separately in a short section.

3.2.1. Wadi As Sirhan and Great Nafud

In the northeast, Wadi As Sirhan is mainly accompanied by young volcanic rocks, and in the southeast and northwest by Eocene rocks. In the basin itself there are Young Tertiary sandstones and marls, as well as Quaternary deposits, with clusters of spring outlets. In relatively large areas there are sabkhah formations owing to evaporation of the shallow groundwater slightly under the surface.

Waters with normal mineral content and higher mineralization are also found in proximity to one another.

The second-largest self-enclosed sand desert (Qe) of the country, the **An Nafud,** takes up the largest amount (some 45,000 km^2) of the area of northern Saudi Arabia covered with Quaternary deposits. Despite the NNW winds prevailing here in the north both summer and winter, seif dunes take up large areas, but alternate with individual sand forms, barchan dunes and dikakah. The latter, with their thorny bushes and arid plants, are evidence of the more frequent winter rains here in the north. An unmistakable abundance of wells in the eastern area of the "Great Nafud" indicates that a measurable percentage of the precipitation seeps into the ground, as we saw before in the Ad Dahna (T. Dincer, 1978). This seepage is positively influenced by the fact that the nocturnal temperatures sink low enough to decrease evaporation considerably.

Quaternary deposits of local occurrance are found in flat basins and wadis. Some of these are partly coarse lag deposits, for example at the southeast border of the Hashemite Kingdom of Jordan, or silt and associated fine sediments, including caliche-like deposits, as well as fine-grained gravels with silt.

3.2.2. Area Between the Great Nafud and the Basement Complex

This area (Table 47, Plate V, see insertion at back cover) is of great hydrogeological significance; it is built up mainly of different Paleocoic and Lower Mesozoic sandstone formations. The area extends from about 29°35′ N to 30°25′ N and 39°50′ E to 40°30′ E. In generally the sandstones have good aquifer characteristics.

In the vicinity to the immediate east there are thousands of square kilometers of Upper Cretaceous **Aruma Formation;** this extends on down to the southeast and has been covered in the description of the cuesta landscape. The Aruma Formation here also dips under the extensive areas of northeast-trending **Umm Er Radhuma** layers, whose significance as an aquifer has already been indicated by the numerous wells. The Umm Er Radhuma layers in places extend into Iraq, where they are often covered by deposits from the Dammam Formation.

In the southeast, the Great Nafud is limited by Tabuk Formation and Saq Sandstones (Paleozoic). Both sandstone formations trend far to the southeast.

The aquifers were studied in more detail by N. F. AL-WATBAN (1976).

The **Tabuk Formation** (Plate V, see also Table 47) is composed of alternating shale and sandstone units. The sandstone units are important aquifers. In the Qusayba area (some 80 km northwest of Buraydah), the Middle Tabuk Member is some 240 meters thick and, separated from it by shale layers, the Lower Tabuk Member is 220 meters thick (N. F. AL-WATBAN, 1976, p. 16). The **Saq Sandstone** underlies and conforms with the Tabuk Formation; it lies directly upon the basement complex. The Saq Sandstone is Cambrian in age; it is red-brown cross-bedded sandstone.

According to the profile made by R. A. BRAMKAMP et al. (1963) the layers dip toward the northeast, and decrease in thickness toward the south. In the Qusayba area (26°50′ N, 43°10′ E), the Saq Sandstone is some 1,000 meters thick and is covered by 500 m of Tabuk sediments.

It may be assumed that the aquifers of the Lower Tabuk Sandstone and the thick Saq Sandstone are hydraulically interconnected. Owing to their thickness

Table 47. *Paleozoic stratigraphic sequence in Northern Saudi Arabia.* Modified after N. F. AL-WATBAN (1976)

Age			Formation		Generalized Lithologic Description	Aquifer Characteristics
Paleozoic	Permian	Upper	Khuff		Grey shale, dolomite and limestone with some calcarenite, and evaporite at top. Lowest strata of Sakaka sandstone.	Poor
		Lower	Pre-Khuff		Grey, red, occasionally yellow, in part silty shale and fine to medium grained, occasionally coarse, micaceous, sandstone.	Aquiclude
	Carboniferous	Upper				
		Lower	Berwath		Grey and varicoloured shale in part pyritic and micoceous and fine to medium grained sandstone.	Poor
	Devonian	Upper	Jauf		Poorly sorted, micaceous sandstone with interlets of varicolored pyritic, micaceous shale and Impure dolomitic limestione beds Toward bottom.	Limited Extent
		Middle				
		Lower	Tawil	Tabuk	Fine to medium grained, micaceous sandstone alternating with gray and gray-green, micaceous, silty shale.	Yields moderate to high, considered prolific aquifer
	Silurian		Qusay Bah			
			Middle Tabuk			
	Ordovician	Upper	Ra'an			
		Middle	Lower Tabuk			
		Lower	Hanadir			
	Cambrian		Saq		Coarse and medium grained sandstone with shale stringers.	Yields moderate to high, prolific aquifer
Precambrian Basement Complex						

and, for their depth, their good water quality (suitable for both domestic and agricultural purposes except in the area of Wadi Ar Rimah; N. F. AL-WATBAN, 1976, p. 214). These sandstone sequences are to be counted among the most important aquifers on the northeast edge of the basement.

The **Umm Sahm and Ram Sandstones** (Table 44) accompany the southern edge of the Great Nafud.

The western and central parts are sparsely settled owing to topology, lack of surface runoff and the deep-lying groundwaters, but in the northeast there are occasional hand-dug wells. The lower layers sometimes reach the Saq Sandstone. No information was available on deep drillings for water through the 1970's.

The zone marginal to the basement is taken up by **Siq Sandstone.** It is massive, with local cross-bedding of local conglomerate deposits of pebbles from the underlying Precambrian complex. It may in part be equivalent to the Saq Sandstone. Umm Sahm, Ram, Quweira, and Siq are members of the **Saq Sandstone.**

Settlements are mainly located on Quaternary deposits in the contact areas of the sandstone formations with the Precambrian basement of the Shield. This is also true of the (mainly sandstone) sequences of the **Tabuk Formation** following the Umm Sahm and Ram Sandstones in the north. Here, particular notice should be taken of the Quaternary depression, which is almost 200 kilometers long and reaches into Jordan, following NW–SE trending tectonic faults. The city of Tabuk is located in the southern part of this depression. Deep wells supplying the city and the most important airport in the northern part of the country indicate that the sandstone, when thick enough, is a potential aquifer.

3.3. The Arabian Shield and Its Coastal Areas on the Red Sea

3.3.1. The Basement

3.3.1.1. General remarks

The basement begins in the north as a relatively narrow area that already shows a considerable altitude (some 2,000 m a. s. l. for example in the Midyan Region). Even here in the north the steeper escarpment to the coastal area of the Red Sea is apparent in comparison to the adjacant sandstone hills following to the east.

Where the Shield is at its widest, the enormous areas of Tertiary and Quaternary basalt flows between 22° and 25° N influence the course of the watersheds (e. g. Wadi Al Hamma northwest of Al Madinah).

But only from the north of At Taif and southward does the escarpment form a distinct orographic feature toward the west, which dominates the geomorphology down to the southern border of the country. This means that in the hinterland of Jeddah in particular, the area between the Red Sea and the escarpment is a more or less independent hydrographic entity, whose groundwater-bearing wadis have won special hydrological significance.

Besides the enormous areas of young effusive rocks and young clastic sediments, in the Precambrian of the Shield we encounter a variety of rocks that must be classified in groups to provide a picture of the hydrogeological situation.

H. TORRENT and G. SAUVEPLANE (1977) divided the large variety of hard rocks into plutonic rocks, and metamorphic and volcanic rocks (Plate V).

3.3.1.2. Plutonic and metamorphic rocks

Following H. TORRENT and S. SAUVEPLANE, we count among the crystalline rocks granite, granodiorite, syenite, gneiss and subgroups (e. g. granite gneiss, etc.), while the "metamorphic and volcanic rocks" include schist, rhyolite, andesite, tuffite, diabase, etc. The young volcanic basalts occupy a special position owing both to their age (Young Tertiary-Quaternary) and their extensive distribution in circumscribed areas (harrats).

This classification may seem oversimplified, but it is a result of the scale of the map, and is intended to provide a general picture.

There are in fact morphological differences, as, for example, the more rapid flattening of larger granite areas as a result of pronounced physical weathering, facilitating the development of higher-lying peneplains, overtopped by monadnock and more resistant relict mountains. Schist is less susceptible to this process (Figs. 109 and 110).

Fig. 109. Landscape in the highlands of granite rocks in southern Hijaz. (Photo: J. G. ZÖTL, 1972.)

Fig. 110. Plateau area in crystalline schists of northern Asir. (Photo: J. G. ZÖTL, 1972.)

3.3.1.3. Quaternary deposits

The varying local influence of physical and chemical weathering determines the composition, thickness and storage of the Quaternary deposits.

In places where the highland is not dissected by oversteepened valleys and gorges, there are hollows and small trough valleys with hundreds of small and minute oases with fields and wells (Fig. 111). The houses and villages are located on the barren slopes. Particularly in the granite areas in the northern highland, the frequent winter and spring showers can create modest local aquifers of feldspar and quartz sand in the small basins.

In the southern Asir, the summer monsoon intensifies chemical weathering and soil formation in the schists. Meadows, trees and cultivated terraces may be found here at altitudes exceeding 2,500 meters a. s. l. (Fig. 119).

The Quaternary deposits in the highland of the Arabian Shield thus are found in basins, or in the uppermost basin-like roots of the main valleys and their tributaries.

With regard to the Quaternary sediments, the map by H. TORRENT and S. SAUVEPLANE (1977) was further simplified in that wadi alluvium, superficial deposits and regs were collected together under one signature, and only the eolian sands (dunes) and the small areas of kabrah deposits were shown separately. The latter are silt and clay in troughs or basins without outlets; they hardly ever carry groundwater, but after heavy rains they may store larger amounts of surface water for months. These natural ponds provide water for grazing animals.

The most important Quaternary aquifers are in the middle and upper courses of the largest wadi systems extending far back into the highland.

Parts of the northernmost of these wadis, known as **Wadi Ar Rimah in its middle and upper course,** were studied in 1974; this expedition also covered the uppermost area near Hulayfah, some 200 kilometers within the Arabian Shield. The description given in Vol. 1 (pp. 173–194) shows, among other things, an episodic surface runoff in the upper part of Wadi Sha'bah, a tributary of Wadi Ar Rimah (p. 179, Fig. 59). The runoff lasts for weeks and feeds groundwater sufficient for oases and settlements (the village and oasis shown in Vol. 1, p. 179, figure 59 are located at about 26° 35′ N and 41° 40′ E). The Quaternary is especially extensive in the upper Wadi Rimah at 26° N and 40° 50′ to 41° 40′ E.

In the south, the wadis Tathlith and Bisha join on the eastern edge of the Shield to form Wadi Ad Dawasir. After preliminary exploration in 1975 (for a description see Vol. 1, pp. 230–252), the uppermost regions of Wadi Bisha were studied in 1977.

The area around the town of Bisha was first studied on behalf of the Ministry of Agriculture and Water by ITALCONSULT (Final Report 1969). It was followed by studies of our group in 1977 and a conceptual hydrochemical model by J. W. LLOYD et al. (1979). All together makes one of the most extensive studies of the hydrogeology of Quaternary deposits within the crystalline basement (cf. chapter 2.5.).

Fig. 111. One of the hundreds of small grain-fields in trough areas of Northern Asir peneplain in an altitude of about 2500 meters. Irrigation by handdug wells, depths ca. 5 to 7 meters. (Photo: J. G. ZÖTL, 1972.)

Fig. 112. Oases of Najran (17° N, 44° E). At present under rapid development. (Photo: J. G. ZÖTL, 1972.)

Fig. 113. Basaltflow erupted in the 13th century near Al'Madinah. (Photo: J. G. ZÖTL, 1976.)

The most important hydrological development in the southeastern Shield occurs the area of Najran (Fig. 112), which has a great development at present.

3.3.1.4. Harrats and basalt flows

The basalt sheets (harrats) and basalt flows cover tremendous areas, and there have been eruptions during recorded history (e. g. the basalt flow near Al Madinah in 1256 AD, Fig. 113).

A large amount of the water from heavy rains seeps into the ground in places where the basalt fields have been broken into debris, or where there are deep cracks and crevices. Brief showers evaporate when they hit the sun-baked, hot stones.

As far as we know, the percent of seepage has not yet been measured. If, however, the young basalt flows cover clastic sediments in higher-lying wadis and smaller basins, then groundwater could well be formed in these deposits. A flight over the border mountains northeast of Jeddah in 1976 showed flowing channels in the upper, inaccessible parts of some wadis, but they vanished behind unreachable steep steps. We could not see that these streams originated in the bed subjacent to the basalt flows, but that possibility is not to be excluded.

G. DUROZOY (1972) provided a closer hydrogeological study of the Harrat Rahat, one of the largest basalt areas. The basalt plateau is located at 23° N and 40° E along a NW–SE line, upon which there are a number of centers of eruption. The lava flows cover an old, S–N oriented valley systems and large depressions filled with detrital sediments. Within these basalts, which in places are porous and clefted, a groundwater current circulates mainly from south to north (Fig. 114).

G. DUROZOY assumes that this basalt groundwater current in the northeast is

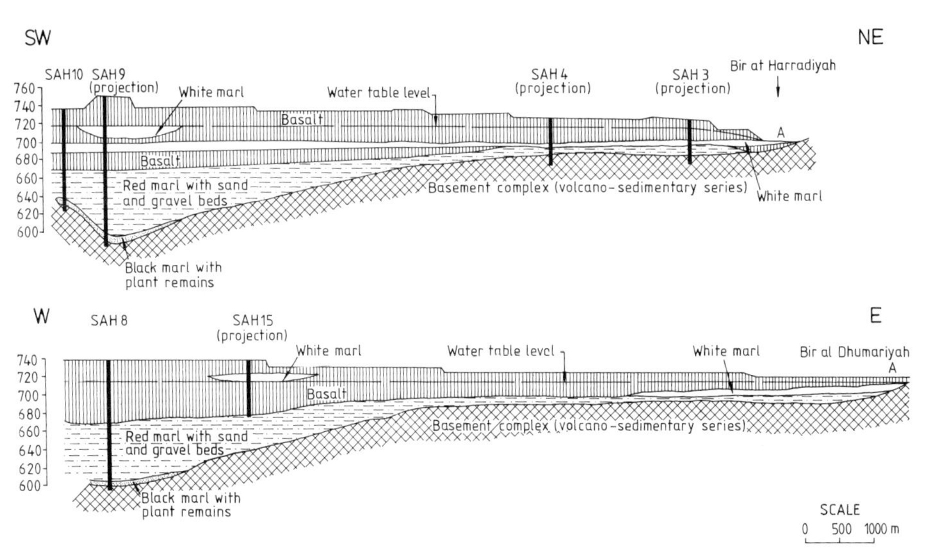

Fig. 114. Confined aquifer; overlying stratum basalt flow. SAH = boreholes in the Harrat Rahat. (After SOGREAH, 1970.)

fed by flood waters from the eastern Wadi Al Aqiq and it seems that subsurface waters flow under the recent basalts in the direction of Al Madinah.

Finally, the large amounts of water taken in Al Medinah indicate a circulation in the basalts in the northernmost part of the Harrat Rahat from SE to NW. Currently, 120 l/sec are being taken from a drilling in the subbasaltic valley fillings above an old well. The groundwater flow of the alluvial aquifer draining the basalts amounts to about 2.9 cbm/sec.

An idea of the extension of the basalt flows gives Fig. 115 showing only a small part of Harrat Rahan (ca 27° N, 37° E about 100 km SSE Tabuk).

Fig. 115. Part of Harrat Rahan, ca. 100 km SSE Tabuk. (Photo: J. G. ZÖTL, 1978.)

3.3.2. The Coastal Areas on the Red Sea

3.3.2.1. The area between the Gulf of Aqaba and Al Lith

The hydrogeological situation along the Red Sea is varied and hard to classify. Distinctions can most readily be made where there is a decisive change in climatic conditions. This is the case where the influence of the monsoon, coming from the Indian Ocean, begins to be felt. Although it is a matter of a zonal transition, the area around Al Lith is assumed to be the northern limit of the monsoon effect.

Along the **Gulf of Aqaba,** the basement often forms the coast and, besides the mouths of smaller wadis, the **Ifal Depression** makes a first deep, almost triangular cut in the Midyan Region. Although the accumulation surface amounts to some

1,200 square kilometers and the Quaternary (90 m in channels) and most especially the Tertiary (2,000 m, SOGREAH, 1970) sediment bodies are very deep. The groundwater situation was disappointing because of high mineralization owing to numerous anhydrite layers present even at slight depths. Only a limited amount of groundwater is available to supply the locality of Al Bad' and its small cultivated surfaces (palm gardens).

The seaboard is interrupted by the alluvial fans of small to medium wadis and generally narrows toward the southeast; between 28°30′ N and 26°00′ N the coast is mainly made up of basement rocks.

The individual wadis here in the north have sufficient water for small settlements and limited agriculture. The natural water supply does not, however, allow any substantial increase in output without danger of damage to the entire resource. A suitable example is the agricultural settlement of Ash Sharmah (28°05′ N, 35°16′ E) in the wadi of the same name south of the Ifal Depression. The palms were not irrigated and grew naturally (total surface area of the grove 50 hectares, SOGREA, 1970). In the meanwhile, bores have been sunk to irrigate further agriculture but the salt content of the lower palm grove water has already risen above 1,000 ppm.

Larger settlements along the coast also have their watersupply resources in the wadis in the hinterland. This was, for example, the situation in the small coastal city of Al Wajh before the desalination plant went into operation.

Between Al Wajh in the north (26°15′ N, 36°25′ E) and Umm Lajj in the south (25°05′ N, 37°12′ E) there is a bay up to 40 km wide and dominated morphologically by Quaternary gravel terraces with wadis that only episodically carry water. In general, this area can be designated as the delta of **Wadi Al Hamdh,** which must have borne considerable quantities of water during the moister phases of the Pleistocene. Archeological findings show that the region was inhabited in the past. Currently, the wells contain generally brackish water and, with the exception of the fishing village of Hanak, one finds only sheep and shepherds here.

In the southern part of the Quaternary bay there is the fishing village of **Umm Lajj.** A barrier of basement rocks running along the coast here protects a trough lying further back from sea-water intrusion. There is high-quality groundwater with only slight mineralization in the sediment filling of the basin. The SOGREA report (1970, p. 419) says that the wells here supply the town of Umm Lajj with "the best water on the whole coast, and Umm Lajj is the only town which is in such a favourable situation". As if to demonstrate the contrasts in the country in the very smallest space, just 10 km southeast of the well area there is the sand desert area of the Nafud Al Murawin.

The coast from Umm Lajj to Yanbu is desert with a high content of gypeum and anhydrite rocks in the underground, partly even on the surface.

Yanbu Al Bahr used to be a fishing village with a small harbor, and drew its water supply from the oasis area of Yanbu An Nakhl 50 kilometers in the hinterland. Increasing demand soon was at the cost of agriculture in the oasis. As Yanbu Al Bahr developed into an industrial city, the only satisfactory solution was to install a desalination plant. The development of a water supply in Yanbu Al Bahr accomplished in a short time what was done stepwise in Jeddah.

The old harbor city of **Jeddah** was able for a long time to satisfy its water demands with the tremendous groundwater reserves from the wadis in the hinterland (cf. chapter 2.3., Wadi Fatimah). But the rapid development of the city in recent decades and the building of the gigantic airport placed excessive demand on the groundwater reserves in the entire region. This problem was again solved by the building of a desalination plant. It may thus be hoped that this will lead to at least a partial regeneration of the natural groundwaters in the wadis, which would be important should an emergency supply at some time be required (cf. the oil catastrophe in the Arabian Gulf).

From Jeddah, an unsettled coastal strip of desert-like nature runs more than 200 kilometers to the southeast, until basement rocks again approach the coast at Al Lith.

3.3.2.2. The northern and southern Tihama and the basalt plateaus of Al Birk

Wadi Al Lith forms the approximate climatic northern border of the area in southwestern Saudi Arabia that is influenced by monsoons. This effect may be seen in precipitation with a yearly average along the foothills around 100 mm, and around 300 mm at 1,000 meters a. s. l. Also of note are the hot springs (49–88° C) some 50 kilometers inland in Wadi Al Lith; it was surprising that tritium measurements showed a water age of only 20 years (SOGREA, 1970, p. 233); both the variations in temperature and the isotope values suggest a mixture of hot deep water and recent seepage water. The measured flow rates vary between 30 and 50 l/sec.

The monsoon rains, which increase from here in the northern Tihama toward the south, are also responsible for the larger number of wadi channels running from Al Lith to the coast. Settlements are more numerous here and provide a very different picture for the coastal areas from Jeddah to Al Lith, and from Al Lith (20°02′ N, 40°20′ E) to the basalt plateau of Al Birk (18°14′ N, 41°32′ E). As far as precipitation is concerned, this is a transitional zone, and both the depth of rainfall and the groundwater enrichment increase remarkably from north to south.

Quaternary basalt flows cover the entire **Al Birk** plateau. On the coast, the basalt runs from 18°30′ N to 17°45′ N. This is an inhospitable and water-poor barrier diagonal to the coastal plain that forms the northern edge of the southern Tihama, which is the coastal area richest in water in Saudi Arabia; in the south this area extends to the Yemeni border.

The hydrogeology of the **southern Tihama** was described in chapter 2.4. Confined and semiconfined groundwaters permit year-round irrigation and continuous cultivation of crops. The Ministry of Agriculture and Water has an Agricultural Experimental Station doing work with non-native plants; the results are passed on to the larger oases to provide alternatives to monocultures. The monsoon rain becomes noticeable here and allows both storage of surface waters at higher altitudes (the Maloki Dam in the Jizan hinterland) and terrace cultivation and storage of drinking water (at Abha, or at altitudes exceeding 3,000 meters a. s. l., Fig. 119).

4. Aspects of Geomorphological Evolution; Paleosols and Dunes in Saudi Arabia

(D. ANTON)

4.1. Introduction

The present work has the purpose of using available resources to improve our understanding of the geomorphological evolution of the Arabian Peninsula and Saudi Arabia. Not too many studies related with the geomorphology of the peninsula have been carried out, being in most of the cases the indirect result of geological investigations. Nevertheless, a number of papers and reports have been prepared during the last years which provide a considerable amount of useful information and a suitable background for this paper.

In 1951 R. A. BAGNOLD studied the sand formations in Southern Arabia and in 1953 D. A. HOLM published a paper on the "Dome Shaped Dunes of the Central Najd". A few years later D. A. HOLM (1960) prepared this "Desert Geomorphology in the Arabian Peninsula". The same year G. F. BROWN published "Geomorphology of Western and Central Saudi Arabia". In 1967 the firm ITALCONSULT was retained to study the development for groundwater in the Central Arabian Sedimentary Basin and Western Shield including a geomorphological interpretation and classification of these areas. In 1971 R. W. CHAPMAN studied the "Climatic Changes and Evolution of Landforms in the Eastern Province of Saudi Arabia". In 1977 the firm MACLAREN Ltd. carried out hydrogeological studies in the Arabian Shield including a new geomorphological classification made partially through Landsat interpretation. In 1978 a first volume of the "Quaternary Period in Saudi Arabia" was published jointly by the University of Petroleum and Minerals of Dhahran and the Austrian Academy of Sciences. This volume included a detail review of general and regional subjects on the geomorphological and geological evolution during this period.

In addition several general studies or reviews on the geology of the peninsula including geomorphological interpretations were prepared and published by R. KARPOFF (1957), G. F. BROWN and R. D. JACKSON (1960), R. W. POWERS et al. (1966), F. GEUKENS (1966), Z. R. BEYDOUN (1966), J. E. GREENWOOD and D. BLEACKLEY (1967), D. I. MILTON (1967) and K. M. NAQIB (1967).

General studies on more restricted fields with some relationship with the geomorphological evolution were also carried out by R. W. POWERS et al. (1966) (sedimentary geology), by G. F. BROWN (1972) (tectonism), and by D. L. SCHMIDT (1973) (stratigraphy and tectonism).

Regional or local studies and/or mapping were carried out in many areas of the Kingdom and they have been consulted in many cases to obtain or confirm data or to assist in the interpretation of known phenomena.

The 1 : 2,000,000 scale geological map (1963) compiled by ARAMCO and USGS and the 1 : 500,000 geological maps published by the USGS were also consulted and some geological boundaries were included with only minor changes when required for the purpose of the maps accompanying this paper. Standard Landsat Imagery was used to trace boundaries of eolian accumulation units and the hypothetical or actual paleo and/or present hydrographic system.

Due to insufficient double coverage, stereoscopic analysis was used only in small strips near the edge of the images.

Eolian units were analyzed in Landsat Imagery and longitudinal features were measured (spacing, length, definition, direction). Alluvial basins were also approximately reconstructed and regional interpretations were performed.

In selected areas, black and white panchromatic and infra-red aerial photography was also used for description and/or interpretation purposes. Field checking and surveys were carried out in several areas of the Kingdom, particularly in the Eastern Province, Western Shield, Western Rub Al Khali and Central Shelf Platform. In these areas selected exposures and outcrops were described and sampled. The samples were also described and compared and the obtained data were integrated with the rest of the information for interpretation purposes.

The main results of this study are a preliminary inventory of eolian units, a reconstruction of present and paleo-drainage systems expressed on maps 1 : 2,000,000, several geomorphological maps of selected areas (Wadis Bishah, Tathlith, Habawnah, Sahba, and Dawasir and Jafurah sand fields), a few cross sections and sketches, and an interpretative reconstruction of the evolution of the landforms in the Kingdom of Saudi Arabia during the Neogene and Quaternary.

4.2. Structural Background and Tectonism

In a general way it can be said that any geomorphological evolution depends on 2 main complex factors:

a) Tectonic behaviour of structural units.

b) Characteristics of the succession of surficial processes.

Both factors inter-relate and combine during the different stages of the geological history producing as a final result a superimposition of features that can be analyzed and interpreted for a better understanding of the actual evolutive history of the landscape.

The structural units of the Arabian Peninsula have been described by R. W. Powers et al. (1966) as follows:

a) Arabian Shield.

b) Arabian Shelf (including the interior homocline, the interior platform and the basins).

c) Mobile belt (Zagros and Oman mountains and respective forelands).

We are not concerned in this study with all the tectonic processes that took place during the recognizable geological history but nevertheless we need to mention a few that gave rise to the present structural configuration.

The Shield, composed of Pre-cambrian rocks, was stabilized and levelled by the end of the Paleozoic. In some areas Paleozoic sedimentary formations had filled existing depressions (as for instance, in the southwest, the Wajid Sandstone sedimentary basin).

Paleo-shorelines moved back and forth from the Early Paleozoic to the Cretaceous accumulating mainly epicontinental sediments on the lower edges of the Shield.

During the Tertiary, the Alpine orogenia affected the rocks of the Zagros and Taurus area folding and thrusting them to form mountain chains. During this same period the Rift system was formed and the Red Sea graben fully developed. The western edge of Shield rose and a steep scarp formed along the peripheral faults (Mountain Scarp geomorphological region). The rise of the Shield was stronger in the south and weaker in the north and this explains the different present elevation of the southern and northern Arabian Shield.

The Shelf area has experienced almost continuous subsidence from the Paleozoic to relatively recent times, especially in the deeper eastern parts of the sedimentary basins. In the interior (central) homocline of the Shelf this subsidence was interrupted in several moments of the geological history and unconformities, disconformities and alternating marine and continental rocks are observed.

A few areas of the sedimentary region had stronger subsidence activity during different moments of the history and collected a thicker sedimentary load. These are the "basins" units (Rub Al Khali, Sirhan – Turayf, Northern Arabian Gulf and Dibdibba basins). This structural background was more or less in place during the Early Tertiary and it is on this paleo-landscape that the surficial (and internal) processes were going to leave their trace.

4.3. Landscape Evolution During the Tertiary

The first direct indications of an area emergent in the Paleogene are found in the southwest of the Kingdom between Abha and Najran where a lateritic paleosol lies between the trap basalts and the basement complex. When the basalts were extruded this area had been emergent for a considerable time. These laterites are mainly found on gneisses, schistes or granites and are very weakly developed where remnants of the Wajid Formation (sandstone) cover the crystalline bedrocks. The laterites are usually composed of a medium- and coarse-grained, angular sand with red or yellow clay cement (probably of kaolinitic – gibbsitic nature) and their thickness may vary considerably reaching in some places up to 15 meters.

These laterites indicate the existence of a rainy tropical forest before the basalts were extruded somewhere between the Eocene and the Early Miocene.

The trap basalts in the southern region of the Saudi Arabian Shield are related to the trap series in Yemen and as such, they seem to be Lower or Early Tertiary in age (no dating was obtained yet and current estimates vary from the Late Cretaceous to Pliocene). These basalts are dissected and occupy the top of mesa-type reliefs in the Asir region.

The Middle Tertiary in Arabia appears to have been a transition period between the humid Early Tertiary climates and the arid Late Tertiary and Quaternary environments.

Early Tertiary sedimentary formations (Umm Er Radhuma, Rus and Dammam formations) seem to have been almost entirely deposited in marine (shallow epicontinental) environments and as such they give little information on continental surficial processes and geomorphic conditions.

The first well defined continental deposits found in the Kingdom are related with the Hadrukh Formation in the Eastern Platform area. The Hadrukh sediments are considered to be Early Miocene due to the depositional continuity with the younger Dam marine deposits (chronostratigraphically located in the Middle Miocene). The sediments of the Hadrukh Formation are marine along the eastern edge (near the Gulf) and continental in the rest of the area. They are mainly composed of marly sandstones, sandy marls, sandy clays and sandy limestones with chert and less frequently gypsum suggesting alluvial and lacustrine environments. Loaded streams seem to have deposited calcareous sandy and fine grained fans along the Platform forelands near a shallow epicontinental sea. They seem to correspond to the above mentioned transitional climatic period and they are the result of strong erosion of ancient soils in the headwaters. We do not know if some eolian action was also present but it is likely that it could have existed as lateral facies of the alluvial and lacustrine beds.

Younger lacustrine and alluvial deposits are found overlying (locally) the Miocene Dam Formation. They belong to the Hofuf and Kharj formations. The type section of the Hofuf Formation (R. W. POWERS, et al., 1966) starts with a basal conglomerate 20 meters thick including boulders and pebbles of limestone in a quartz matrix. This Hofuf basal conglomerate seems to define a clear alluvial episode probably in later Miocene or Early Pliocene, related with increasing humidity or tectonic dynamics.

Even considering that the sedimentary cover on the Shield was much more extensive than today, the absence of crystalline pebbles indicate a relatively near source. Correlative deposits including crystalline fragments would be most probably associated with active subsident basins (as g. e. Rub Al Khali, Dibdibba and Sirhan – Turayf basins). Limestones (probably lacustrine) and thick argillaceous sandstones are found on top of the basal conglomerate.

A second clearly identified alluvial episode is seen on the upper part of the type section consisting of a conglomerate composed of limestone pebbles. In other areas these upper conglomerates include quartz showing an expansion of the headwaters and/or an erosional exhumation of the crystalline Shield along the edges of the sedimentary Shelf. These conglomerates may indicate an increase on regional rainfall and/or an increase of tectonic activity.

The Hofuf Formation has been eroded in subsequent geological times and extensive platforms covered with widespread gravel pavements were formed. In the eastern part of the outcropping area, a high escarpment showing typical sections of this formation is also found, probably related with an ancient higher sea level.

4.4. Acting Processes and Geomorphic Systems

4.4.1. Introduction

During the last periods of the geological history, the geomorphological dynamics of the peninsula have been mainly the result of the inter-relation and interference of several surficial processes.

These inter-relations and interferences are integrated in different geomorphic systems and actual landscapes are the result of the historical succession of these acting systems.

The main geomorphic systems identified as having acted in the region of Arabia are the arid, semi-arid and semi-humid systems, in the sense explained in sections 4.4.2. to 4.4.4.

4.4.2. The Arid System

When the rainfall decreases below a certain level most of the water is spent in evaporation, dehumidification of ground surface and decreasing infiltration. Runoff and erosion decrease and very little fluvial flow is observed.

The ground remains dry during long periods and the loose grains become available to be removed by the wind (deflation) and transported to variable distances according to the strength (and other characteristics) of the airflow, and the grain size (and other characteristics) of the particles.

Eolian accumulations are the local or regional result of this system activity.

4.4.3. The Semi-Arid System

Given a certain distribution of the rainfall beyound at a certain level the runoff on the slopes reaches a maximum. This happens when soil humidity is not enough to support a dense perennial vegetation cover and rainfall is frequent and concentrated, providing more water than the amount that can be evaporated or infiltrated. In these cases runoff and soil erosion reach a maximum and the greatest peak volumes of fluvial flow are obtained. As a consequence thick coarse alluvial deposits may be accumulated.

Eroding headwaters and alluvial plains and fans are normal features in the semi-arid systems.

4.4.4. The Semi-Humid System

If rainfall increases above a certain level (given a certain annual distribution) vegetation density and foliar cover also increase. Evapotranspiration rises, run-off decreases, and soil erosion is reduced to a much lower value. Fluvial peaks are not so high (or at least so concentrated) and rivers flow in a more regular way with less instantaneous efficiency. Sedimentary accumulations become finer and less important and deeper soils are formed on the slopes.

Slope erosion and flooding may still occur but they are not an every-year phenomenon. For the purpose of this study, this geomorphic system will be called semi-humid.

Due to the fact that sub-humid systems did not act extensively or long enough during the last geological epochs (Pliocene – Pleistocene – Holocene) they are not described in this text.

4.5. Geomorphic Systems and Accumulations During the Late Pliocene and Quaternary

4.5.1. Background

Identification of present geomorphic systems mainly requires field observations and meteorological data. Active wadis and dune fields, presence of gullies or ravines, and observation of the vegetation cover, complemented with data on rainfall, wind direction and speed, humidity and temperature can also provide a good understanding of the current geomorphic system. Identification of older systems requires a different approach: sedimentary accumulations, landforms and paleo-pedogenetic processes must be described, analyzed and integrated in order to obtain the necessary knowledge to reconstruct the paleo-environments.

Taking into account these elements six main climatic phases are proposed in the present study for the Late Pliocene – Quaternary. During each phase the distribution of geomorphic systems in the Tertiary changed. During "humid" phases arid systems became semi-arid, and semi-arid systems became semi-humid. During the "arid" phases the opposite phenomenon occured (semi-humid systems became semi-arid and semi-arid systems became arid).

In the following sections these phases are described and analyzed.

4.5.2. The Late Pliocene – Early Pleistocene Humid Phase

4.5.2.1. Introduction

This phase has been identified mainly because of the existence of extensive ancient gravel deposits associated with old or present drainage systems.

Red soils in the mountains and weathering on shield rocks and basalts seem to confirm the existance of this humid phase during the Late Cenozoic.

In Wadi Ranyah, gravel deposits belonging to these periods were found between two basalt layers. Both basalts were dated using K-Ar method and the ages found were as follows: Upper basalt: 1.1±0.3 million years; Lower basalt: 3.5±0.3 million years (H. Hötzl et al., 1978). These gravels were therefore accumulated during the Late Pliocene – Early Pleistocene subepochs. Due to sedimentological and geomorphological similarities it is reasonable to assume that other gravelly accumulations and deep weathering in the shield area also occured during the same period. In the following sections a brief description of each hydrographic basin is presented including an analysis of accumulations, soils and landforms belonging to this first humid phase.

4.5.2.2. In Wadi Ad Dawasir basin

Wadi Ad Dawasir is the name used for the lowest identifiable active wadi channel in a big basin covering the southern Shield and some areas of the

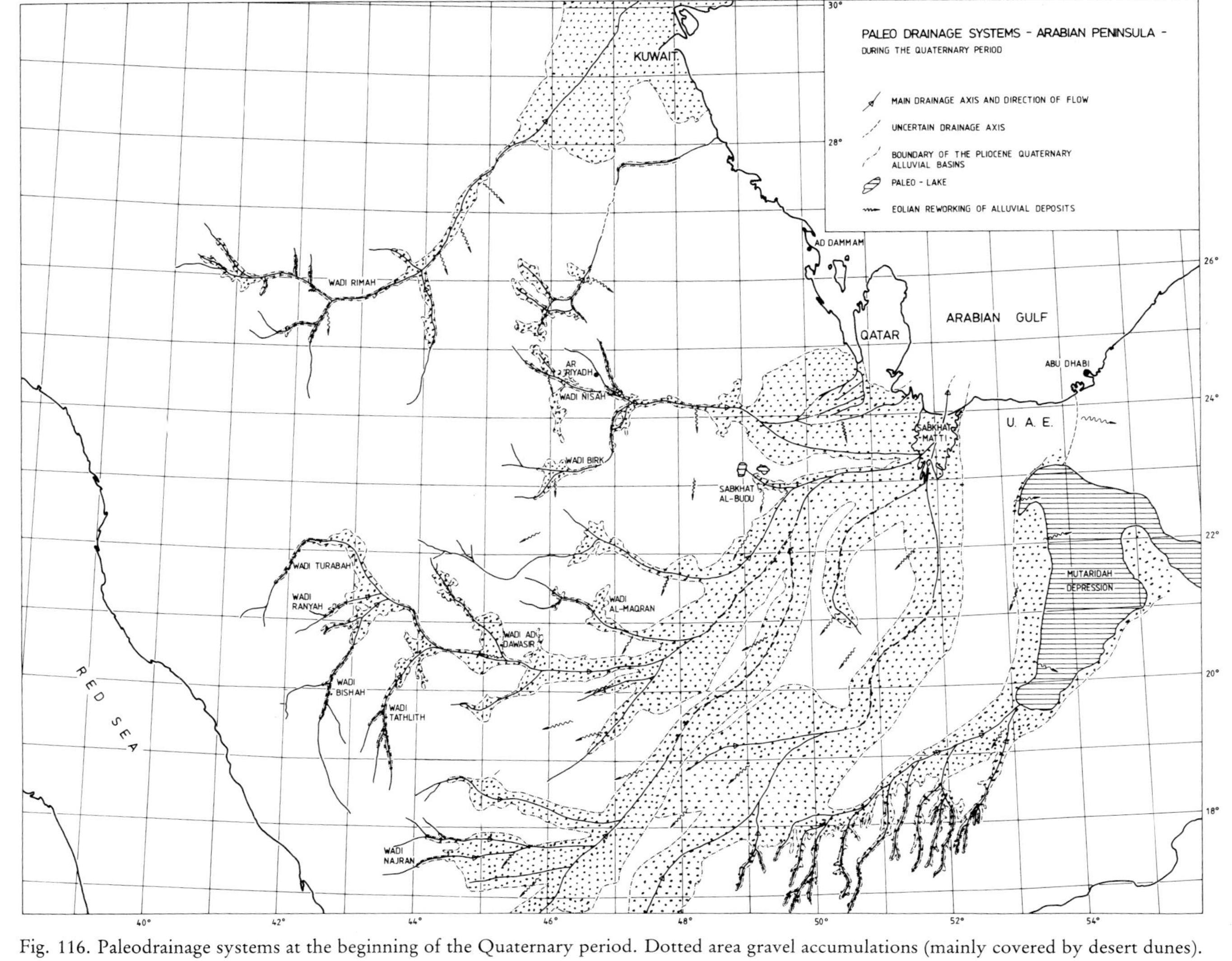

Fig. 116. Paleodrainage systems at the beginning of the Quaternary period. Dotted area gravel accumulations (mainly covered by desert dunes).

sedimentary homocline. The basin is formed by several sub-basins among which the most important are as follows: (a) Wadi Bishah, (b) Wadi Tathlith, (c) Wadi Ranyah and (d) Wadi Turabah (Fig. 116).

In all these wadis (and normally following present drainage lines) the valley depressions are filled with thick gravel accumulations seldom appearing as lateral fluvial terraces. They are composed of coarse sand and fine gravel in Wadi Bishah, coarse and fine gravels and coarse sand in Wadi Tathlith, coarse sand in Wadi Ranyah, coarse sand and fine gravel in Wadi Turabah and coarse sand and gravels in Wadi Ad Dawasir. In all the cases gravels are polymictic with a variety of crystalline rocks and in some places sedimentary pebbles (g. e. Wadi Ad Dawasir limestone pebbles).

Thickness is usually around 20 meters in the middle course of the wadis but may reach up to 100 meters in the Wadi Ad Dawasir area east of the Sulayyil gorge.

They are overlying unconformably the eroded bedrock and are covered with extensive silt deposits (often forming terraces). East of As Sulayyil (20°19′ N, 45°32′ E) Wadi Ad Dawasir forms extensive alluvial fans of silty gravel and downstream the plain is gradually covered by eolian sand. Nevertheless, in some places the ancient gravels can be seen and followed to the northeast, down to the Sabkhat Matti lowlands for about 700 kilometers. There are usually related, well sorted gravel pavements, resulting from the residual deflation of silt and sand from the otherwise poorly sorted original alluvial formation. Studies of samples taken from the alluvial fan show a very poorly sorted sandy and gravelly silt in which sand grains are about 85% quartzic and 12% are mafic minerals. Grains show iron coatings in about 10–20% of the individuals, roundness is high (60% subrounded, 10% rounded) in the larger fractions (500–1,000 mm) and lower (mainly subangular) in the smaller fractions.

Pockets of red soils are commonly found in the upper reaches of the basin (Asir region) and they seem to correlate with the peak of humidity during this period.

All these sediments and soils seem to correspond to a climatic period with strong slope erosion and fluvial transportation (semi-arid system) with soil formation in the southwestern highlands (semihumid). Soil erosion probably happened later during the first stages of the aridification phase.

4.5.2.3. In Wadi Sahba basin

In Wadi Sahba basin the valleys are filled with a basal coarse gravelly deposit underlying a finer formation (e. f. maps I-207A, I-208A [1 : 500,000], and I-270A [1 : 2,000,000]).

In the upper reaches of the basin several larger alluvial valleys following more or less the present drainage pattern are found. They are Wadi Hanifah, Wadi Al Luhy, Sha'ib Nisah, Sha'ib Al Awsat, Sha'ib Al Ayn and Wadi Birk.

In Wadis Hanifah and Al Luhy there is a lower gravelly formation (underlying grey clayey silts) probably correlative with the older gravels in Ad Dawasir, corresponding to the widespread semi-arid system existing in most of the Kingdom during the Pliocene – Pleistocene transition. This lower coarse formation can be also identified in Wadi Birk where it is composed of gravel sand with lenses of pebbles and clay.

Downstream of the Al Kharj basin a single valley is observed running eastwards through the Ad Dahna eolian corridor and down to the Jafurah sand fields. Along this valley, terraces composed of gravelly deposits (and often covered with a well defined calcicrust) are encountered. These terraces are normally dissected along the axis of the valley and in a lesser degree along the edges, next to the bedrock outcropping areas.

The elevation of the terraces reaches a maximum in the high gradient section west of Haradh station and decreases in height eastwards with their flat tops merging with the surface of an extensive alluvial plain (actually a true alluvial fan) with probable similar age. The fan is partially covered with eolian sand and has been dissected by a later channel but can be easily traced on Landsat images and observed in the field. Extensive gravel pavements (consisting of quartz pebbles) are observed in many places. The fan has been considered a delta by H. HÖTZL et al. (1978) corresponding to a high sea level existing during the Pliocene – Pleistocene transitional phase.

4.5.2.4. Wadi Ar Rimah – Wadi Batin basin

In the Wadi Ar Rimah – Wadi Batin basin the older gravels are not clearly expressed in the upper basin. H. HÖTZL et al. (1978) identified a flat pediplain covered with gravels presenting a terrace morphology near the wadis. The fact that this pediment has been covered by Late Tertiary lava flows in the Harrat Khaybar region suggests a probable Miocene–Pliocene age, simultaneous with the accumulation of Hadrukh – Hofuf continental complex. It is not known if a formation correlative with the older gravels of Ad Dawasir lies filling the bottom of the wadi valleys in the Wadi Ar Rimah area, but in any case coarse formations showing the actual existence of an old semi-arid environment in the watershed are seen covering an extensive area north of the As Summam Plateau in the Dibdibba sedimentary basin along the Wadi Batin course. These sediments are composed mainly of coarse poorly sorted materials, and gravels are found paving the surface in many places. The surficial distribution of this unit shows a fan type appearance and the proximity to the Gulf shoreline suggests a possible deltaic accumulation dynamics (at least along the northeastern edge).

This formation is strongly dissected by Wadi Batin which forms a very well defined valley only partially re-filled by younger alluvium. The older gravels surface remain as high terrace, while the slopes of the Wadi Batin valley go deep into the Tertiary bedrock. More studies will be able to bring more light on this matter.

What is certain, however, is that the grain size and the extension of this alluvium requires much stronger rainfall than today, probably with dry periods (seasons) long enough to reduce the vegetation density and allow heavy run off, strong erosion, high peak flows and long distance transportation of large rock fragments as explained in Section 5.4.

This semi-arid system seems to have been dominant until some time in the Early Pleistocene when silts became dominant and peak flows had decreased in volume.

4.5.2.5. In other basins

Deposits correlative with the Ad Dawasir older gravels can be found in several other basins. In Wadi Habawnah and Wadi Najran the wadi valleys are mainly excavated in the shield rocks and to a lesser degree in the Wajid Sandstone. The bottom of the valleys has been partially filled up (probably during the Late Pliocene – Early Pleistocene time) with a gravelly formation, which may reach a thickness of 30 meters in some areas. This formation has been covered by a silty unit similar to the silty formation found in the Ad Dawasir basin. The terraces along the wadis only show the lower gravels in a few places, as for instance in Wadi Thur near the confluence with Wadi Habawnah where the older gravels appear with minor consolidation underlying a silty material less than 4 meters thick.

Alluvial forms presenting widespread deposits of gravels probably related with the older gravels are also found in several other basins. Among them Wadi Al Maqran, Wadi Al Jadwal, Wadi Idimah and Wadi Al Atk.

4.5.2.6. Tentative correlation

A number of formations which can be correlated with this First Humid Climatic Phase are found in neighbouring countries.

A humid period was identified in Sudanese and Egyptian Nuba (A. J. WHITEMAN, 1971) during the Plio-Pleistocene expressed by three terraces (50, 65 and 100 m high) along the Nile valley. Similar accumulations are described by K. W. BUTZER (1971) assigned to the Early to mid-Pleistocene in the Nile and wadis of southern Egypt. In Jordan, a fluviatile lacustrine formation including coarse conglomerates was observed in the eastern side of the Jordan valley (F. BENDER, 1974) and according to its fossil content was tentatively assigned to the Late Pliocene – Early Pleistocene, corresponding with a humid climatic phase.

In the Afar region (Djibouti and Ethiopia) large lakes developed during the Plio-Pleistocene (F. GASSE et al., 1971) in depressions which are presently dry, confirming the occurrence of a more humid period near the Tertiary-Quaternary boundary.

4.5.3. The Middle Pleistocene Arid Phase

There are several evidences showing that the Middle Pleistocene was not as humid as the Early Pleistocene and that this trend continued until Late Pleistocene times. The younger basalts of the Harrats are not strongly weathered and no deep valleys have been dissected in them, and this can be interpreted as due to the absence of long humid periods during the rest of the Quaternary.

The deposits that may correspond with this Pleistocene subepoch are fine and rich in $CaCO_3$ or gypsum suggesting a semi-arid to arid climate with little run off, and limited fluvial flow. Even though some eolian sediments are found belonging to this period, aridity during this period does not seem to have reached the threshold needed for widespread eolian dynamics. A moderately dense xerophytic vegetation cover would have been enough to produce this type of environment. We assume that during the first part of the Middle-Late Pleistocene period, fine gravels and sand intercalated with silty lenses were still being accumulated in the valleys and gradually the finer lenses become dominant (probably a few hundred thousand years ago) giving the common silty formations (today frequently well terraced) that are found in the southern half of the Shield. Correlatible deposits are also found in the other wadi systems (e. g. in Wadi Hanifah terraces).

The silty terraces fill almost completely the valleys and often present rests of fresh water molluscus and mammal bones. It is not excluded that part of the silt could be an alluvial reworking of loessic material deflated during the dry season in more arid regions and retained by the vegetation in the Shield area.

The silts of the Shield area seem to be related with the silt phase (Pleistocene, Lower Sebilian) identified in southern Nuba and Egypt by SANDFORD and ARKELL (1933), with the Ubeidiya and Naharayim formations of Jordan (F. BENDER, 1974) and with the wide pediments of the Afar region (F. GASSE et al., 1971).

4.5.4. The Late Pleistocene Humid Phase

4.5.4.1. General background

Humid phases have been identified in the Quaternary of the peninsula, but they have not been as durable nor as extensive as the widespread increase in rainfall that produced the accumulation of the older gravels in the alluvial plains and fans of the country. A few accumulations, however, are observed in some areas which give elements to define a general increase in humidity by the later stages of the Pleistocene. These deposits are composed mainly of gravels and coarse sand and they are found covering the silty terraces in most of the wadis of the southern Shield.

The dissection of Wadi Sahba and Wadi Batin alluvial fans seems to be mainly related to eustatic changes of the sea level, but the fact that these valleys could have been formed show that fluvial flow was available much more frequently and in a bigger volume than today.

Soils were formed on fans, terraces and dunes. They have been identified near Al-Kharj, along the Dahna sand fields and in several other places. They seem to be related with a semi-arid environment with local areas of stronger pedogenetical activity (due to favourable topographic position, dense vegetation cover, etc.).

Lakes in the Rub' Al Khali showed high levels at least in two different moments, from 36,000 to 17,000 B. P. and from 9,000 to 6,000 B. P. (H. A. McCLURE, 1978). The first of those high lacustrine levels seems to be also associated with this humid phase.

4.5.4.2. The younger gravels of the southern Shield

They are found overlying the silty terraces or filling the wadi valleys under the recent sediments. In some wadis, the silty terraces appear to transgress conformably to a thin layer of gravels (often less than one meter thick) which constitutes the surface of much of the high terraces in the area. These gravel layers usually extend over a major portion of the terraces. In some areas, however, only scattered gravel lenses occur.

In Wadi Tathlith they are normally less than one meter thick and only found along the wadi very near the present valley. In Wadi Habawnah important accumulations of gravels and coarse sand can be observed occupying the bottom of the wadi channel near the confluence of Wadi Thar and on the top of a relatively low terrace west of this confluence.

In Wadi Bishah the correlative deposits are mainly sandy with occasional gravel lenses.

4.5.4.3. Paleosols on eolian deposits

In several areas, present sand dunes are covering paleosols developed on ancient eolian sands. An examples is found in the Ad Dahna sand fields west of the town of Khurays.

There, the present dunes are covering two paleosols developed on eolian sand. In the upper paleosol, calcified root channels can be seen near the boundary between the paleo-solum and the paleo-C. The thickness of this paleosol (only including the solum) is about 0.2–0.3 meter. The lower paleosol is better developed and may show a thickness of 0.3–0.4 meter.

We assume that the lower paleosol can be chronologically correlated with this Late Pleistocene humid phase showing that dunes were stabilized by a denser vegetation cover (much denser than the present cover in the dikaka areas).

4.5.4.4. Dissection on older alluvial plains and fans

Strong dissection of older alluvium can be observed in many places of the Kingdom. The main dissection valleys are found in Wadi Sahba and Wadi Batin alluvial fans. In other wadis, the Late Pleistocene dissection is not found or is only weakly developed. In Wadi Sahba the dissected valley was partially or totally filled with sediments and later covered with eolian sand. However, Landsat Imagery allows a very accurate reconstruction of the old wadi channel from the As Summam Plateau gorge to the proximity of Sabkhat Matti in the Emirates.

In Wadi Batin the dissection was deep enough to uncover the underlying Tertiary formations and during a later stage, alluvial deposits, filled the bottom of the valley forming a new alluvial plain at a lower level.

Dissection is also observed in the Shield in Wadi Tathlith (probably this phase produced the gravel deposits and a second one the dissection of the channel forming the present terraces), in Wadi Habawnah and in lesser degree in Wadi Ad Dawasir alluvial fan where not too well defined terraces are observed in the Landsat imagery and in the field.

4.5.4.5. Lake levels

Due to the present aridity of the area, no lakes are found in the region (only scattered sabkhahs in the Eastern sector).

In the less arid periods of the Quaternary more water was available and lakes formed in the depressions. These depressions were related with the old alluvial (pre-dune) surface or with interdunar areas. Lacustrine deposits are found in some areas and some of these sediments were studied and dated by H. A.

McClure (1978). The older lakes (36,000–17,000 B. P.) do not show any fish remains, but they do present a relatively rich mammal fauna including bovids and hyppopotamus indicating a Savannah grassland (semi-arid to arid). These high lake levels can be easily correlated to the above-mentioned coarse alluvial accumulations, dissection features and the older paleosol on dunes of the Ad Dahna sand fields.

4.5.4.6. Correlation

This humid phase seems to correspond with the aggradational sequence of the Masmas Formation on the river Nile (K. W. Butzer, 1971); and with the high lake levels (Abhe II and Abhe III) in the Central Afar (F. Gasse, 1977).

4.5.5. The Late Pleistocene – Early Holocene Arid Phase

During Late Pleistocene times a strong increase in aridity, seems to have happened (it probably became more arid than ever before). Dunes systems started to form in the vast expanses of the Rub' Al Khali and Nafud (probably also in Jafurah and Ad Dahna) and fluvial flow decreased to a minimum. No alluvial accumulations occurred with the probable exception of evaporites or some local fine materials. Lacustrine sediments stratigraphically located above and below these eolian deposits were dated in the Rub' Al Khali (H. A. McClure, 1978). According to these dates this arid phase seems to have started about 17,000 B. P. ending 9,000 years ago.

This arid phase was identified in the Nile valley (dune invasions in some flood plains, K. W. Butzer, 1971), in the Afar and Ethiopian rift lakes (from 17,000 to 12,000 B. P. with a period of high aridity; F. Gasse et al., 1971) and in most of the Saharan and Sahelian lakes (F. Alayne Street and A. T. Grove, 1979).

4.5.6. The Early Holocene Humid Phase

A later increase in humidity coincided with the beginning of the Holocene (probably about 12–10,000 years ago). Shallow lakes started to form again in the Rub' Al Khali (dated 9,000–6,000 year B. P.; H. A. McClure, 1971), soils developed on the Ad Dahna sand fields and probably part of the younger coarser alluvial deposits of wadis As Sha'ba and Batin are related with this stage.

In the Shield the gravel terraces were dissected and thin coarse accumulations developed in the wadi valleys. The short duration of this period did not allow deep weathering and only a weak pedogenesis is observed. The younger paleosols in the Ad Dahna sand fields (which seem to correspond with this humid phase) are not deep (about 20 cms) and poorly developed. The younger paleosols in the Rub' Al Khali also show a weak pedogenetic history probably related with a steppe vegetation cover. This humid period has been confirmed by a great number of evidences for many lakes in the Sahara, Eastern Africa and Indian Subcontinent. According to a survey of lake levels in about 30 different lacustrine basins during the 9,000–8,000 B. P. period (F. Alayne Street and A. T. Grove, 1979) all recorded bodies of water had a high water level during the above mentioned period (for details see chapter 5).

4.5.7. The Recent Arid Phase

4.5.7.1. Introduction

The younger dates obtained by H. A. McClure (1979) in the shallow lakes of the Rub' Al Khali show a decrease in humidity starting about 6,000 years ago.

F. Alayne Street's and A. T. Grove's survey of lake levels shows a general decrease of water level after 6,000 B. P. reaching a minimum in historic times.

During this phase the main sand fields of Saudi Arabia were activated (most of them, reactivated) and a widespread arid system became dominant in the peninsula. Probably, this aridity was less intense in the first 3,000–4,000 years but human action (overgrazing, cutting and burning trees and bushes, etc.) seem to have accelerated the aridification trend.

In the following pages we will describe the main active sand fields of Saudi Arabia according to available geological, geomorphological, mineralogical and climatological information. Landsat imagery was mainly used to identify, and analyze eolian features in 130 eolian territorial units.

4.5.7.2. Jafurah sand fields

The Jafurah sand fields are located between 27° and 24° latitude, and east of the meridian 49°30′ in the coastal lowlands (under the 200 m contour line) along the Arabian Gulf littoral. They are a narrow band in the north that widens southwards merging with the Rub' Al Khali sand fields.

In spite of a relatively high humidity (associated with the proximity of the Gulf) and a not negligible winter rainfall of nearly 80 mm in the northern edge, the geodynamics belong to the arid type. The vegetation cover is not dense, but density may be substantially higher than the one observed in other more typical arid areas.

A number of facts should be considered in this region:

a) Due to low topographic position, surficial aquifers are commonly very shallow.

b) There has been a long history of migration of water (through artesian and non artesian sources and wells) from deep to shallow aquifers raising still more the water levels and locally improving the quality of the surficial groundwater.

c) Recently this trend has been substantially increased by new agricultural developments and extensive drilling communicating otherwise isolated aquifers.

d) The proximity of the Gulf increases the average humidity decreasing evapotranspiration.

e) The particular geological and topographical position due to which, sandy alluvial and littoral deposits are found in great amounts determinating a specific geomorphic behaviour of the area.

As a result of these facts the area has an extensive eolian morphology with very scarce runoff features. However, the vegetation may be moderately dense in some areas with low intensity of sand movement. These areas are usually related with shallow relatively fresh surficial groundwater. In other areas, the presence of a shallow saline surficial aquifer or the excessive depth of the water level may stop altogether plant growth. In the first case the final result is a sabkhah plain and in the second a field of active dunes. Therefore, in the Jafurah region arid conditions produce dune fields only in some appropriate areas.

These dune fields are generally a local phenomenon and the size of observed eolian features is relatively small. Due to a nearly constant wind direction and very frequent high wind speed the typical eolian landforms are barchans with associated barchanoids ridges and parabolic dune fields.

Barchans travel lonely paths on the flat sabkhah surfaces or merge into barchan fields. On the sabkhahs, dunes loose the bottom structures retained by moisture coming from the shallow water table and salt cementation (stopping deflation under a certain level) and the top due to increased wind velocity associated with smaller surface roughness. In the sand fields, dunes remain active and it is there that the most typical eolian arid dynamics are found in the region. Landsat imagery does not show clearly the eolian features in Jafurah because of their relatively small size often beyond the resolution of the satellite imagery.

However, in some areas, eolian features are observed with a spacing of 0.4 kilometer between longitudinal features and a length averaging around two kilometers. Direction of ridges is about N 55° E resulting from an approximate wind direction N 35 W).

Information from a network of azimuthal sand traps in north Jafurah has shown that eolian sand is mainly transported from a NNW direction. The small difference between both observed directions is perhaps due to the fact that in Landsat we are observing eolian features requiring tens or hundreds of years to respond to changes in wind direction, and in the field we are observing the actual direction in the present (recent) storm(s).

The sedimentology of the sand in the Jafurah area was studied in several samples taken in various geomorphological positions in the sand fields.

Sorting is usually medium with grain size ranging from fine to coarse sand (100–1,200 μ) and averaging 300 to 400 μ.

Quartz varies between 90 to 99%, with lower values in the smaller fractions.

The colour of the sand is normally around the 10 YR 7/3 in the Munsell Chart (very pale brown near light gray) probably due to the fact that few yellow grains and no iron coatings are found.

Rounded grains represent between 20 and 30% of the total but angular grains may also be frequently found in some samples.

4.5.7.3. The Rub' Al Khali

Ar Rub' Al Khali is the largest single region where an arid system is acting in present times. More than 550,000 square kilometers in the southern half of the country are covered with more or less continuous eolian accumulations seldom interrupted by eroded remnants of older reliefs (mainly near the edges) or by uncovered gravel pavements in some interduna areas.

Present rainfall in the region seems to be less than 50 mm/y although accurate meteorological information is not available for most of the area. When it rains, most of the water only humects the surface grains promptly evaporating or (in the case of strong rains) it infiltrates through the thick sandy cover only reaching the water table in some areas with shallow surficial aquifers. Very sporadically, a little runoff flows down the slopes of the duna ridges and sorting of different grains size classes of sand takes place in the micro-pediments sloping towards the interduna depressions. No concentration of runoff happens in the whole area, and no geomorphic water erosion features are seen.

Nevertheless the Rub' Al Khali is really a complex area with a long story of climatic (and geomorphic dynamics) changes. Two main parameters have influenced the evolution of the landscape in the area. In first place, the amount of rainfall (and associated with it the density and permanence of the vegetation cover) and in second place the changes of wind direction. The changes in rainfall volume have determined the ruling ecological system and of course the presence or absence of fluvial flow, rise or drop in lake levels, soil formation or erosion, salt accumulation and of course eolian reworking of existing sand.

The changes in wind direction have mainly affected the duna morphology, old ridges are being rebuilt following a new wind direction, complex dunes are being formed, and sand sources change from some areas to other places.

The Rub' Al Khali arid system is relatively young. Pliocene and Pleistocene sediments are mainly alluvial related with the semi-arid systems already mentioned in section 4.5.2.

Two main duna systems are found. The oldest one overlies the Pliocene and Pleistocene alluvial surfaces, and locally the Late Pleistocene lacustrine deposits. These old dunes developed a paleosol during Early Holocene times and were covered by new dunes in a later stage. The paleosols on the dunes are composed of root and stem encrustations (H. A. McClure, 1978) and seem to have been contemporary with the younger lakes in interduna depressions being presently covered by the present active arid environment. The arid systems of the Rub' Al Khali started 17,000 years ago, and lasted until about 9,000 B. P. when a more humid period developed. 3,000 years later a new arid period (the present one) started to develop, probably accelerated during the last millenia by human intervention (due to excessive grazing activities, etc.) Wind directions seem to have been approximately the same during both arid periods, although in some areas specially in the east, two or more wind directions are observed, with one or more older duna patterns in the process of being the present active eolian system. The Rub' Al Khali desert was already studied using Landsat by C. S. Breed et al. and a map 1 : 500,000 of the area was included in "A study of global sand seas" (McKee, 1980). In general terms, the area was divided according to the following main types and units: linear dunes, crescentic dunes, star dunes, complex dunes, compound dunes, sand sheets and streaks.

Fifty units were identified in a study using Landsat imagery in the Rub' Al Khali area, among which 12 have undefined eolian features (nonexistant or beyond Landsat resolution limit) and four have variable characterictics (difficult to quantify). In 34 units spacing of eolian longitudinal features range from 0.3 to 5.5 kilometers (average 1.8 km), length from 0.3 to 200 kilometers (average 20 km), definition from 0 to 9 with an average of 4, and directions are concentrated between N45 and N90 (57%) and N and N45 (33%). N90 to N180 directions are uncommon (about 10%).

Due to their size or geomorphological importance a few selected units deserve a special reference. They are:

– An unit of linear dunes, with ridge spacing averaging two km and length 200 kms in the South West region. Definition is very good probably due to the fact that the alluvial – lacustrine floor between dunes often appears much darker than the sand ridges.
– Two oblique wind directions giving a N60 fairly constant resultant seem to be the reason for this unusual development of long ridges. Sand supply, on the other hand is probably relatively scarce leaving bare a considerable extension of the non-eolian floor.
– A well defined other unit of barchanoid ridges is found in the Central region. Spacing is about 0.8 km and features length 6 km. Direction is around N45. The fact that this unit's western and southern boundaries approximately coincide with the 350–360 contour lines suggests an older alluvial and/or lacustrine unit being marked by present (recent) eolian processes.
– **Al-Mutaridah.** This unit is extremely complex including "crescentic ridges with star dunes superimposed" (C. S. Breed et al., 1979) and a great number of small sabkhahs occupying the interduna depressions. Spacing varies between 1.5 km (north) and 2.5 km (south), length between 20 km (north) and 50 km (south) and direction from N75 (north) to N100 (south) with a good and very good definition.

Selected samples were obtained and studied from the western and south-western areas. Grain size may vary considerably depending on the geomorphic position of the sampled sediment. The typical dune sand to range from 0.1 to 0.4 mm (L) averaging around 200–220 μ with a good to very good sorting.

Quartz represents often more than 97% of the sample, with feldspars varying from 0.5 to 2% and mafic minerals being less than 3%.

Grains are subrounded and subangular (rounded grains being around 25% of the total number) and grain colour ranges from yellow to light yellow, with dull surfaces and remnants of iron coating in about 10–20% of the grains. The general colour of the sand varies between 5 YR 7/6 to 7/8 and 7.5 YR 7/6 to 7/8 of the Munsell chart corresponding to the reddish yellow color definition.

These results are different from other sand fields allowing the identification of the Rub' Al Khali sediments among other eolian samples. The main characteristics of these samples compared with the other sand fields are the amount of iron hydroxides present in grain interstices (much higher than in Jafurah, much lower than in Dahna), good sorting and yellowish colour (not found in Jafurah), percent of quartz (similar to Ad Dahna, much higher than in Jafurah) percent of angular grains (less than 10% in Rub' Al Khali, much higher in Jafurah) and general colour of the sample (lighter than in Ad Dahna, more red than in Jafurah).

4.5.7.4. The Ad Dahna sand fields

The Ad Dahna sand fields form a long and narrow strip of eolian sand from 20°30′ to 28°30′ latitude north. The width of this strip does not exceed 40 kilometers but the length reaches more than 1,100 kilometers.

The Ad Dahna sand fields are mainly formed of several relatively parallel ridges including complex dune systems. In Landsat imagery the pattern does not always appear well defined. In the south and with a low moderate definition N20 to N oriented strips are seen. The spacing varies between 0.8 and 1.2 kilometers and the length rangs from 10 to 20 kilometers. Farther north the strips are not clearly observed, but two directions can be defined (N80 and N170). In the northern part, directions change from N100 to N120 and definition becomes moderate or moderate-low. Spacing is much larger (3–9 km) and length ranges from 10 kilometers to 100 kilometers.

The fact that the Ad Dahna sand fields occupy a relatively low position and approximately follow the curved geological structures, suggests a genesis related with nearby sand sources in the contiguous watersheds.

In the area west of Khurays five strips with interduna vegetated sand plains (sand sheets) are observed (Geol. Map 1 : 500,000). Eolian activity is more or less reduced to the duna ridges. Under the ridges two paleosols developed in eolian sand are interesting evidence of the existence of (at least) **two humid periods** interrupting the eolian dynamics in the regions. We can quite safely correlate these two paleosols with the humid periods identified by H. A. McClure in the Rub' Al Khali. Based on these elements we can assume that the arid eolian system ruling the Ad Dahna sand field dynamics has probably started around 40,000 years ago with two interruptions (one in the Late Pleistocene, around 30–20,000 years B. P. and the other in the Early Holocene) during which the system became semi-arid, a denser vegetation cover developed and pedogenetic processes started to act allowing the formation of the above mentioned paleosols.

The Ad Dahna eolian sediments have well defined characteristics. They are strongly reddish (Munsell 7.5 YR 6/8 falling in the reddish-yellow zone), quartz represents about 99% of the total samples and grains are mainly subrounded, yellow, and iron coatings in approximately 75–80% of the cases.

Along the Ad Dahna low corridor, small and medium sized alluvial fans are observed, in which relatively reddish soils have developed. These alluvial fans (with probable Pleistocene age) seem to have been the source for the iron coatings in the grains found in the younger formations, supporting the idea that the sand fields of Ad Dahna are mainly related with the regional structures and local geological and geomorphological formations and only in a lesser degree with a climatic type.

4.5.7.5. The Nafud sand field

The Nafud or Great Nafud sand fields are located in the north of the Kingdom occupying a relatively depressed area between latitudes 27°20′ N and 29°45′ N and longitudes 38°20′ E and 42°40′ E with an approximate length of 300 kilometers and a width of 250 kilometers. The Nafud sand fields according to R. W. Chapman (1978) are composed by longitudinal (elongated sand sheets undulations) and Uruq (parallel ridges) eolian sand terrains.

In the Landsat imagery about 22 units were differenciated. In eight of them no defined features are seen. In the rest, definition is mainly moderate with spacing between ridges ranging from one to two kilometers. Length may reach up to 35 kilometers but it is usually much less (average length: 12 km). Directions are usually ENE corresponding to NNW wind directions (features are perpendicular to wind flow).

Genetically, the Nafud sand field seems to be related to the eolian reworking (under arid conditions) of alluvial deposits brought in by wadis originated in the surrounding watersheds (wadis Fajr, Nayyat, etc.). The presence of sandstones outcropping in the southeast of the region suggests an important source of sand from these sandstones (the Ordovician Saq Sandstone and equivalent units). It is not known if the same climatic and geomorphological phases observed in the Rub' Al Khali and Ad Dahna sand fields are applicable to the Nafud fields.

More information will be necessary to reconstruct the evolution of the eolian landscape in the northern sand fields.

4.5.7.6. Other dune fields

A number of smaller dune fields are found in different areas of the Kingdom. Some of them are geographically related with the four largest fields mentioned before, as for instance the Ramlat Dahm and Al-Arid located on the western edge of the Rub' Al Khali or Ramlat Al Ikrish not far from the Jafurah fields.

Some dune fields follow the general curved structures in the Central Arabian sedimentary homocline (as the Ad Dahna) further west, not far from the shield boundary.

These "Dahna type" sand fields are discontinuous and receive different names from north to south as follows: Nafud Ath Thuwayrat, Nafud Al Ghamis, Nafud Ash Shugayyigah, Nafud As Sirr, Nafud Qunayfidhah, Nafud Ad Dahi and Nafud Al Wadi (s. Geol. Maps 1 : 500,000).

Some of these "Nafuds" are also associated with main alluvial plains as in Nafud Al Ghamis (located along Wadi Rimah lower valley) and Nafud Al Wadi (along the Wadi Dawasir plain). A few other sand fields are found on the Shield among which the largest ones are the Al Uruq sand field (south of W. Rimah), Irq Subay (northeast of Wadi Turabah), the Nafud Hanjaran (associated with the lower alluvial plain of Wadi Bishah) and the Nafud As Surrah (northeast of Irq Subay).

From a geomorphological point of view these fields have an "intrazonal" behaviour (they are more the product of local conditions than the result of regional equilibrium with the existing climate). However, all these fields show an arid geomorphic dynamics, with very little or no runoff, strong eolian activity, and reduced vegetation cover.

Due to the varied location these fields have different characteristics. Observed on the Landsat imagery some of them show very elongated features like Ramlat Dahm with ridges 150 kilometers long, or Nafud Ad Dahi with shorter ridges about 20 kilometers long. In some cases large fields of pyramidal dunes are observed, and in many other cases eolian features appear poorly defined on the Landsat images. From a sedimentological point of view, and due to the different location of these fields, varied materials are found. Nevertheless a few samples studied in these fields seem to show a certain similarity between some of them and the Ad Dahna in many sedimentological parameters (grain size, sorting, roundness, percent of quartz and iron coatings). The sand fields seeming to present this relatively homogenous characteristics are the Nafud Ath Thuwayrat, Nafud

Qunayfidhah, Nafud Al Ghamis and a few smaller ones north of Al Riyadh (Irq Banban, etc.).

Other sand fields have sedimentological features similar to the Rub' Al Khali type (Al-Arid, Dahm) or the Jafurah type (Nafud Al Ikrish). In a few other cases they cannot be grouped with any of the large fields (Uruq Subay, Nafud Hanjaran, Nafud Al Wadi, etc.).

4.6. Conclusion

Saudi Arabia presents a complex geomorphology developed from a geological structure composed of a raised crystalline Shield descending westwards to a deep tectonic graben through a steep scarp, and a low eastwards dipping Shelf whose subsidence originated the development of several sedimentary basins. Cuesta type reliefs are found following the outcropping harder layers and/or formations. Along the coast, flat plains, frequently sandy (sabkhahs) are found. In several areas extensive sand fields can be observed.

The historical geology of the region shows a general trend of aridification starting from humid climates. During the Paleogene a rain forest vegetation developed on the Asir region leaving thick laterites in the area east of Abha and Khamis Mushayt. The Neogene seems to have been a period of aridification in which mainly semi-arid climates gave rise to a wide spectrum of continental units. The Hadrukh and Hofuf formations were almost completely deposited on emerged lands including coarse as well as fine sediments and even evaporites. These two formations seem to be respectively Miocene and Pliocene.

About 3.3 million years ago a very well defined formation started to fill up the bottom of the wadi valleys. This formation is composed of coarse materials and is well expressed in the main basins of the country. The paleohydrography can be reconstructed using the geological map and the Landsat imagery (Fig. 116).

Extensive alluvial fans are observed in the lower courses of the important wadis such as Wadi Ad Dawasir, Wadi Sahba and Wadi Batin. The floor of the Rub' Al Khali is also formed of the upper layer of these older gravels (more or less transformed in desert pavements and usually covered by extensive dune accumulations). A semi-arid climate (in most of the country) probably can explain the required erosion and necessary fluvial transportation competence to accumulate these coarse materials over such large areas.

There are evidences that the southwest was simultaneously experiencing a semi-humid climate (pockets of red soils in Asir, weathering of some crystalline rocks, etc.). This period finished probably about one million years B. P., but less humid semi-arid climates continued to exist in most of the country. Silts were accumulated in the Shield valleys possibly due to the erosion of the semi-humid soils and perhaps with some contribution of loess particles carried westward by the wind from the eastern deflation areas and retained by the vegetation in the relatively dense western steppes.

An increase in humidity is observed in the Rub' Al Khali where a more humid period was dated (36,000–17,000 years B. P.) in the lakes that were formed in the depressions of the older alluvial topography.

Gravels in the Dawasir basins, and an older paleosol in the Ad Dahna sand fields seem to have been formed during this same period. Strong dissection in the older alluvial fans is (clearly observed in the Landsat imagery) in Wadi As Sah'ba valley (partially covered by the dunes of Jafurah) and in Wadi Batin. An arid climate with active eolian dynamics was identified starting after 17,000 years B. P. in the Rub'al Khali. This period seem to have been simultaneous with the eolian sand covering the old paleosol in Dahna. Some eolian sand ridges formed during this period may still be conserved in some areas.

A new increase in humidity about 10,000 years ago (9,000 in the Rub' Al Khali, probably sometime before in the Northern areas) brought again a semi-arid environment to most of the country. Shallow lakes in the interduna depressions of the Rub' Al Khali soils on dunes in the Ad Dahna and Ar Rub' Al Khali,

Table 48. *Correlation of climatic phases, geomorphological evolution and geological dynamics* (D. ANTON)

	Chronology	Time scale	Climatic phase	Continental accumulations	Landforms	Dynamics	Soil	Vegetation
Quaternary	Holocene	6,000	Arid	Eolian sands	Dunes	Eolian	Soils covered by dunes	Mainly steppe and desert
		11,000	Semi-arid	Gravels & sands in wadi valleys, lacustrine deposits	Low alluvial plains	Locally torrential erosion	Shallow soil	Mainly steppe and savannah
	Late Pleistocene	17,000	Arid	Eolian sands	Dunes	Eolian	Soils covered by dunes	Steppe and desert
		35,000	Semi-arid	Gravels, lacustrine deposits	Some terraces in the west, dissection in the eastern fans	Locally torrential erosion	Soils on dunes	Steppe and savannah
	Middle Pleistocene	1,100,000	Arid to semi-arid	Alluvial silts	Terraces cover on fans	Erosion on slopes	Soil erosion	Mainly steppe
	Early Pleistocene	3,500,000	Semi-arid to semi-humid	Alluvial gravels	Large fans, filling of wadi valleys	Torrential erosion and alluvial accumulation	Red soils	Savannah and forest (s. w.)
Tertiary	Pliocene		Semi-arid	Alluvial silts, marls, sands & gravels	Mainly old fans, pediplains in plateau position	erosion on slopes	soil erosion	
	Miocene		Semi-humid					
	Oligocene	25,000,000?	Humid				Latosols (Laterites)	Rain, tropical forest

alluvium in the dissected valleys of wadis Sahba and Batin and the dissection of the silty gravel terraces in the upper Dawasir sub-basins seem to be associated with this "more humid" Early Holocene period. About 6,000 years ago a new phase started, eolian activity intensified and new dunes started to form. In some areas, man action has undoubtedly helped to accelerate the aridification process due to overgrazing and use of the vegetation by man (cutting, burning acitivities, etc.).

During all the successive climatic and environmental changes, sea levels also changed considerably. From the high sea levels (130 m above present s. l.) suggested for the Pliocene–Pleistocene boundary (based among other things in the scarp near Al-Hasa) to the low regressive levels estimated for the Würm glaci-eustatic minimum (−100 to −130) situating the shoreline outside the Arabian Gulf, a number of old shorelines have been proposed.

Among them two shoreline associated features seem to have been clearly identified. The first one related with the interstatial Würm (30,000–40,000 years ago, up to 10 m above present s. l.) and the second one is associated with the Dunkirk transgression (4,000–7,000 years ago, two m above present s. l.). In several places the areas previously occupied by these transgressions evolved afterwards into sabkhahs (surfaces in deflational equilibrium with a shallow water table).

5. Geochronology and Climate of the Quaternary

5.1. The Current Climate

(J. G. Zötl)

In Vol. 1, **Quaternary Period in Saudi Arabia** (1978, 31–44) E. Schyfsma provided general information on the number and position of the meteorological stations on the Arabian Peninsula, temperature and precipitation measurements (generally from 1966 to 1974), relative humidity, prevailing winds and solar irradiation. The data available were presented in figures showing the locations of meteorological stations in Saudi Arabia, summer and winter isotherms, isohyetes of the average yearly precipitation, and the yearly average of the actual evaporation (op. cit., figs. 3–7).

Additions to the descriptions of the current climate will be made here only as far as required by the subjects at hand of climatic variations and neighboring areas.

The network of meteorological stations is rather limited in view of the tremendous area shown in Vol. 1, fig. 3, and observation periods seldom cover as much as ten years. This information, together with personal impressions and experience, nonetheless made it possible to sketch a valid general picture for Saudi Arabia.

The attempt to isolate current climatic provinces is necessary, among other reasons, for purposes of comparison in the presentation of effects of Quaternary climatic variations.

With the exception of the southwestern highland, the Kingdom of Saudi Arabia at present lies in climates that according to W. Köppen (1931 and 1939) are to be classified as dry climates (W. Köppen, „B-CLIMATES"). The **desert climates** (highly or extremely arid) are to be divided into four large units: the Rub'Al Khali; the mainly inland deserts and desert steppes; the basalt deserts; and, the coastal deserts.

The **Rub'Al Khali** takes up the greater part of the area shown in Fig. 120 as extremely arid (between 45° and 55° E). The area falls from SW to NE from some 700 meters a. s. l., to less than 100 meters a. s. l.; in the NE, it runs directly into the coastal desert, and extends in the east in a somewhat different variable form to the coast. The core of this area is avoided even by the bedouins. With its reddish sand dunes reaching heights of hundreds of meters and the hard-as-cement salt floors between them, it may rightly be described as an "empty quarter" (cf. Vol. 1, 281, fig. 96 and D. Anton, chapter 4).

Also extremely arid are the immediate **coastal areas** of the Red Sea and the Arabian Gulf (Vol. 1, fig. 3). Unlike the Rub'Al Khali and the intercontinental deserts, these seaboards of varying width south of the tropic of cancer are exceptionally hot with a high (relative) humidity above 80%, and show only slight daily and yearly variations in temperature (annual average for Jizan: January 26.1° C, July 33.1° C). With the exception of episodically or periodically waterbearing wadis (Tihama and Yemen), they are nonetheless desert areas and, north of the tropic, actually extremely arid.

The **inland desert** is highly arid, but episodic cloudbursts can sometimes cause flood-like runoffs in the wadis. These include the sand deserts (northern and southern Nafud, Ad Dhana) and the remarkable vegetation-free plateau area of As Summan (sandstones, marl, lime; cf. Vol. 1, 163ff, fig. 54) as well as parts of the central coastal landscape.

The daily and yearly temperature variations are significant; they increase from south to north, and decrease with altitude: As Sulayyil: 20°28′ N, 45°40′ E (612 m a. s. l.) annual average January 15.5.° C, July 33.7° C, Turayf 31°41′ N, 38°40′ E (824 m a. s. l.) annual average January 4.5° C, July 28.1° C, Khamis Mushait 18°18′ N, 42°48′ E (2,057 m a. s. l.) annual averages January 13.7° C, July 23.2° C.

The **basalt deserts** in the mountains of western Saudi Arabia are, beyond doubt, of great climatologic interest. These are (with altitudes of 800 to 1 000 m a. s. l.), from south to north, the areas Harrat Al Bugum and Harrat Nawasif (approximately 20°30′ N to 22° N, center 42° E); Harrat Hadan and Harrat Al Kishb (21°20′ N to 23°30′ N, center 41°30′ E); Harrat Rahat (approximately 21°40′ N to 24°30′ N, center 40°30′ E); Harrat Khaybar (approximately 25° N to 27° N, center 40° E), Harrat Uwayrid (approximately 26°30′ N to 28° N, center 37° E) and Al Harrah (30°20′ to 32°30′ N). Each of these Harrats forming flat plateaulike highlands is built up by a series of young basalt flows and has an area of at least 10,000 square kilometers; the largest, Harrat Khaybar, has some 22,000 square kilometers. The Harrats are total deserts. After thunderstorms (and especially after repeated short rains in the spring), the sand dunes will sprout a fuzz of green grasses, but even in the north the black basalt blocks reflect the oven-like heat during the day, only to turn frigid at night. Although no measurements are available – the central block fields sometimes can only be reached by helicopter – it may be assumed that it is here where the greatest diurnal temperature variations are to be found.

At the edges of the inner-Arabian deserts there is in the direction of both the highland and the central-Arabian cuesta an imprecise transition from desert to desert steppe and highland steppe. There are many small oases on the terraces of the numerous wadis of the cuesta. The capital, Ar Riyadh, rose upon one such spot. Average annual precipitation is 100–200 millimeters. The daily and yearly temperature variations can be significant: the annual average in Ar Riyadh is 13.8° C in January and 34.7° C in July; the record low to date was −7° C! In comparison to these small oasis the large oases between Ar Riyadh and the Arabian Gulf (Al Hasa, Al Qatif, see Vol. 1, chapters 2.1.3., 2.1.6–2.1.8.) did not develop as a result of the present climatic conditions.

The climate in the western highlands shows a classical differentiation caused by the longitudinal extension over almost 2,000 kilometers reaching from the Mediterranean to the edge of the humid tropic areas. While in the far north under Mediterranean influence, scant **winter rains** occur, there is a transition from desert steppe to bush steppe in the Hijaz highland from north to south. In the trough shaped valleys, there are numerous small oases with agriculture on the flat alluvial floor. Between At Taif and Abha the mountain rise to 3,000 meters and above, precipitation rises beyond 500 millimeters; it is the transition to the area of **summer rains.** The same is also true for the south and southeast. In between lies the Yemen high mountain regions where (between 1,800 and 3,700 m a. s. l.) winter and especially summer precipitation of 1,000 to 2,500 millimeters occur. In winter time during the night there are temperatures below zero, and snowfall is not uncommon. Measurements for these are scarce and are taken only occasionally. It is reported that there was 100 millimeters snow in 1949 in Abha (E. GABRIEL, 1968, vol. 1, p. 204). We witnessed ourselves to drive through an area east of At Taif in February 1962 which was covered with hail (Fig. 117).

It is apparent that the Arabian Peninsula as well as North Africa lie within the zone of Passatwinds with its characteristic dry hot innercontinental climate. But some facts influences the regional climatic conditions. In North Africa the Atlas Mountains which run east-west provide a rather sharp limit in the west to the Mediterranean climate (lee side of Atlas Mountains northernmost part approximately 30° N). In the east, however, sometimes Mediterranean rains pass through the low elevations between Sinai and Lebanon over Jordan and Syria towards the southeast where they are blocked by the Persian border mountains.

Fig. 117. Hail storm east of At Taif, March 1972. (Photo: J. G. ZÖTL, 1972.)

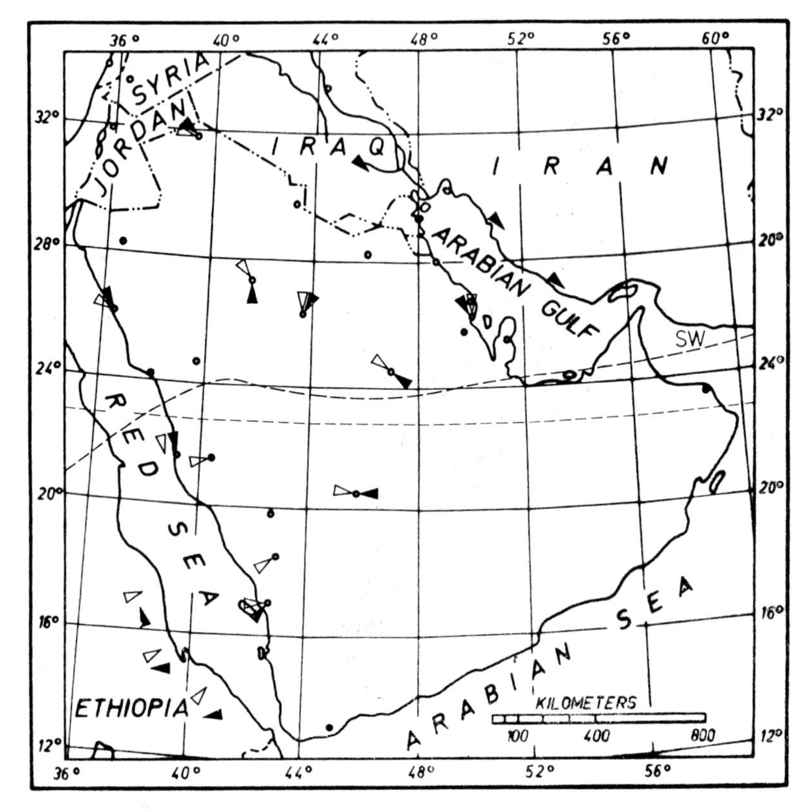

Fig. 118. Prevailing wind directions. (Data: Ministry of Agriculture and Water, Ar Riyadh.) White arrows = summer, black arrows = winter season.

In the west of the Arabian Shield the north and northwest winds extend into the area of Jeddah, having lost generally their rains but have still high relative humidity.

In the area of the Arabian Shield, the west slope which increases in altitude from north to south forms a north-south barriere. It rises from the southern Hijaz to the highlands of Asir to 3,700 meters altitudes of Yemen. Figure 118 shows that the winds coming from the Ethiopian highlands sweep towards the east, cross the Red Sea, ascend the west side of these mountains moisture-laden and break, monsoon-fashion, over the highlands south of Abha (altitude somewhat more than 3,000 m) (Fig. 119).

In North Africa, the climatic belts cover the continent from west to east in a fairly unbroken manner. The Sahara (W. KÖPPEN, 1931, BWh-climate), seen as a whole, thus takes up a strip between 15° W and 30° E, undisturbed by the Ahaggar and Tibesti massifs, as well as the Egyptian-Nubian coastal range. Even though it is of course divided into a variety of subgroups, this is the classical type of dry region in the tropical desert zone reaching from the Atlantic to the Red Sea (approximately 6,000 km). In the west, this desert zone is accompanied by a strip

Fig. 119. Monsoon clouds in southern Asir at 3000 meters a. s. l. (Photo: J. G. ZÖTL, 1977.)

of salt steppe, and in the south by the thornbush steppe of the Sahel Zone (both BSh according to KÖPPEN). The wind system is relatively uniformly the northeast Passat (trade wind); its front nearer the Equator is at about the latitude of Bilma (22–18° N) in the summer, and at the latitude of the Niger curve and the north edge of the Trehad (approximately 15° N) in winter. This relatively simple wind system becomes more complicated on the Arabian Peninsula, as shown by the wind directions given in Fig. 118.

All this shows, most importantly, that in the southwest Arabia there is already a pronounced monsoon climate with rainy seasons. This makes climatic comparisons with North Africa difficult, although the deserts in the interior, no doubt, also belong to the trade-wind zone and cover about the same latitudes (15–30° N) as the African desert belt.

5.2. Climatic Fluctuations in the Holocene

(H. HÖTZL, A. R. JADO, H. MOSER, W. RAUERT, J. G. ZÖTL)

A statistical compilation of climatic variations based on measurement data can at best be based on a period of 100 years, from the beginning of systematic meteorological and oceanic observation. Fluctuations doubtless occur, for example, the glaciation in the Alps in 1850, and the increase in the average water temperature (ocean surface 0.6–0.7° C) south of Greenland from 1910–1940. Since 1950, however, there has again been a falling tendency and the causes of even these present fluctuations have not yet been explained; consideration is given to changes in solar constants, atmospheric CO_2, ozone or water-vapor content, and the effect of worldwide changes in volcanic activity. Cycles related to sunspot cycles are the subject of conjecture (11 years, or 22 and 33 years; 35 years according to E. BRÜCKNER).

The observation of small and short fluctuations, as important as it is for meteorologists, is not a suitable basis for our problems (cf. H. FLOHN, 1957, and 1960). The question of climatic changes in the Quaternary involves those climatic periods which led to a worldwide and longer-lasting shift or significant change in climatic zones in the Holocene or Pleistocene.

As far as the terms used in this text for the current climate and climatic zones are concerned, the reader is referred to the standard works on general climatology by W. KÖPPEN (1931), W. KÖPPEN and R. GEIGER (1930–1939); J. BLÜTHGEN (1964), and H. OESCHGER, B. MESSERLI and M. SVILAR, (eds., 1980).

5.2.1. Euro-African Examples

The best-studied areas for postglacial climatic fluctuations are without a doubt Great Britain, Scandinavia, the Baltic States, the North German lowland, the Alps and the Mediterranean coast from France to Yugoslavia. The retreat phases of the northland glaciation of the last Ice Age and the related tectonic (glacioisostatic) movements, the stepwise retreat of the Alpine glaciers, and the youngest coastal terraces which can be followed along virtually the entire Mediterranean coast, are all owing to their youth, very well preserved locally and allow both a chronological fixation and an estimation of the extent and duration of the climatic fluctuations required to produce these geological phenomena. Methods for climatic research are making rapid advancement at present and, owing to the nature of the data these methods generate, the main concern is with the last Ice Age and the Holocene.

The data available include the present, and also the short (from the geological point of view) changes in climate and their effects (cf. H. FLOHN, 1959). Within this extent, the question arises which criteria can be used for climatic fluctuations seen in broader terms. The basic requirements are occurrence of an event on more than one continent, an extensiveness of time and change, and the availability of various comparable criteria. Anthropogenic climatic changes may be a subject of worldwide concern, but they do not fit into this context, although the burning-off of larger volumes of natural gas in the Iranian-Arabian area (Gulf region) has reached a point where it merits climatological discussion. The same is true for the so-called "Little Ice Age" (CH. PFISTER, 1980; F. F. MATTES, 1939) on the one hand, and a warming phase on the other (H. H. LAMB, 1977; H. FLOHN, 1980); the latter is concerned with the Norwegian emigration to Iceland and south Greenland (unhindered passage for ships through the then ice-free Denmark strait, which only iced up again in 1320 A. D. (H. FLOHN, 1980, p. 10).

The youngest phenomenon of a climatic fluctuation expressed in climatic curves representing a variety of criteria (temperature, glaciology, variations in sea level, climatic zones) is the period designated as a post-Ice Age "climatic optimum" by European scientists (W. W. KELLOG, 1977; H. FLOHN, 1977; H. FLOHN and S. NICHOLSON, 1979). W. W. KELLOG (1980, p. 26) quite rightly asks, optimal for whom, and calls this fluctuation "altithermal" (as a part of the hypsithermal), or also "Atlantic". For Saudi Arabia, in Vol. 1, we used the term Atlanticum (Altithermal) paralleling it with the "Neolithic Pluvial," which is applied by other authors (e. g. B. K. KAISER et al., 1973). But the term "Pluvial" is misleading for the only semiarid conditions of the Arabian Peninsula at that time.

In the chronological classification on the "Neolithic Pluvial", closer study shows that there are considerable fluctuations in temperature, and an alternation of wet and dry periods.

M. SCHWARZBACH (1974) differentiates, in the Late Glacial Period in Northern Europe after the Bölling interstadial epoch (approximately 18,000–12,700 years B. P.) a last high point in the Late Glacial, the so called **Alleröd-Period** (approximately 12,700–12,300 years B. P.) with average temperature in Middle Europe in July 4° C colder than today and a more pronounced worsening in the Younger Dryas Period (Younger Tundra Period 11,000–10,000 years B. P.) with temperatures in Germany 7–8° C colder than today (Alpine Gschnitz and Daun are Stages of glacial retreat).

The Holocene finally comes after a last rapid retreat of the ice in Scandinavia; in the post-glacial warm period of the Holocene (climatic optimum, Atlantic or Altithermal) from about 7,000 to 5,000 years B. P., the average annual temperature was 2°–3° warmer than today. This post-glacial climatic optimum (Altithermal) has also been demonstrated in North America, though it was less pronounced[1].

M. SCHWARZBACH (1974, p. 248) doubts an eustatic increase in sea level related to the Altithermal.

This brief discussion of the European classification permits us to use well-known research data on North Africa to interpret the observations in Saudi Arabia with a relatively high degree of certainty.

Even a first comparison between Middle Europe and North Africa shows, for example, that cold periods can be either cold and humid or cold and dry. More will be said about the problem of the pluvials in the discussion of the Würm and the Pleistocene.

In North Africa and the mountains of the Sahara, the "Holocene wet period" also began with increasing dryness at first (around 12,000 years B. P., B. MESSERLI, 1980, p. 77). Later there were moister periods from the equatorial region to the north of today's arid zones (B. MESSERLI, 1980, p. 78). Soil formations and limnetic deposits were found, especially in Tibesti and Hoggar; lake datings of 8,530 years B. P. and soil formations dated 6,600 years B. P. indicate "morphodynamic stability phases".

According to the Paleocene soils and lake deposits studied so far, M. MESSERLI assumes that the period of maximum warmth in the post-glacial period in the arid zones in Africa occurred from 6,000 to 5,500 years B. P. corresponding to the

[1] The data of H. V. RUDOLF (1980) agree generally with M. SCHWARZBACH. H. V. RUDOLF places the beginning of the Holocene at about 10,000 years B. P. and the climatic optimum at 7,500 years B. P. His average annual temperatures for 6,700–4,500 years B. P. are 1°–2° C higher, and the summer temperatures 2°–3° C higher than today.

Holocene Climatic Optimum of Europe. Shorter moist periods from 3,500 to 3,000, and 2,500 to 1,500 years B. P. etc. (cf. B. MESSERLI, 1980) do not appear to us to be climatic periods on the basis of their geographical extent or duration (for details see 5.2.2.3.).

It is only these Euro-african research data that permit a general classification of the data collected in Saudi Arabia during field work at selected sites.

5.2.2. The Altithermal in the Arabian Subcontinent

5.2.2.1. Fluctuations in sea level

In the coastal areas of the Arabian Peninsula, the most important evidence of a super regional nature is provided by traces of sea-level fluctuations with static periods long enough to permit the development of morphological forms (beach terraces, wave-cut notches) and the deposition of marine fossils (mussels, corals, etc.). They are of particular significance in that the absolute age of the calcareous shells can be established with carbon-14 dating.

Volume 1 (1978, p. 65) has already given examples of Holocene beach marks demonstrated on the Arabian Gulf coast. In the meanwhile, H. A. MCCLURE and C. VITA-FINZI (1981), and A. P. RIDLEY and M. W. SEELEY (1979) have made further datings. What with the older studies by J. C. M. TAYLOR and L. V. ILLING (1969) in Qatar and G. EVANS et al. (1969) in Abu Dhabi, the west coast of the Arabian Gulf is now one of the best-studied areas as far as Holocene coastal development is concerned (cf. Tables 49, 50, 51). The tables do not show the morphological Holocene marine-cut terraces (cf. Vol. 1, Fig. 11).

Besides the data given in Table 49 and 50 for the Saudi Arabian Gulf coast, data are also available for the shoreline of Qatar and the Trucial coast (G. EVANS et al., 1969). J. C. TAYLOR and L. V. ILLING (1969) describe three carbon-14 measurements of shells (Cerithies) from sand barriers in Qatar approximately 2–3 meters above high sea-water level (a. HW) with an age of 3,930±130, 4,200±200 and 4,340±180 years B. P. (Table 51).

Studies on Holocene sediments of coastal sabkhahs in the Abu Dhabi area were made by G. EVANS et al. (1969). They worked out a sedimentological profile, whereby they differentiated four main lithological units and gave their interpretation of the geological history of the sabkhahs sediments based on 34 ^{14}C measurements. The two upper horizons are characterized by fossils of different ages, meaning that the sediments were reworked by waves. The sixteen samples of pelecypods and gastropods with ^{14}C ages of 3,100 to 6,500 years B. P. come from the very flat sabkhah plain with subsurface depths of 0.4–2.7 meters.

G. EVANS et al. (1969) conclude that the Late Holocene transgression began approximately 7,000 years ago and continued until some 4,000 years B. P. The transgression inundated a dune-covered area. Since the end of the transgression phase, while subaqueous sedimentation took place in a lagoon, the intertidal and supertidal zones progressed seaward.

If we compare data and finding places (Fig. 120) the question at first remains open as to whether beach terraces presently located 1–3 meters above recent high water marks (HW) on the Arabian side of the Gulf are the result of a decrease in sea level or a tectonic uplift, as A. P. RIDLEY and M. W. SEELEY (1979) assumed.

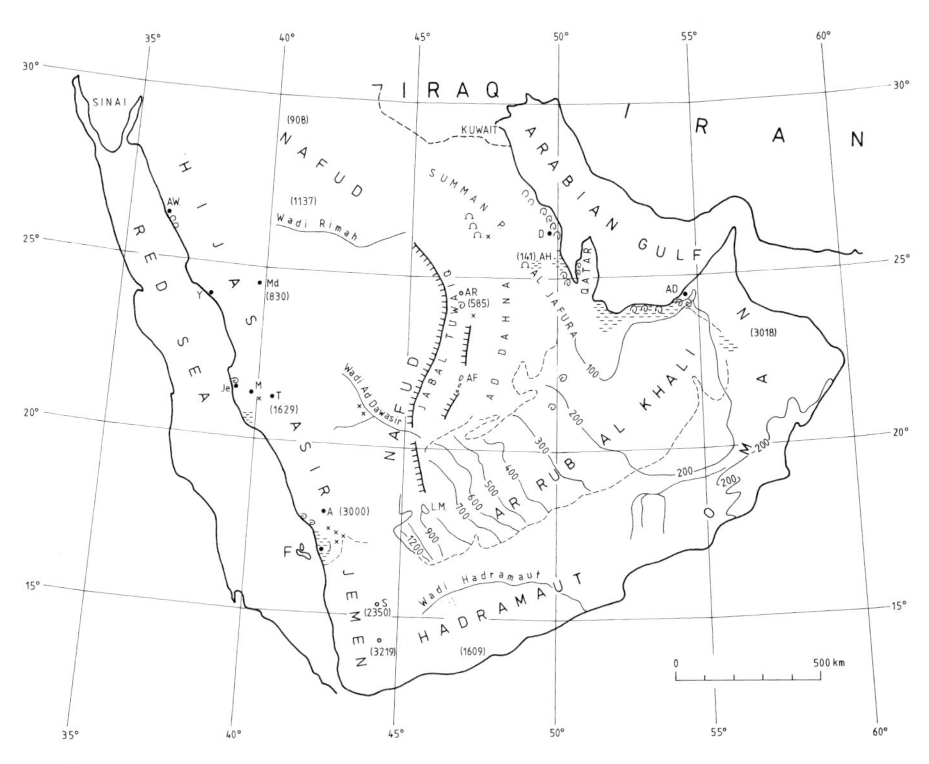

Fig. 120. Location of fossils dated by ^{14}C-measurements as altithermal. (ƏƏ shells or snales, *x* calcareous sinter, *AW* Al Wajh, *Je* Jeddah, *J* Jizan, *D* Dhahran, *AD* Abu Dhabi, *AR* Ar Riyadh, near *AH* Al Hasa, *AF* Aflaj, *L. M.* Lake Mundafan, *A* Abha, *M* Mecca, *Md* Al Madinah, *T* At Taif, *S* Sana, *F* Farasan; Ω caves.)

This is contrary to our opinion. H. A. McClure and C. Vita-Finzi (1981) studied this question very carefully and reached the following conclusions: Their finds were usually in depressions removed from the coast and there were dunes as high as 10 meters between these sabkhah areas (see below). There are also considerable discrepancies in the elevations of findings and in measurement data given by A. P. Ridley and M. W. Seeley (1979). This means that it is not permissible to apply the uplift tendencies still effective in the Zagros chain to the coast of the Arabian Peninsula[1]. This is ultimately demonstrated by a sea level of 1–3 meters a. HW shown by dating (fossils with ages between 3,600 and 6,700 years B. P.).

We believe on the basis of these similar conditions that we can draw the further-reaching conclusions that in the Holocene period limited by dating there were no important tectonic movements in the entire Arabian Peninsula, with the possible exception of weak uplifts in the area of the salt diapirs in the Gulf and in

[1] C. Vita-Finzi (1980) dates seawater fossils from the Iranian coast between 3 and 28.6 meters a. HW at 4,625±115 and 6,110±95 years, indicating recent tectonic mobility of the Zagros Mountains. The southern coast of the Arabian Platform was not studied.

the Red Sea areas, and the very slow subsidence tendencies in the area of the lower Euphrates and Tigris.

The fossils found by H. A. McCLURE and C. VITA-FINZI, increase the significance of the **coastal sabkhahs** as regards the extent of the sea during the Holocene high water level.

Sabkhah research in general clearly shows three sorts of sabkhahs for Saudi Arabia. These are: 1) coastal sabkhahs resulting from marine transgression; 2) inland sabkhahs resulting from high groundwater levels and deflation of originally sand covering (D. H. JOHNSON et al., Vol. 1, 1978), and 3) the "wadi sabkhah sediments" introduced by H. HÖTZL (Vol. 1, p. 272) as a comprehensive term for fine sandy, silty, clayey and evaporitic sediment sequences accumulated in pan-like wadi sections owing to the interaction of eolian and episodic fluviatile processes.

Table 49. *Results of carbon-14 measurements on the west coast of the Arabian Gulf* (from Quaternary Period in Saudi Arabia Vol. 1, 1978)

No.	Height above HW (m)	Coordinates	Lab No.	Age (yrs. B. P.)	Species
1	ca. 3 m	lat 26°53′ N long 49°56′ E	IRM-3649	4,670±190	*Cardies* and *Pectes*
2	ca. 2.5 m	lat 26°50′ N long 50°00′ E	IRM-3650	3,380±180	*Cardies* and *Pectes*
3	ca. 2.5 m	lat 26°30′ N long 49°50′ E	VRI-406	3,990±90	Cemented shells of *Cardies* and *Pectes*
4	ca. 1.5 m	lat 26°30′ N	VRI-383	1,090±80	*Oysters*

Table 50. *Results of carbon-14 measurements on the west coast of the Arabian Gulf* (from McCLURE and VITA-FINZI, 1981)

No.	Height above HW (m)	Coordinates	Lab No.	Age (yrs. B. P.)	Species
1	1.0 m	24°53′ N 50°43′ E	BETA-2679	4,585±60	*Punctada margaritifera*
2	1.8 m	25°53′ N 50°07′ E	BETA-2681	3,695±50	*Punctada margaritifera*
3	2.0 m	26°01′ N 49°59′ E	BETA-2682	4,460±60	*Punctada margaritifera*
4	1.2 m	26°13′ N 50°07′ E	BETA-2533	6,020±80	*Circe arabica*
5	3.0 m	27°32′ N 49°12′ E	BETA-2532	4,205±70	*Circe arabica Pectes*

HW = High Sea Water level.
B. P. = before present.

Table 51. *^{14}C-ages of witnesses of a higher mean sea level of ca. 2–3 m above high sea water level in Qatar and the Red Sea Coast*

No.	Coordinates	Age (yrs. B. P.)	Material	Remarks
1	ca. 25°00′ N 50°45′ E	3,930±130	*Cerithies*	Qatar coastal terrace (TAILOR & ILLING, 1969)
2	– do –	4,200±200	– do –	– do –
3	– do –	4,340±180	– do –	– do –
4	ca. 18°12′ N 41°33′ E	5,400±200	Sinter with shells and gastropodes	Red Sea coast; IRM 7623
5	– do –	4,700±400	sinterlimestone	Red Sea coast; IRM 7622
6	ca. 16°46′ N 42°00′ E	4,700±400	Shells	Farasan Kabir NW IRM 7616

B. P. = before present.

Figure 120 provides a general survey of the most important findings of the samples for ^{14}C-measurements in the coastal areas and the interior, as well as of the coastal sabkhahs.

5.2.2.2. Facts regarding the moister period

As far as climatic changes are concerned, the demonstration of phases of altered **humidity and temperature** is more important than fluctuations in sea level. Clarification may be provided by the following: characteristic changes of wadi fillings, demonstration of previous inland lakes and their alteration, and datings of residues of organic carbon as well as calcareons material from shells and precipitates. We were not able to perform palynological studies.

Both in the wadis of the cuesta landscape (Wadi Hanifah, Wadi Al Luhy, Vol. 1, p. 202ff) and those of the shield it is apparent that they are currently undergoing an erosion phase in their middle and upper courses. The terraces are thus remnants of a previous accumulations (see Vol. 1, fig. 64, p. 204).

The most interesting example in the central-Arabian cuesta landscape is doubtless that of the terraces of the wadis **Hanifah** and **Al Luhy** (Vol. 1, fig. 63, p. 203). Wadi Hanifah flows along the northern Tuwayq Mountains (geology, see Vol. 1, p. 202f). The erosion caused by recent episodic floods has formed a vertical edge on the remnants of the youngest accumulation or terrace. The height of the terrace above the recent valley floor – i. e. the work of the current erosion – increases from about 0.3 meter in the uppermost course of Wadi Hanifah to about three meters downstream west of Ar Riyadh to about six meters at Al Ha'ir, where Wadi Al Luhy flows into Wadi Hanifah (lat. 24°20′ N, long. 46°40′ E).

The aggradation of the accumulation that today is only present in terraces in Wadi Hanifah must have taken place rather quickly, as it led to damming-up of the water in Wadi Al Luhy, a right tributary of Wadi Hanifah. This backup brought about a longer-lasting deposition of gray calcareous stillwater silts. Gastropod shells from these stillwater sediments were collected for ^{14}C dating (see Vol. 1, Paleontological description of *Gastropoda*, p. 205). The deposition of the stillwater sediments and gastropods on the lower part of the terrace and their later overcovering with wadi sediments indicate that they developed at the beginning of the considerably moister period. That the stillwater sediments were covered over by the terrace sediments mentioned above at the same terrace surface level for both wadis indicates a breakthrough of the natural dam before the end of the accumulation phase.

^{14}C dating of the gastropod shells showed an age of 8,400±140 years B. P. (Vol. 1, sample VRI-384).

The **Wadi Birk** system, formed in the pre-Pleistocene and rising on the eastern slope of the Shield, crosses the southern Tuwayq Mountains. The breakthroughs show terrace systems of the same sort as those in Wadi Hanifah. Although we were unable to find datable stillwater sediments, we did find a calcite-sinter gravel with a ^{14}C age of 6,880±290 years in Wadi Al Hawtah, the most important tributary of Wadi Birk in the Tuwayq Mountains. These data and ^{14}C measurements to be discussed below also increase the significance of the find in Wadi Al Hawtah (see Table 52).

Wadi Ad Dawasir and its largest tributary, Wadi Ranyah, also belong to the large pre-Pleistocene river systems that cross the entire Arabian Platform from west to east (Vol. 1, p. 230ff). In Wadi Ranyah, although it is still in the eastern slope of the Shield (between Al Amlah and Rawdah, see Vol. 1, p. 234f), there are tongues of two basalt flows to which we shall return in the discussion of the early Pleistocene owing to their different degrees of weathering. What is significant for Holocene climatic fluctuations, however, is the fact that various kinds of sinter troughs and channels are still relatively well preserved in the bay-like foreland of these two basalt flows. The ^{14}C ages for two calcite sinter samples are 6,700±280 and 6,110±250 years B. P. (Lab. Nos. IRM-4235 and IRM-4366, lat. 21°15′ N, long. 42°44′ E, Table 52).

A river system rose during the Pliocene/Pleistocene transition in the upper Wadi Ad Dawasir that crosses the Arabian Platform to the Arabian Gulf (see Fig. 116, and Vol. 1, p. 310). Today the northern Rub'Al Khali occupies this space on the Shelf. Surprisingly, what today is an extreme sand desert (H. A. McClure, 1976, and 1978) shows remnants of Holocene playas or "mud lakes" from the edge of the basement – the eastern flank of the crystalline rocks of the Shield – over nearly 800 kilometers of the Rub'Al Khali. Radio-carbon dating of their dried-up sediments placed them unambiguously in a Holocene series (9,000 to 6,000 years P. B. table 52). The most important starting point is the fossil **Lake Mundafan** (Fig. 120) that extends for 150 kilometers in a north-south direction along the Al Arid escarpment and today is partially covered by dunes. H. A. McClure (1976) divides the roughly 24-meter thick lake deposit into three series, of which series B covers ten ^{14}C datings between 6,100±70 and 8,800±90 years B. P. There are two additional finds 700 and 750 kilometers farther NE in the center of the Rub'al Khali aged 6,520±115 and 7,160±115 years B. P. (Fig. 120).

Regarding fossil lakes, it must surely be the case that systematic palynological

Table 52. *Holocene ^{14}C ages of inland fossils from various locations in Saudi Arabia* (see Fig. 120)

No.	Coordinates	Age (yrs. B. P.)	Material	Remarks
1	23°40′ N, 46°35′ E	6,880±290	Sinter gravel	Wadi Birk; IRM 4241
2	24°20′ N, 46°40′ E	8,400±140	Gastropoda	Wadi Al Luhy; VRI-384
3	26°25′ N, 47°20′ E	5,060±250	Calcareous tufa	As Sulb Plateau; IRM-3660
4	21°15′ N, 42°44′ E	6,700±250	Calcite sinter	Wadi Ad Dawasir; IRM-4235
5	21°15′ N, 42°44′ E	6,110±250	Calcite sinter	Wadi Ad Dawasir; IRM-4366
6	18°30′ N, 45°05′ E	6,100± 70	Algal encrustation	Lake Mundafan; McClure 1976, MF 12
7	18°30′ N, 45°05′ E	7,040±115	Diatomaceous marl	Lake Mundafan; McClure 1976, MF 13
8	18°30′ N, 45°05′ E	7,190± 85	marl	Lake Mundafan; McClure 1976, MF 14
9	18°30′ N, 45°05′ E	7,265± 80	Diatomaceous marl	Lake Mundafan; McClure 1976, MF 15
10	18°30′ N, 45°05′ E	7,400±210	shells	Lake Mundafan; McClure 1976, MF 16
11	18°30′ N, 45°05′ E	7,770± 90	Diatomaceous marl	Lake Mundafan; McClure 1976, MF 17
12	18°30′ N, 45°05′ E	8,060± 95	marl	Lake Mundafan; McClure 1976, MF 18
13	18°30′ N, 45°05′ E	8,155± 85	marl	Lake Mundafan; McClure 1976, MF 19
14	18°30′ N, 45°05′ E	8,565±110	shells	Lake Mundafan; McClure 1976, MF 20
15	18°30′ N, 45°05′ E	8,800± 90	charcoal/ash	Lake Mundafan; McClure 1976, MF 21
16	21°55′ N, 49°45′ E	6,520±115	shells	Rub Al Khali; McClure 1976, Site 4
17	23°15′ N, 50°20′ E	7,160±115	shells	Rub Al Khali; McClure 1976, Site 5
18	16°55′ N, 42°50′ E	8,070±340	Calcareous crust	S Jabal At Tirf; IRM SA-77/18
19	16°55′ N, 42°50′ E	7,160±340	roots	S Jabal At Tirf; IRM SA-77/20
20	17°05′ N, 42°44′ E	8,080±270	Calcareous sinter	Wadi Jizan below Maloki dam; IRM SA-77/4
21	17°05′ N, 42°44′ E	3,140±357	Snail shells in dust deposits	Wadi Jizan below Maloki dam; IRM SA-77/5
22	17°10′ N, 42°50′ E	6,220±280	Calcareous sinter	Wadi Jizan west edge of basalt; IRM SA-77/9

Some more dates of Holocene ^{14}C ages contain Tables 18 and 54.

studies in the **Al Hasa** area, the largest oasis in Saudi Arabia, would provide valuable information. Neolithic spears and fishhooks found in higherlying island-like areas prove that a good deal of fishing was done during the Middle Holocene (H. A. McClure, 1971).

The **area of Aflaj,** which we call the "Central-Arabian Lake District," is unusual (Fig. 120). This consists of actual lakes located in a desert like area south of Layla near the As Sulayyil Road to Ar Riyadh (approximately 22°10′ N, 46°37′ E). What the larger of the open waters in a NNE–SSW trending chain some 500 meters long do not show, may be seen immediately in the smaller ones: they are solution collapses within the Hith anhydrite formation. These are groundwater lakes of considerable depth (more than 20 meters). This water was once used, as is shown by the remnants of two kanat systems left uncovered by the dunes). More recently (perhaps during the Ottoman Empire), construction of a deep, open canal was begun, but never finished. The dune-covered kanat systems differ visibly in age. The largest lake doubtless developed from a number of collapses; its steep eastern bank has broken down. A worn intermediate terrace could indicate that the groundwater level used to be higher. It was not possible to date these topographic forms.

We also find the same positive criteria for a wetter Holocene phase (Table 52) such as in the east on the **Red Sea coast** and in the wadis rising in the escarpment of the Shield. Here as well, ^{14}C age measurements have been made from samples both from the hinterland (Table 52) and from the seacoast. The extensive coastal sabkhahs, especially north and south of Jizan, also provide distinct indications that the water level of the Red Sea was some meters higher in the "Altithermal" of the Holocene.

The northernmost limit for a comparison with neighboring areas would be the **Damascus Basin** with the Quaternary studies done there (center 33°30′ N, 36°30′ E).

Although the very detailed studies of K. Kaiser et al. (1973), which were based mostly on palynology, resulted mainly in an emphasis on the northern pluvials in relation to the Pleistocene cold periods (see K. Kaiser, 1973, p. 278), carbonatic sinter 70–80 meters thick is mentioned for the upper Baranda valley (1973, p. 351) with a lignite band in its upper layer dated 3,580 years B. P. The authors agree on a "Middle Holocene pluvial of possibly 4,000–5,000 years duration" for the underlying carbonate precipitates; this pluvial separated the Holocene dry periods "in the area of the Damascus Basin and its vicinity" (ibid., p. 352).

5.2.2.3. Conclusions

The available ^{14}C datings and related information are entirely sufficient to demonstrate a Middle-Young Holocene eustatic rise in the seas bordering the Arabian peninsula in the east and west, as well as the occurrence of a more humid phase.

The same is true for the conclusion that there were no significant tectonic movements in this entire area in the Holocene.

A number of other problems, however, remain to be discussed.

With the absolute age determinations it is noteworthy that for the **eustatic increase** in sea level the oldest datings begin with 7,000 years for only one sample of coastal sabkhah, while other samples are dated mainly between 6,000 and 3,500 years B. P. (and this from more than 50 samples, some considerably younger). Of the **fossils found inland,** the twenty-two samples shown in Table 52 demonstrated, with two exceptions, ^{14}C ages between 9,000 and 6,000 years B. P. The indisputably longer moist phase of this period is shown most clearly by McClure's data on a closed series of ten samples from Lake Mundafan aged approximately 9,000–6,000 years B. P. (Table 52), but also by the gastropods aged about 8,000 years B. P. from Wadi Al Luhy, the sinter in Wadi Ad Dawasir, and not least the small, formerly fresh-water lakes in the central Rub'al Khali. Younger

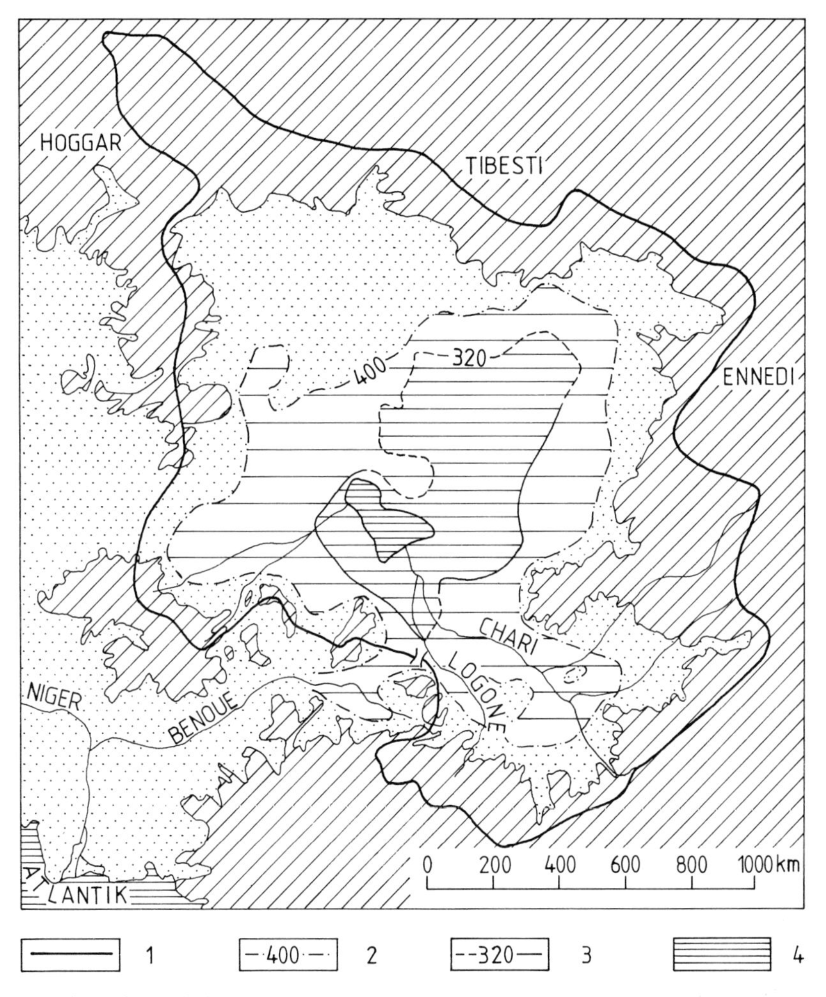

Fig. 121. Chronology of the water table of Lake Chad (H. Oeschger, et al. 1980) (*1* border of recharge area; *2* Paleo-Chad niveau ca. 400 meters. a. s. l. [30,000–22,000 yrs. B. P.]; *3* Paleo-Chad niveau ca. 320 meters a. s. l. [ca. 5,400 yrs. B. P.], *4* recent Chad niveau changing between 281–283 meters a. s. l. [B. Messerli, 1980]).

fossils from Lake Mundafan are missing entirely. McClure's assumption (1976, and 1978) of a sudden **hyperarid** period around 6,000 B. P. seems plausible.

In Africa, the development of Lake Chad shows unmistakable parallels with Lake Mundafan (cf. H. Oeschger et al., 1980, p. 80). The so-called paleo-Chad level of 320 meters above sea level is placed here (with interruption) in the period of 9,200–5,000 years B. P. A second phase of interrupted sinking at 320 meters above sea level is given for the period of 3,200–1,800 years B. P.; the current water level is 281–283 meters above sea level on the average for 1908–1974 (Fig. 121).

The moist period in all of Africa from the Equator to the Mediterranean is assumed to have reached its peak around 8,000 years B. P. (H. Oeschger et al., 1962, and fig. 121); the region of summer rains reached to the southern foot of the Atlas Mountains.

An interruption by the abrupt onset of an arid phase is also assumed for the period from 7,300 to 6,300 for the Sahara during "this first Holocene moisture maximum" (H. Oeschger et al., 1980, p. 81); other measurements, however, place it between about 6,500 and 5,700 years B. P.

Complete profiles of ^{14}C age determinations of fluviatile deposits from the Tibesti Mountains are also available for the period of about 9,000 to 7,000 years B. P. (M. Geyh et al., 1974, p. 112f).

The numerous parallels between the extensive literature on North Africa and our own work in Saudi Arabia show that the climatic history of these areas is by all means comparable. This is of great help in the processing of the Quaternary geology of Saudi Arabia, in-so-far as the adjacent Iranian highland in the east puts an abrupt end to the Afroarabian climatic belt.

Observation of the zone of inner tropic convergence alone (Fig. 122; cf. H. Blüthgen, 1964, p. 464) shows that the Iranian highland is outside of the climatic belt under study. H. Bobek, 1963, emphasized that an important finding of his work is "that Iran occupies a very distinct position in the Near East in that its vast interior plateau (at least) did not experience pluvial periods in any way comparable with those of other Near Eastern countries (or subtropical countries in general) (p. 403). H. Bobek is particularly concerned with questions of the zones of glaciation, nivation, soliflucation, etc., and their changes in the Quaternary, but he does not go into absolute dating.

The attempt to quantify Holocene climatic fluctuations is also different.

The temperature increase in the Altithermal is assumed to be 1–1.5° C for this entire area (H. Oeschger et al., 1980, p. 78). With regard to the effect on the climate, this temperature increase is far less important than the changes in **amount** of precipitation and its distribution. There is no doubt that the lower-lying regions of Saudi Arabia were not hyperarid, as today, but were semiarid, especially between 9,000 and 6,000 years B. P., although potential evaporation was also higher than the precipitation level in this phase.

We thus have to depend on estimates of the precipitation levels at the peak of the Altithermal; these, however, are not entirely unfounded.

Based on the open inland waters shown by dated fossils and the duration of their existence, it may be assumed that during the Altithermal, the area of Saudi Arabia that today is hyperarid showed conditions of moisture rather like those

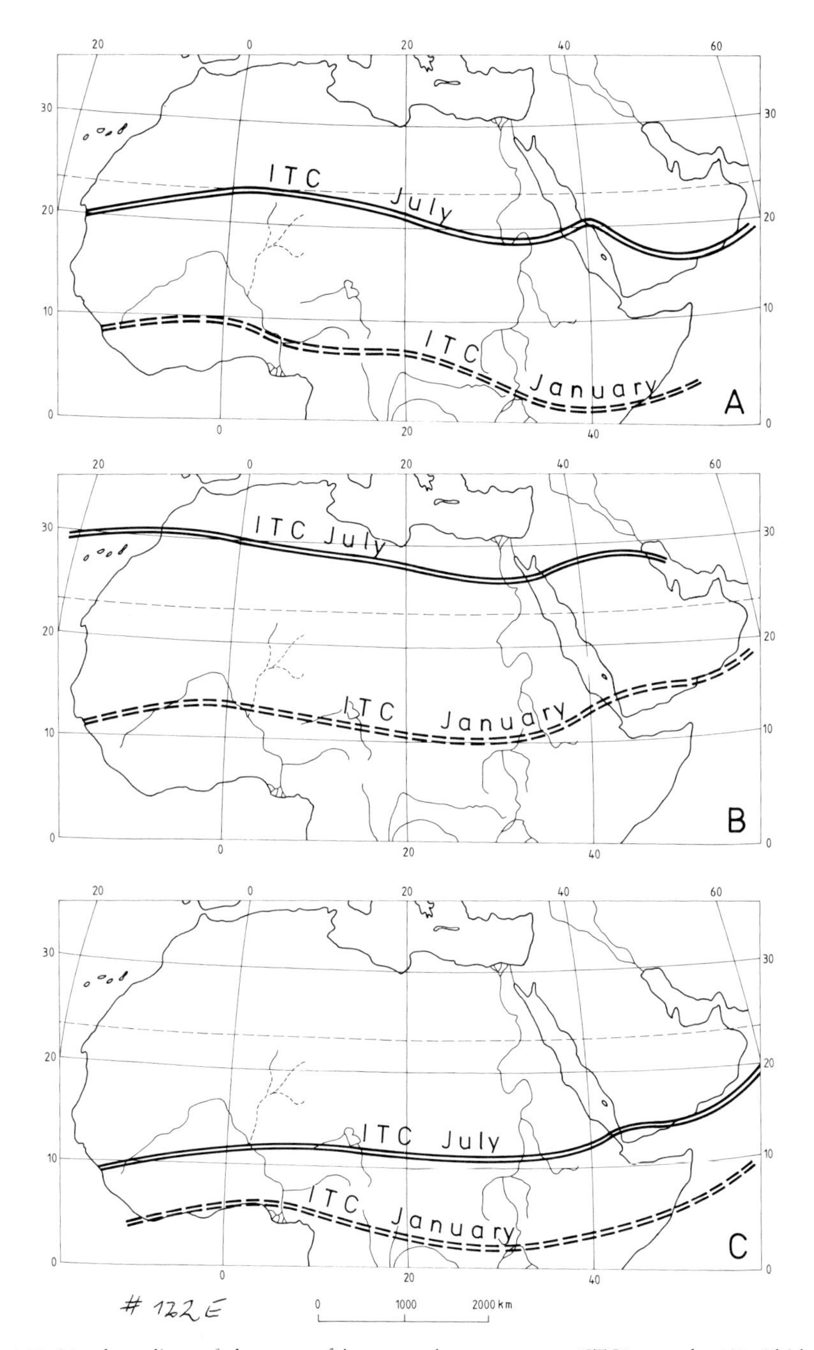

Fig. 122. Northern lines of the zone of inner tropic convergence *(ITC)* recently *(A)*, Altithermal *(B:* ca. 8,000 yrs. B. P.); an Youngest Pleistocene ca. 18,000 yrs. B. P.

currently prevailing in the Sahel zone south of the Sahara. Decreasing from south to north, annual precipitation of 400–100 millimeters more than the present may be assumed to be within correct limits. This would also explain why Lake Mundafan and a few small standing waters on the edge of or within the Rub'Al Khali – owing to the different climatic conditions – could exist during Pleistocene and the Holocene phases, but not under current conditions (cf. Fig. 122). These figures and figure 140 in J. BLÜTHGEN (loc. cit.), clearly show that shifts in atmospheric circulation and thus in climatic conditions in the Afro-Arabian area. Figure 122 B shows that the region of equatorial summer rains in the Holocene pluvial reached the northern boundary of Saudi Arabia.

The question is still open as to the late occurrence, or the proof, of the eustatic increase in the sea level in the Holocene. Here it is necessary to take a look at Europe. The increase in sea level is not (or is scarcely) due to increased precipitation, but to the melting of still rather large glacial masses in the high mountains of Eurasia and North America, and polar ice. It may be assumed that this did not occur directly and to its full extent at the beginning of the European "climatic optipmum" (Altithermal), but rather in gradual steps, with the highwater mark at the end of the Altithermal. Fluctuations and a slow sinking of the sea level to the present have been confirmed by ^{14}C ages for the whole period from 6,000 years B. P. to date.

5.3. The Youngest Pleistocene

5.3.1. General Considerations

(H. HÖTZL, A. R. JADO, H. MOSER, W. RAUERT, J. G. ZÖTL)

Today, the Pleistocene-Holocene transition is quite uniformly placed at the end of the younger Tundra period, about 10,000 year B. P. This is the beginning of the retreat in Northern Europe of the ice of the "third Salpausselka" in Finland (P. WOLDSTEDT, 1958, p. 140), the last of the North European Ice cap which was up to 3,000 meters thick and covered not only Scandinavia, but also North and Middle Russia, the Baltic Sea, North Germany and the British Isles[1].

In the Alps, the end of the Würm is placed at about the high-lying Gschnitz and Daun retreat stages (snow line some 300–400 meters below the present line; also about 10,000 years B. P.).

The course of the Würm, as the last of the great Ice Ages, has been studied extensively – as the youngest of the large units it left the best-preserved traces behind – but there are numerous uncertainities, or even contrary opinions, as to the details (Plate VI, see insertion at back cover). Put very simply and according to the majority of opinions, the Würm began about 70,000 years B. P. It covered Glacial A (with small interstadials from 55,000–40,000 B. P.), the Major Interstadial Paudorf, 40,000–30,000 B. P. and High Glacial 30,000 (25,000)–15,000

[1] In North Germany, the last Ice Age (alpine "Würm") is called "Weichsel", in North America "Wisconsin" glaciation.

B. P.; Late Glacial with glacier retreat occurred in Europe, Asia and North America until 10,000 years B. P.

The climatic curves suggested by well-known authorities presented for purposes of comparison in Plate VI (see insertion at back cover) seem at first glance to show little agreement, and not only because of the different areas they cover and the differing working methods. However, more generalized curves (as for example, the reproduction according to B. FRENZEL et al. (1967), B. FRENZEL (1980), D. NEEV and K. O. EMERY (1967) and, for the last 25,000 years, G. F. MITCHELL) shows surprisingly good correlation of the time units for the climatic periods.

If we start with the fact that, in contrast to the Altithermal, we are concerned with a climatic change coming from the north as far as the inland waters of the North Arabian area are concerned, this leads us back to the Damascus Basin. K. KAISER (1973) supports his powerful "north pluvial" with a pollen diagram of Young Pleistocene lake deposits from the Sahl Aâdra in the Damascus Basin (1973, fig. 5), which is also supported by ^{14}C age determinations by M. A. GEYH. This, however, only covers a period from about 24,000 to 19,000 years B. P. There is agreement with Middle European information when K. KAISER concludes that the phase of maximal cold occurred after 20,000 years B. P. in the Syrian-Lebanese Levant.

The most important deposit sequence for the effects of the Würm in the North Arabian-South Levantine area is the so-called Lisan facies found on the slopes going down to the Dead Sea, the lowest point on the major tectonic line of the Jordan graben and Gulf of Aqaba.

The beginning of the Würm must have put a relatively rapid end to the Riss-Würm Interglacial (Ems interglacial or Sangamon).

If we again start by considering the European area, we are first confronted with the fact that all of northern and northwestern Europe was covered with an ice cap up to 3,000 meters thick which either drastically limited and brought an end to plant and animal life.

With such a large amount of water tied up in the North European and North American inland ice and in the glacial masses in the mountains, the sea level dropped by about 100 meters. This has not only been studied carefully in the Mediterranean; a special study of the Arabian Gulf also showed that it dried up in the Würm (M. SARNTHEIN, 1972; cf. Vol. 1, fig. 9, p. 53).

While there is little question as to the cooling of the areas south of the glacier regions, the amount of precipitation is still quite open to discussion (cf. R. W. FAIRBRIDGE, 1964). The high seasonal runoff of the European rivers and the related aggradation of extensive gravel terraces is mainly to be attributed to the very considerable melting of the glacier zones in the summer months during the Ice Age. Today, maximal moisture (i. e. cold and moist conditions) in the waxing first half of the Ice Age and a cold, dry climate in the waxing Ice Age are generally assumed for the glacial and periglacial areas (B. FRENZEL, 1980).

There is little uniform opinion on non-European areas. Here, an important step could be a comparison of lakes without outflow that during the Ice Age as well had no open outlet into a river system or the sea, or despite such an outflow were very subject to climatic influences.

5.3.2. Results of Research in Anterior Asia and Africa

If we begin with the fact that in contrast to the "Neolithic Pluvial" we are here concerned with a climatic change coming from the north with inland seas in the North Arabian area, then we again come to the **Damascus Basin.** K. KAISER (1973) bases his proof of a strong "north pluvial" on a pollen diagram of Young Pleistocene marine deposits from the Sahl Aâdra in the Damascus Basin (1973, fig. 5), which is supported by ^{14}C measurements made by M. A. GEYH. This is, however, only a period of about 24,000–19,000 years B. P. There is agreement with Middle-European informations when K. KAISER concludes that the phase of maximal cold in the Syrian-Lebanese Levant was after 20,000 years B. P.; he only has difficulty providing evidence of maximal moisture, as many of his palynological findings indicate the opposite showing a dry climate during the extreme cold (Würm maximum from 21,000–20,000 years B. P. according to K. KAISER).

The most important deposit series for the effects of the Würm in area of northern Arabia and the southern Levant is the so-called Lisan facies; it is located on the slopes of the Dead Sea, the lowest point of the major tectonic line of the Jordanian region to the Gulf of Aqaba.

The **Dead Sea** lies in the graben zone extending from East Africa through the Red Sea and the Gulf of Aqaba to Syria.

D. NEEV and K. O. EMERY (1967, cf. also 1966) assume that the graben began to subside in the Late Tertiary or at the beginning of the Quaternary (p. 24). The first sediment series contains Plio-Pleistocene marine-brackish fossils. D. NEEV and K. O. EMERY suppose "that a connection with the ocean persisted until early or middle Pleistocene times" (op. cit., 24). During this time, ocean water flowed into the trough of the graben. These authors' comparison with the Gulf of Kara-Bugaz is at first impressive, but transient early Pleistocene downwarps of the Mediterranean could also very well have contributed to extreme conditions of evaporation at times. Pluvial episodes are suggested by deep canyons eroded into the cretaceous dolomites and limestones bordering the graben. They call this period the "Salt Unit Stage", which "ended with a change from a dry to a more humid climate".

According to these authors, a transitional phase is followed by the so-called "Lisan Stage"; remnants of the typical Lisan marl are also well preserved in its highest regions. The highest level of Lisan Lake was 180 meters below s. l., and it was at least 190 meters deep. It was 220 kilometers long and probably less than 17 kilometers wide. Including the transitional period from the "Salt Unit Stage", the approximate period of 100,000 to 20,000 years B. P. is assumed for the "Lisan Stage". A schematic section of the Dead Sea from north to south shows that the Lisan formation, especially in the deeper northern basin of the sea (the present bottom is at 790 m b. s. l.), later was subject to strong tectonic downwarping (D. NEEV and K. O. EMERY, 1966, p. 6).

According to the climate curve (cf. Plate VI), there was low precipitation in the transitional period from 100,000 to 70,000 years B. P. The level of Lake Lisan reached a maximum level of 180 meters below m. s. l. at about 25,000 years B. P. D. NEEV and W. O. EMERY suppose a more or less humid phase between 50,000 and 10,000 years B. P. Around 20,000 years B. P. the climate became drier and

Lake Lisan shrank to form the Dead Sea (D. NEEV and K. O. EMERY, 1967, p. 26). This is surely mainly due to tectonic subsidence of the northern basin, but climatic conditions also doubtless play a part, as the course of the curve in more recent time shows. Stages with a lower sea level show terraces and cliffs in the Lisan formation and the erosion of canyons. D. NEEV and K. O. EMERY assume that both the northern and southern basins subsided near the end of the Lisan Stage.

By the end of the Lisan Stage the waters of the graben became restricted to isolated depressions. The north and south basins became divided at some time during the post-Lisan Stage (e. g. at the time of the Roman Empire). It is difficult to differentiate climatic and tectonic effects. It is interesting that the climate curve from 5,500 to 4,500 years B. P. again shows a pronounced, though short moist period, which certainly is in accordance with the altithermal described.

B. MESSERLI (1980) provides the newest and most comprehensive summary of climatic history in the high mountains of middle and **northern Africa;** he is particularly concerned with the effects of the last Ice Age. As far as the problems in Saudi Arabia are concerned, the climatic changes in the mountains of the Sahel Zone and the Sahara on the one hand, and certain relatively well-known lakes (Turkana or Rudolf Lake, lakes in the Ethiopian Danakil Desert and Lake Chad) on the other, are of interest.

In the Semien Mountains of northern Ethiopia, which B. MESSERLI describes as the mountains of the "Sahel Zone" (p. 73), there were no monsoons in the high and late Würm Age, so that the snow line was higher than in the southern equatorial mountains. It was a dry zone between areas with summer monsoons from the south and occasional cold waves from the north.

The central mountains of the Sahara are quite another matter. Old nivation and periglacial forms may be seen there at heights far below 2,000 meters a. s. l. The Young Pleistocene periglacial forms suggest winter temperatures 10°–14° C colder than today; for the summer temperatures B. MESSERLI (loc. cit.) assumes a lesser value not exceeding 6°–8° C. The extensive periglacial developments, however, would have required considerably higher humidity than is available today. This was provided in the form of cold waves from the north in the winter months.

All in all, B. MESSERLI believes that the **Southern Sahara** was extremely dry during the period of maximal cold (dune formation up to 10° N Lat., i. e. south of Lake Chad), while northern Africa had a moister climate than today, with heavier hibernal precipitation and generally dry summers. R. W. FAIRBRIDGE (1964) also supposes that the Sahara dunes were at least 1,000 km farther south.

D. JAEKEL (1977) concludes from terrace studies that these precipitation periods only reached the central Sahara after the maximum cold (about 16,000 years B. P.); this would, however, again open the question as to the time of optimal development of the nivation and periglacial forms.

The question is also open as to what happened during the climatic transition from the north pluvial of the Youngest Pleistocene to the quite effective summer rains at the beginning of the Holocene moist phase. When we consider Lake Mundafan during the last Ice Age we will again encounter this problem in Saudi Arabia.

Comparisons with the best-studied steppe lake in northern Africa, Lake Chad, are inevitable.

The **Chad Basin** only began to subside at the end of the Tertiary; as far as climate is concerned, it varies from total desert in the north to alluvial landscape with plentiful rain in the south. Lake Chad is thus a steppe lake, and its nature and size underwent both tectonic and climatic influences in the Quaternary.

In the last 60 years, Lake Chad which takes up the deepest part of the basin, has changed with regard to the size, form and distribution of its reed belt and water surface by thousands of square kilometers; this is mainly due to the influence of the Chari River system coming from the southeast[1].

Figure 121 shows the water levels of Lake Chad during the last Pleistocene in the Neolithic pluvial (Altithermal) and at present (B. MESSERLI, 1980, p. 80, simplified). B. MESSERLI does not describe his measurements, but he is certainly correct when he speaks of "water-level variations"; his contours of the paleo-Chad water level shown in Fig. 121, however, only indicate that there has been a step-wise retreat of the lake surface from the last Ice Age to the present. That the "pluvial periods" could at most have been static phases is due to the fact that annual evaporation exceeded annual precipitation which in most years was not compensated for by the perennial flows from the south.

R. W. FAIRBRIDGE (1964 p. 404ff) also brings the silting and erosional phases of the Nile into the discussion. First, there are the terraces around Wadi Halfa; their silt masses are placed in the Late Würm (so-called "sebil sedimentation"). Before the sebil sedimentation, R. W. FAIRBRIDGE assumes a wide erosion valley in the mid-Nile section incised down into the rock bed and up to 2–3 kilometers wide, which at certain times "took up very large water masses, perhaps five times as much as the present maximum amount of water". After the period from 25,000–20,000 there were profound changes in the river's regime. Above all, R. W. FAIRBRIDGE protests vehemently the – long discarded – classical pluvial-glacial correlation, whereby it is only possible to agree with him completely.

R. W. FAIRBRIDGE can not, however, give absolutely precise dates for runoff variations for the Nile; for one thing, because the Nile is 6,670 km long, the catchment areas of the White and Blue Nile vary considerably in altitude and the river flows through different climatic zones in every climatic situation.

5.3.3. Climatic Conditions Inland in Saudi Arabia

What used to be Mundafan Lake at the beginning of the Arabian Shelf some 300 kilometers east of Abha in Saudi Arabia is dry today and the area it formerly covered is obscured by dunes.

At 18°30′ N, the former lake is at a geographical latitude between Lake Chad (about 13° N) and the Tibesti Mountains (about 20° N).

H. A. MCCLURE's systematic samples show that Mundafan Lake is a true product of Quaternary climatic fluctuations.

As was the case with the Altithermal, radio-carbon dating of MCCLURE's samples shows an uninterrupted sequence of the last high glacial (Table 53).

---

[1] Cf. Westermann, Lexikon der Geographie, vol. 4, p. 679, 1908, 1904, 1967.

Concerning the problematic of age determination based on very low content of ^{14}C the reader is referred to chapter 5.4.

Table 53. *Carbon-14 ages of marl and shells in samples from Lake Mundafan and the western Rub' Al Khali* (after H. A. McClure, 1976, Table 1)

	Site, author, sample No.		Age, yrs, B. P.	Material	coordinates
1	L. Mundafan, McClure	MF 1	17,460±245	marl, ca.	18°30′ N, 45°15′ E
2	L. Mundafan, McClure	MF 2	21,090±420	marl, ca.	18°30′ N, 45°15′ E
3	L. Mundafan, McClure	MF 3	21,280±275	marl, ca.	18°30′ N, 45°15′ E
4	Ar Rub'Al Khali McClure	Site 1	21,400±450	marl, ca.	10°20′ N, 46°20′ E
5	L. Mundafan, McClure	MF 4	22,345±415	marl, ca.	18°30′ N, 45°15′ E
6	L. Mundafan, McClure	MF 5	22,965±390	marl, ca.	18°30′ N, 45°15′ E
7	L. Mundafan, McClure	MF 6	23,075±425	marl, ca.	18°30′ N, 45°15′ E
8	L. Mundafan, McClure	MF 7	24,145±400	marl, ca.	18°30′ N, 45°15′ E
9	L. Mundafan, McClure	MF 8	25,660±800	algal limestone	
10	Ar Rub'Al Khali McClure	Site 2	27,160±940	marl, ca.	21° N, 49°10′ E
11	L. Mundafan, McClure	MF 9	28,750±615	marl, ca.	18°30′ N, 45°15′ E
12	L. Mundafan, McClure	MF 10	29,595±780	shells ca.	18°30′ N, 45°15′ E
13	Ar Rub'Al Khali McClure	Site 3	29,660±1 400	shells ca.	18°20′ N, 47°55′ E
14	L. Mundafan, McClure	MF 11	36,300±2,400	shells ca.	18°30′ N, 45°15′ E
15	L. Mundafan, McClure	MF 22	11,465±115	calcareous siltstone	
16	L. Mundafan, McClure	MF 23	14,965±195	marl, ca.	18°30′ N, 45°15′ E

These data seem to indicate that Mundafan Lake existed in the Youngest Pleistocene parallel to the Würm Cold period and that Saudi Arabia was subjected to its effects.

H. A. McClure (1976, 755–756) believes that the pile-up of data between 30,000 and 21,000 years B. P. indicates a high-water mark for the lake at that time. That this lake had a higher water level than it did in the neolithic pluvial is shown by the fossils and deposits in the exposed layers some 24 m thick at its higher levels.

H. A. McClure (1976) sees the catchment area in the backslope of the eastward-dipping Wajid Plateau and the foreslope of the Jabal Tuwaiq.

The remnants of surface-water deposits found far to the east in the Rub' Al Khali, however, indicate a larger catchment area and runoffs of longer duration.

During the Würm period, the highland on the Shelf also had open outlets to the east. For the highest-lying areas (up to 3,700 m a. s. l. in Yemen), it may be assumed that there were glaciers at the maximum of the last Ice Age (nocturnal frosts are currently increasing at altitudes of 1,800 to 3,700 m). Old periglacial and nivation forms (slope displacement, etc.) may be seen in the area southeast of Abha (above 3,000 m a. s. l.).

The wadis of the Asir Highland still bear water from fracture springs. It is virtually certain that at the snow melt the lower wadi region from Wadi Bisha to Wadi Najran sometimes carried plentiful water and contributed to the existence of desert lakes in the western Rub' Al Khali. There was and is (?) underflow of dune areas in surfaces beds covered by dunes. It is still assumed that these groundwater flows from the Yemenite eastern slope of the Shield replenish the groundwater in Wadi Hadramaut and the Rub' Al Khali.

The lower part of Wadi Bisha (before it joins Wadi Tathlith to form Wadi Ad Dawasir) and Wadi Ad Dawasir were described in Volume 1. Wadi Ranyah is a tributary on the left-hand side of Wadi Bisha. Two samples of weathered limy material overlying basalt at 0.7 and 1.0 meter below surface were collected from the Mugabil loam pit in Wadi Ranyah; ^{14}C measurements showed ages of 26,400±1 970 and 29,840±2 600 years B. P., respectively (see Vol. 1, Table 44).

Table 54 shows that, insofar as it was possible to take samples from the Shield area and the transition zone to the Shelf platform many of the sinter and crust formations, which are dependent upon heavier precipitation, show ^{14}C ages between 33,000 and 9,000 years B. P. The agreement of many samples with McClure's ^{14}C data for Mundafan Lake can not be overlooked; this would confirm the assumption of a moist period in the Youngest Pleistocene for the area of both the Shelf and the Shield, but concerning a possible contamination of the ^{14}C material by a younger precipitation the reader again is referred to chapter 5.4.

The possibility is not to be excluded that at this time Wadi Ad Dawasir with its tributaries had a "semi-superficial runoff" into the then dry Arabian Gulf between Qatar and Oman. What is meant by this is a freshwater runoff from the upper Najran – western Rub'Al Khali area (approx. 1,200 m a. s. l.), first flowing along alluvial dunes and then alternately under the giant dunes of the middle and northeastern Rub'Al Khali and on the surface between them. This assumption is supported by a number of facts. First, is the fact that there is still flow under the dunes from the runoff from the evaporation pans of the Al Hasa Oasis drainage system (Vol. 1, 61, fig. 12) to the coast of the Arabian Gulf between Dhahran and Qatar under the intermediate dune sediments (see Fig. 123). The wind-sorted

Fig. 123. Open outflow of Al Hasa seepage water from evaporation pans flowing as underground river ca. 70 kilometers under Jafurah dunes into the Arabian Gulf. (Photo: J. G. Zötl, 1974.)

Table 54. *Results of ^{13}C and ^{14}C measurements on samples from the Wadi As Sirhan, the Hijaz and the Asir highlands. Analysis performed by GSF – Institut für Radiohydrometrie, Munich–Neuherberg, Federal Republic of Germany*

Locality	Geographic position		Material	IRM Lab. No.	δ ^{13}C (‰ PDB)	^{14}C-content (% mod.)	Assumed initial ^{14}C content (% mod.)	^{14}C age uncorr. (years B. P.)
Wadi As Sirhan, Manwa 25 km E of Qurayyat	lat long	31°23′ 37°35′	layered sinter limestone from sinter basin	7666	+1.1	2.3±0.5	85	$29,000^{+2,000}_{-1,600}$
Wadi As Sirhan Al Qargar 23 km E of Qurayyat	lat long	31°22′ 37°06′	pisolite from a sinter layer	7671	+1.2	<1.3	85	>33,800
Wadi As Sirhan Al Jufayrat 40 km SE of Qurayyat	lat long	31°13′ 37°43′	sinter limestone from a small bay of the basalt occurrence	7672	−0.1	0.9	85	>36,800
Hulayfa Upper Wadi Ar Rimah	lat long	26°00′ 40°46′	calcrete from a weathering profile of a basalt	5907	+1.0	<1.1	100	>36,500
Wadi Al Gharas Bir Al Afariyah 24 km S of Khaybar	lat long	25°31′ 39°18′	calcrete from weathering pockets in the surface of the basalt	5910	0.0	<0.5	100	>42.900
Hanakiyah 100 km ENE Al Madinah	lat long	24°50′ 40°01′	calcrete from the alluvial wadi filling 8.5 m below surface	5908 b	+1.2	1.6±0.4	100	$33,400^{+2,300}_{-1,800}$
Hanakiyah 100 km ENE Al Madinah	lat long	24°50′ 40°00′	calcrete from a thin weathering layer on top of the basalt	5909 a	+2.1	1.8±0.5	100	$32,500^{+2,400}_{-1,900}$
Harrat Al Kishb Southern appendix near Bir Khuwarah	lat long	22°25′ 41°19′	sinter layer from the wadi floor, cut into a basalt flow	5914	+1.6	2.3±1.0	100	$30,300^{+4,600}_{-2,900}$
Harrat Nawasif western margin Wadi Jakrah 5 km SE Turabah	lat long	21°13′ 41°41′	calcrete from the top of a flat river terrace	5915	−2.3	5.3±0.8	100	$23,600^{+1,300}_{-1,100}$
Harrat As Sarat western margin 4 km SE Usran	lat long	18°02′ 43°06′	calcrete from a small channel on the basalt slope	7627	−2.1	28.5±1.3	85 100	8,800±400 10,100±400

dune sediments differ considerably in their permeability from the less porous underground. Further, even in the current period of extreme aridity, there is the probability of a groundwater stream through the Rub'Al Khali, running in the direction of the old wadi system. Evidence for this includes both wells (Bir Ad Dagma, Bir Umm Al Hadid, Farayah, Kawr Mahyubah, Qalamat Al Juhaysh and others; see map 1:2,000,000) and artificial tappings of water just below the dune floor whose capillary ascent and evaporation make possible the formation of the cemented salt-clay soils.

5.3.4. Sea-level Changes and Shorelines of Saudi Arabia During the Würm

As was the case with the Altithermal, the Youngest Pleistocene also raises the question as to eustatic fluctuations in sea level.

As was shown in Volume 1 (p. 53, fig. 9, after M. SARNTHEIN, 1972), the Arabian Gulf was dry at the time of the high glacial and the 100-meters isobath reached to just above the Strait of Hormuz. Today, the 50-meters isobath is a morphologically distinct step, but we do not yet know whether it indicates a static phase before a rise or fall in sea level, or an interstadial epoch. A tectonic effect is also not to be excluded.

The ^{14}C ages of the corals and shell banks sampled along the coast of the Red Sea provide a definite indication of tectonic movement. Coral banks aged between >43,000 and 30,000 years B. P. (cf. Table 4 and 18) are found above the surface of the Red Sea from its southernmost to northernmost reaches. If we do not view this as an indication of an upheaval after this time, then the level of the Red Sea would have had to have increased during the Würm; this was hardly the case during this entire period of time, even if we assume a powerful interstadial epoch.

5.3.5. Establishing the Age of Underground Water with Isotope Studies

Tritium (^{3}H) and carbon-14 (^{14}C) measurements were used for direct age determination of water samples from the area of study in Saudi Arabia.

From the concentration of the radioactive heavy hydrogen isotopes in groundwater (**tritium content,** see Vol. 1, 156ff) that had been formed from infiltrated rain water, water ages up to about 50 years can be determined depending on the short half-life of ^{3}H, the ^{3}H input concentration, and the detection limit of the analytical technique applied, and using the piston-flow model for interpretation. This can be a valuable help for water-balance calculations in humid areas, but deeper and older groundwaters no longer contain any tritium.

For decades attempts have been made to use **carbon-14** measurements (see Vol. 1, 159ff) to assess longer residence times of underground water. Although these methods have been considerably improved in the course of time, unlike measurements of organic material the results are often difficult to interprete and only given as ^{14}C content in % modern, and not as an age in years B. P. The uncertainty stems, among other things, from the difficulties connected with the determination of the initial ^{14}C content of the groundwater to be dated, and from mixture with other waters. Whilst, for instance, the measurement data from waters flowing through noncarbonate rocks approximate the real ages, the uptake of "dead" carbon by the limestone solution during water transport through the aquifer usually cannot be determined correctly. The measurements available nonetheless give times ranges that provide at least approximate information.

Back in 1961, ARAMCO ordered ^{14}C measurements of water samples (L. THATCHER et al., 1961); Table 55 shows the ARAMCO data and Table 56 C. JOB's summary (1973) of C-14 measurements for the oases Al Qatif and Al Hasa.

Table 55. *^{14}C data of groundwater. Samples of ARAMCO (1961)*

Sample No.	Water well	Depth of sample (ft)	Age of aquifer	Min. distance to outcrop (km)	Water Temp. (°F)	Water Age (yr)
W-904	Town well, Buraydah	1,250	Cambrian and Ordovician	24	100	20,400±500
W-889	Riyadh, WW 180	3,647 to 3,974	Triassic or Jurassic	60	126	24,630±500
W-897	Khurays, WW 8	1,490 to 1,693	Cretaceous	70	80	20,760±500
W-894	Abqaiq, WW 32	3,003 to 3,402	Cretaceous	250	134	22,500±500
W-888	St. WW 7	1,617 to 3,035	Jurassic and Cretaceous	75	100	>33,000
W-887	St. WW 13	3,435 to 3,506	Permian	200	98	>33,000

Table 56. *^{14}C data of confined groundwater. Samples of C. JOB (1973)*

Place or No. of sample	^{14}C content[1] (% modern)	^{14}C age[2] (years B. P., uncorrected)	$\delta^{13}C$[3] (‰)	Tritium content (T. U.)
'Ayn Al Labaniyah I (Al Qatif)	<5.9	>22,000	− 9.0	<0.8
'Ayn Al Labaniyah II (Al Qatif)	<1.2	>34,500	− 8.8	<2.7
No. 32 (Al Hasa)	<1.4	>33,000	−10.3	<0.9
'Ayn Mansur (Al Hasa)	<1.4	>33,000	−10.0	<0.9

[1] Sampling through carbonate precipitation by C. JOB, 1973, and isotope analyses performed by GSF – Institut für Radiohydrometrie, Munich–Neuherberg.
[2] An initial ^{14}C content of 85% modern was assumed.
[3] The ^{13}C content is given as the relative per mill deviation from the limestone standard PDB.

If we look at Thatcher's data, we cannot find a dependence of the ^{14}C age on the depth of the bore, or on the distance to the outcrop of the aquifer drilled into. Samples W-897 and W-894 came from rather different locations, but they probably come closest to the actual age data.

Unlike Thatcher's samples, those published by C. JOB in 1973 came not from bores but from free-flowing springs in the Al Qatif and Al Hasa oases (cf. Vol. 1,

p. 160). While it may be assumed that the Ayn Al Labaniyah I spring contains younger waters from another aquifer (owing to the content of stable isotopes), the water in the samples from Al Hasa came from the so-called Umm Er Radhuma Formation, whose north-south trending outcrops and recharge area are some 150–200 kilometers west of the springs. This aquifer has the most abundant confined water in the cuesta landscape of the Shelf.

The springs of Al Hasa are the largest on the Saudi-Arabian Platform and are artesian springs from the aquifer lying at a depth of some 280–600 m. The Ministry of Agriculture in Ar Riyadh has a collection of ^{14}C data from these wells with a total flow of some 12 cbm/sec, which agree generally with the data in Table 56. With all due consideration of all the open questions regarding age determination of water with ^{14}C measurement, it is possible that the waters from the Umm Er Radhuma Formation are precipitation infiltrates from the Würm Age; we can not say, however, whether they belong to moist phases of the Early or Late Würm, or both. But there can also be no doubt that large amounts of rainwater infiltrated, as borings were also made in the Umm Er Radhuma Aquifer farther to the south and the large bore holes in the King Faisal Project alone in Haradh provide additional and considerable amounts of water (52 wells; the Umm Er Radhuna formation in this area is between 87 and 223 m subsurface).

Measurement of the stable isotope shows that it is unlikely that the stored underground waters did not come from the Youngest Pleistocene.

A large number of water samples from all the areas studied was examined for content of the stable hydrogen and oxygen isotopes **deuterium and oxygen-18.**

Preliminary research results in Austria showed that deuterium content in the form of a temperature effect can express itself not only as an altitudinal or continental effect, but in accordance with the ^{14}C-data on confined groundwater, also as a climatic effect. Without going into unnecessary detail, fig. 50, vol. 1 shows that this climatic effect is apparently involved in the deep waters of the Arabian Platform.

Looking at the $\delta D/\delta^{18}O$ relationship (cf. Vol. 1, p. 155), we see that the values of the Al Hasa and Al Qatif waters lie distincly below the line for $\delta D = 8\delta^{18}O+10$, valid for temperate climates, e. g. for Middle Europe. There can be no doubt that the equation line for the Al Hasa and Al Qatif water samples is based upon a climatic effect. The waters currently discharging there came from precipitation that evaporated from the sea under climatic conditions other than those prevailing today. The low t-values lead to the conclusion that the water vapor of this condensate was in better equilibrium with sea water than is currently the case. The evaporation rate was evidently determined by a cooler and more humid climate than the present one. This conclusion is confirmed by the carbon-14 measurement described above, and the absence of tritium. All in all, these facts definitely indicate that these waters came from the Youngest Pleistocene, and that here as well, this period was certainly cooler and moister, although the distribution and overlapping of temperate and humid phases has still not been determined.

5.4. Problems Involved in ^{14}C Age Determinations in Carbonates

(H. HÖTZL, H. MOSER, W. RAUERT, M. WOLF, J. G. ZÖTL)

In arid regions, carbonate relicts of organic and inorganic origin are often the only materials available for dating, and the ^{14}C method is that used to establish their ages. Organic material can be dated relatively precisely, but there are still unsolved problems and uncertainties with measurements of carbon precipitated from **water** samples, and also the ^{14}C model age **for inorganic material** may vary considerably from the actual age. This is generally because the initial ^{14}C content is insufficiently known, and/or because of unrecognized contamination; this will be dealt with in the following. For more detailed information on the methods and problems involved in ^{14}C age determination, the reader is referred to such works as M. A. GEYH (1971) and H. N. MICHAEL and E. K. RALPH (1971).

5.4.1. ^{14}C Model Age

The given uncorrected ^{14}C model ages A (in years before present) were calculated as agreed with a ^{14}C half-life of 5,568 years according to formula (1), which is a result of the law of radioactive decay:

$$A = 8033 \cdot \ln (C_o/C). \qquad (1)$$

Here, C_o is the initial ^{14}C concentration, and C the measured ^{14}C concentration in the sample in percent modern[1]. To simplify matters, depending on their origin, samples were assumed to have $C_o = 100\%$ modern or $C_o = 85\%$ modern. Using the more precise ^{14}C half-life of 5,730 years increases the ^{14}C model ages by approximately 3%[1].

^{13}C corrections were not made. The uncorrected ^{14}C model age is given together with the double standard deviation.

5.4.2. Initial ^{14}C Content

The initial ^{14}C content required to calculate the ^{14}C model age can be altered by various influences. It is primarily dependent on the ^{14}C concentration in atmospheric CO_2, which has been reconstructed for the last 8,000 years by precision measurements of the ^{14}C content of dendrochronologically dated tree rings (H. E. SUESS, 1980). The given uncorrected ^{14}C model ages for the last 8,000 years can vary from the actual age by up to 1,000 years. Larger fluctuations in the atmospheric ^{14}C concentration may have occurred in the last 50,000 years; M. BARBETTI (1980) has estimated them from changes in geomagnetic field strength. In this case, the calculated uncorrected ^{14}C model ages could vary from the actual age by as much as some thousands of years.

Physical, chemical and biological processes involving atmospheric CO_2 generally produce isotope fractionations that can be recognized by determination of the $^{13}C/^{12}C$ isotope ratio. On the basis of theoretical considerations, something like the double amount can be derived for the change in the $^{14}C/^{12}C$ isotope ratio.

---

[1] 100% modern $\hat{=}$ 13.56 ^{14}C nuclear decays per minute and gram of carbon; this is a near approximation of the natural ^{14}C concentration of living organisms (after application of various corrections).

The age shifts caused by these effects can be corrected when the respective $\delta^{13}C$ value is known, but as a rule they amount only to some decades or hundreds of years.

The so-called "reservoir effect" can have a further influence on the initial ^{14}C content. What is meant by this is the difference in ^{14}C concentration e. g. in the medium surrounding a living organism (e. g. sea water, lake water), and in atmospheric CO_2. This effect also usually leads to a lower ^{14}C content for the organism living in the medium. Thus, for example, I. U. OLSSON (1974), Y. NOZAKI et al. (1978) and E. M. DRUFFEL (1980) found apparent ^{14}C ages of some hundreds of years for recent shells and corals, as the ^{14}C concentration of the sea water in the area under study was several percents modern lower than the ^{14}C concentration in atmospheric CO_2.

Considerably larger age shifts (up to several thousands of years) can occur with ground- and sea water, whose initial ^{14}C content can basically amount to 50–100% modern. Here, geochemical influences and isotope-exchange processes play an important role. In an environment containing carbonate, the HCO_3^- ions dissolved in water are usually formed by dissolution of usually ^{14}C-free fossil carbonate rocks under the influence of biogenic CO_2. Depending on CO_2 partial pressure, temperature and isotope exchange, there is then a more or less pronounced decrease in the original biogenic ^{14}C concentration. Equation (2) gives an example for dissolution of calcite or aragonit:

$$CaCO_3 + x\,CO_2 + H_2O \rightarrow Ca^{2+} + 2\,HCO_3^- + (x\text{-}1)CO_2. \quad (2)$$

In contrast, in a carbonate-free environment the HCO_3^- ions are formed by weathering of silicate rock. Equation (3) gives an example for the decay of potash feldspar:

$$2\,KAlSi_3O_8 + y\,CO_2 + H_2O \rightarrow$$
$$2\,K^+ + 2\,HCO_3^- + (y-2)\,CO_2 + 4\,H_4SiO_4 + Al_2Si_2O_5(OH)_4. \quad (3)$$

In this case there is, disregarding isotope fractionations, no change in the ^{14}C concentration in the water as compared to biogenic CO_2. Inorganic precipitates (e. g. tufaceous limestone, calc-tufa) from the water contain nearly the same initial ^{14}C concentration as the water sample they came from, and can therefore have an initial ^{14}C content within the 50–100% modern range. As examples, J. LABEYRIE et al. (1967) found an initial ^{14}C content of 65% modern in a stalactite, and D. SRDOČ et al. (1980) an initial ^{14}C content of 85% modern in a tufaceous limestone. In special geological situations, as, for example, under the influence of fossil CO_2, or formation from water with a slight ^{14}C content, the initial ^{14}C content can also be considerably lower (M. A. GEYH, 1970).

5.4.3. Contamination

In the ^{14}C dating of inorganic material, contamination of the sample material by atmospheric or biogenic CO_2, detrital carbonates and cementation with carbonates can play an important role. Depending on the ^{14}C concentration of the contaminating substance, the measured ^{14}C model age can then be greater or lesser than the actual age. This is especially true of older material, where even slight contamination with recent material can lead to considerable age shifts (Fig. 124).

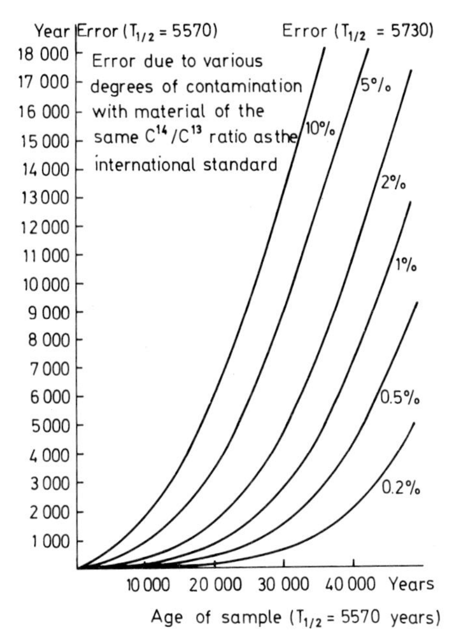

Fig. 124. Age shifts caused by contamination with recent material (ex: I. U. OLSSON et al., 1968.)

Laboratory work by I. U. OLSSON et al. (1968) shows that it is very difficult to obtain reliably datable samples from material that is more than 25,000 years old. The contamination can have been caused by CO_2 and depends on the grain size and storage conditions of the samples. To prevent contamination, it is suggested that samples be stored in sealed glass jars. A comparison of measured ^{14}C model ages of corals and molluscs from New Guinea with the expected ages showed age differences of as much as several thousands of years (J. CHAPPELL, H. A. POLACH, 1972). With virtually constant $\delta^{13}C$ values for the samples, large age shifts were generally correlated with high degrees of recrystallization. In recrystallization, in these cases, aragonite, which under normal conditions is thermodynamically unstable, is changed for example under the effect of H_2O and CO_2 into the thermodynamically stable calcite (equation 4):

$$CaCO_3 \text{ (aragonite)} + H_2O + CO_2 \rightarrow Ca^{2+} + 2\,HCO_3^- \rightarrow CaCO_3 \text{ (calcite)} + H_2O + CO_2. \quad (4)$$

High-Mg calcite, which under normal conditions is also unstable can in a similar manner be transformed into the more stable low-Mg calcite. Depending on whether the recrystallization takes place with preference in an open or closed system, the contamination is greater or lesser and can not therefore be correlated exactly with the degree of recrystallization. This research shows that reliable ^{14}C dating can only be performed on unchanged material. This can be checked by X-ray analysis. According to T. L. GRANT-TAYLOR (1972), ^{14}C datings should only be performed on aragonitic material with a calcite content below 1%; here, both X-ray analysis and infrared analysis can be used to check the results. C. VITA-

FINZI (1982) sometimes also uses the scanning electron microscope (cf. also B. M. WALKER, 1979) and determines the Mg and Sr concentrations in the sample objects. A very recent study by W.-Chr. DULLO (1983) showed beyond doubt diagenetic changes in Red-Sea coral reefs. Here, in addition to vadose diagenesis owing to meteoric seepage water, marine diagenesis is to be found, especially among corals. This leads to cementation with aragonite or Mg calcite and sometimes even to biological carbonate extraction by endolithic organisms, whose effects on ^{14}C content are difficult to estimate. The ^{14}C dating of paleosoils is on the one hand especially problematic owing to a detrital carbonate component that is not easily recognizable (G. E. WILLIAMS, H. A. POLACH, 1971) and can lead to an excess age of 500–7,000 years. These same authors found, on the other hand, ages that were too small owing to contamination with atmospheric CO_2. For the interpretation of ^{14}C analyses for soils as well, precise knowledge of their development and diagenetic changes is necessary.

The problem of young contamination is shown with some selected ^{14}C-results from the Tables 4 and 18 and put now together in Table 57. From our understanding of the rather weak tectonic activity at the Red Sea coast we have to assume that the constant levels of beach terraces over long distances are a consequence of eustatic sea level changes. Therefore the terraces represents interglacial sea water high marks. The age determinations of samples No. 7634, 7631 or 7659 from the higher terraces with values older than 34,000 respectively 37,000 years are in agreement with such a finding. In contrary sample No. 7639 also from the 10 m-terrace where it is covered addionaly by an basalt flow dated with K-Ar-method (0.4±0.2 Mill. years) gives an ^{14}C-age of approximately 30,000 years B. P.

Similar ages are received for a Miocene algal limestone (No. 5911: $35{,}500^{+7{,}000}_{-3{,}700}$ years B. P.) and for middle or older Pleistocene coral reed from Farasan Al Kabir (No. 7619: $35{,}500^{+3{,}500}_{-2{,}400}$ years B. P.). In the latter case a second sample (No. 7618) from the top of this old reef (^{14}C-age: 17,500±500) even gives the false impression of a youngest Pleistocene profile sequence. Here it's obviously that the contamination caused by several possible influences from the surface is larger in the upper layers than in the more protected deeper parts of the profile.

Special attention we have to pay to the possible contamination of soil material as was mentioned above. In this light we have to see the ^{14}C-age of the calcrete sample No. 5908b, where alterations may be caused by precipitation from groundwater. On the other hand a calcrete from the top layer (No. 5910) exposed to the different weathering conditions on the surface, practically does not show any contamination.

But this is not the normal way. In general it seems that infiltration of rain water and subsequent solution as well as precipitation processes is a main source for alterations, which above all effect the upper soil layers. We believe that just the rains of the wetter early Holocene phase with denser vegetation, changing soil forming conditions and higher ground water levels may be responsible for the increase of ^{14}C content in distinct older sediments. The sinterlimestone of sample No. 7622 was formed by the discharge of these Holocene precipitations while the older sinter (No. 7621) with strong recrystallisation may habe received its recent ^{14}C-content by processes caused from the infiltration of the same precipitations.

As far as it could be controlled by the general geologic situation, e. g. pre-Würm age of the older eustatic seaterraces, the ^{14}C contamination is less than 5% modern but may rise up under extrem conditions to more than 10% modern (cf. sample No. 7618, Table 57).

5.4.4. Use of ^{14}C-dating?

The remarks on the Youngest Pleistocene are mainly based on ^{14}C data. The discussion of the problems that this method still causes was thus appended to the chapter on the Würm. Here are a few final notes on the subject.

The main sources of error with these measurements are from inorganic skeletal material (problems of physical, chemical and biological processes) as well as carbonate precipitates from water. The latter are now often limited to data on the ^{14}C content in percent modern and often questionable through mixing problems.

But factors should be mentioned with regard to the doubtfulness of the data from inorganic material that limit the probability of a major deviation of ^{14}C data from the actual age. First of all, there are the **similar results** from large **regions of different geological structure.** There are, for example, similar ^{14}C data of less problematic material for a cool and humid phase lasting for some thousands of years in the period between 20,000 and 35,000 B. P. in the entire Middle East and North Africa. If these deposits then are found in **exact chronological series** (Lake Mundafan, Table 52, McClure, 1976) and ultimately show a completely different phenomenon, i. e. the study of the Ad Dahna dunes (D. Anton, 1980) showing that their movement through semi-arid circumstances in this period was inter rupted by vegetation, then this might be an indication together with other geologic and climatic facts that the ^{14}C datings are more or less correct. It may thus be assumed that the ^{14}C ages assumed for this area correspond to the actual climatic course to about 30,000 to 25,000 years B. P., whereby the Holocene Altithermal is not questioned. When dealing with samples older than about 25,000 years B. P., the contamination problem represents a serious limitation of the ^{14}C dating technique and has to be considered during future investigations in even more detail.

The situation is different for the data shown in Table 53, and the studies of water samples described in Table 54. The samples taken in Al Qatif and Al Hasa show values of the stable isotopes deuterium and ^{18}oxygen indicating cooler air (and water) temperatures for the period of precipitation, or when the water percolated. It can not, however, be said whether this period was in the younger or older Würm. These waters rising mainly from the Umm Er Radhuma Formation can most probably be attributed to the last 100,000 years, but also an older age can not be excluded.

Plate V (see insertion at back cover) shows a comparison of a selection of climatic curves from the beginning of the last Ice Age to the present. In spite of all the differences in details, it is apparent that in general there were related climatic variations occurring from northwestern Europe over Middle Europe, the Mediterranean area and Middle East into southwestern Saudi Arabia.

Table 57. *Selected results of ^{14}C and ^{13}C measurements on samples from the western part of Saudi Arabia. Data mainly extracted from table 4, 17 and 18 for special comparison. Analysis performed by GSF – Institut für Radiohydrometrie, Munich–Neuherberg, Federal Republic of Germany*

Lab. No.	Locality	Sample description	$\delta^{13}C$ (‰ PDB)	^{14}C content (% modern)	Assumed initial ^{14}C content C_0 (% mod.)	^{14}C age uncorr. (years B. P.)	Remarks concerning alteration and contamination
16353	Ash Sheykh Humayd	*Platigyra* from recent corals reef 0.8 m below mean sea level	+1.5	116.2±3.5	100	recent	unaltered recent coral
7634	Al Wajh	massive coral from 3 m-terrace, 300 m N of Al Wajh airport	−2.5	<1.0	100	>37,000	diagenetic recrystallization of aragonite to calcite; contamination possible by infiltration of rain; exposed to weathering; **age: pre-Würm**
7631	Al Wajh	massive corals *(Platigyra)* from 10 m-terrace, 300 m N of Al Wajh airport	−1.4	<1.0	100	>37,000	diagenetic recrystallization of aragonite to calcite; contamination possible by infiltration of rain; exposed to weathering; **age: probably pre-Riss**
7616	Farasan Al Kabir	shell agglomeration of young beach sediments	+3.3	55.3±2.8	100	4,800±400	no or only small recrystallization; exposed to weathering; **age: Flandrian or younger**
7659	Ash Sheykh Humayd	*Tridacna* shell from 12 m-terrace of Gulf of Aqaba 20 km N of Ras Ash Sheikh Humayd	+2.8	<1.4	100	>34,800	little recrystallization, contamination possible by infiltration of rain; exposed to weathering; **age: pre-Würm**
7618	Farasan Al Kabir	massive coral *(Platigyra)* from the top of the upper reef layer of Farasan near Hussein	−2.7	11.0±0.7	100	17,700±500	strong recrystallization (blocky calcite), contamination possible by infiltration of rain; exposed to weathering; **age: middle or older Pleistocene?**

7619	Farasan Al Kabir	massive corals *(Platigyra)* from the base of the conglomeratic lower coral layer of Farasan near Hussein	−0.4	1.5±0.5	100	$33{,}500^{+3{,}500}_{-2{,}400}$	strong recrystallization (blocky calcite), contamination possible by infiltration of rain; but more protected from weathering; **age: middle or older Pleistocene**
5911	Masturah	algal limestone, coastal plain 10 km south of Masturah	−6.0	1.2±0.7	100	$35{,}500^{+7{,}000}_{-3{,}700}$	strong recrystallization (blocky calcite), contamination possible by infiltration of rain; exposed to weathering; **age: Miocene**
7639	Umm Lajj	corals and shells from reef layer below basalt flow, 3 m above m. s. l., coast 3 km north of Umm Lajj	−0.8	2.4±0.6	100	$30{,}000^{+2{,}300}_{-1{,}800}$	diagenetic recrystallization, contamination possible by fresh and salt water; exposed to weathering; age of the superimposed basalt **0.4±0.2 m y.**
7622	Al Birk	young sinterlimestone from the wadi floor, 2 km south of Al Birk	−11.6	47.6±2.5	85 100	4,700±400 6,000±400	little recrystallization, contamination possible by surface water, exposed to weathering; **age: probably Flandrian**
7621	Al Birk	old sinterlimestone from the wadi terrace, 2 km south of Al Birk	−7.9	7.0±0.4	85 100	20,000±500 21,300±500	strong recrystallization (blocky calcite), contamination possible by surface water exposed to weathering; age: wadi terrace of **pre-Würm age (?)**
5908 b	Hanakiyah	calcrete from a layer *8.5 m* below surface	+1.2	1.6±0.4	100	$33{,}400^{+2{,}300}_{-1{,}800}$	possible recrystallization and additional precipitation, contamination possible by infiltration and groundwater; age?
5910	Khaybar	calcrete from top of weathered basalt, 24 km south of Khaybar	0.0	<0.5	100	>42,900	possible recrystallization and additional precipitation, contamination possible by rain and surface water; age?

5.5. Middle and Early Pleistocene

(H. Hötzl, J. G. Zötl)

5.5.1. The Beginning of the Pleistocene

The quantity and quality of information on the Middle and Early Pleistocene is not as good as that for the Youngest Pleistocene and the Holocene.

It makes sense to first consider the question of the beginning of the Pleistocene, or the transition from Tertiary (Pliocene) to Quaternary (Pleistocene) (cf. Vol. 1, p. 306ff).

W. A. Berggren and J. A. Van Couvering (1974) place this question in a larger framework, i. e. the Late Neogene. This widens the perspective. They compare intercontinental areas and fulfill the requirement of covering more than regional events.

These two authors relate the beginning of the Pleistocene not to the first lowland ice-sheet in Europe, but to the temperature drop associated with the Olduvai Normal Event about 1.8 million years ago.

J. D. Hays and W. A. Berggren (1971) summarized studies and theories dealing with the Pliocene/Pleistocene boundaries and correlations up to 1967. On the basis of these studies, the Pliocene/Pleistocene boundary was determined to be at about 1.85 million years B. P., as defined by the "first evolutionary appearance of *Globorotaria trumcatulinoides*" (W. A. Berggren et al., 1967; J. D. Hays, W. A. Berggren, J. A. Van Couvering, 1974, p. 80).

In the series of Pleistocene sea levels in the Mediterranean, this is somewhat under the base of the Calabrian Stage.

A series of further studies (e. g. K-Ar age, studies of volcanics) suggests the Pliocene/Pleistocene border can be narrowed down to a time range of 1.61–1.82 million years before present, placing it in the Olduvai-Gilsa normal magnetic event of the Matuyama inverse magnetization epoch. Vol. 1, Table 51 (p. 309) shows a geochronological correlation of the last three million years B. P. after W. A. Berggren and J. A. Van Couvering (1974).

As far as the beginning of the Pleistocene is concerned, it is important to establish, with the authors cited, that it is incorrect to define the Pliocene/Pleistocene border with the sentence "date the base of the Quaternary and you have dated the earliest glaciation". Suggestions as to where to locate this border range from one to three million years.

It is not impossible that future work could lead to a relocation of the Pliocene/Pleistocene border. What is certain is that the Quaternary Period is characterized by climatic variations with temperature differences leading to glacial and interglacial stages at higher latitudes and altitudes.

5.5.2. Climatic Variations

"Die Alpen im Eiszeitalter" the world-wide known publication by A. Penck and E. Brückner (1901–1909) provides fundamental information on climatic changes in the Quaternary. In this standard work, the authors demonstrate four chronologically separate glaciations in the Alpine foreland and river terraces on the basis of end-moraine dams. The four periods were named for end moraines near the small rivers Günz, Mindel, Riss and Würm in the German Alpine foreland, and they are still so called.

As in the Alpine region, the glacials in northern Europe are named for the rivers in the North German lowlands or in Poland that were reached by the inland ice covering Scandinavia at its greatest extent (Weichsel, Saale, Elster). In North America, the Ice Ages are named for the states in the United States touched by glaciers. Starting at the end moraines, terrace systems were created by melt-water runoff; these systems extended far into the foreland and the cool and moist climate of the higher latitudes helped to preserve them. Numerous local studies produced an abundance of detailed results that left open only the absolute dating of all these periods. The same is true for northern Europe and North America.

The landmarks left by changing sea levels provided supplementary new information on a completely different aspect. The long coasts of the Apennine Peninsula provided fertile ground for research, as indicated by the names Calabria, Emilia, Sicily, Milazzo and Tyrrhenia. Here confirmation was found for the fact that there must have been more than four cold periods in the Pleistocene. Today, research is assisted by such new fields as soil science, isotope measurement, and deep-sea studies.

It is much more difficult to find indications of older climatic changes in **arid areas.**

Expeditions and studies on the **Arabian Shelf platform** at first showed only datable changes in temperature and precipitation during the Holocene Altithermal (Neolithic pluvial) and the Youngest Pleistocene in the main wadi sytems and in beach marks on the Arabian Gulf.

Age determination of volcanic rocks with K-Ar measurements is a large help in the study of portions of Saudi Arabia, as long as the rocks are in contact with Quaternary deposits.

As was reported in Vol. 1, age determinations from two basalt flows and comparison of their morphology provided a limit for the period of transition between Tertiary and Quaternary. The climatic changes before the Youngest Pleistocene remained open.

Although many of the basalt flows are Tertiary in age and other young basalts (Al Madinah, 13th C.) provide little information on morphogenesis or classification for age and climate, there are several basalt flows in the coastal region of the Red Sea and its hinterland which are intercalated or superlaying terrestric and marine sequences. K-Ar dating of some of these flows permit to draw geochronological conclusions of supraregional importance. The results of the K-Ar age determinations are given in the Tables 5 and 19; the consequences for the stratigraphic position of the terraces are discussed in the chapters 2.2.3. and 2.4.2.

D. Anton (chapter 4) attempted to reconstruct the huge river systems on the Shelf that were fully active under the tropical hyper-humid circumstances prevailing during the Pliocene-Pleistocene border (Fig. 116). The beginning Pleistocene brought lower temperatures and precipitation, with desert formation in the lee of the southern monsoon, putting an end to the river system crisscrossing the subcontinent. When moisture returned, the tributaries were even less able than the rivers more than one thousand kilometers long to overcome the dune belt that in the meanwhile had developed. Here, the fact can not be overlooked that a moister climate also makes the dunes firmer (cf. D. ANTON). Wadis in non-desert areas were in fact retained, but conditions did not permit them to regain their fluvial function (episodic lake formation, cf. Vol. 1, p. 205).

Table 58. *Geochronological comparison of the Pleistocene in Saudi Arabia and some other regions of the northern hemisphere*

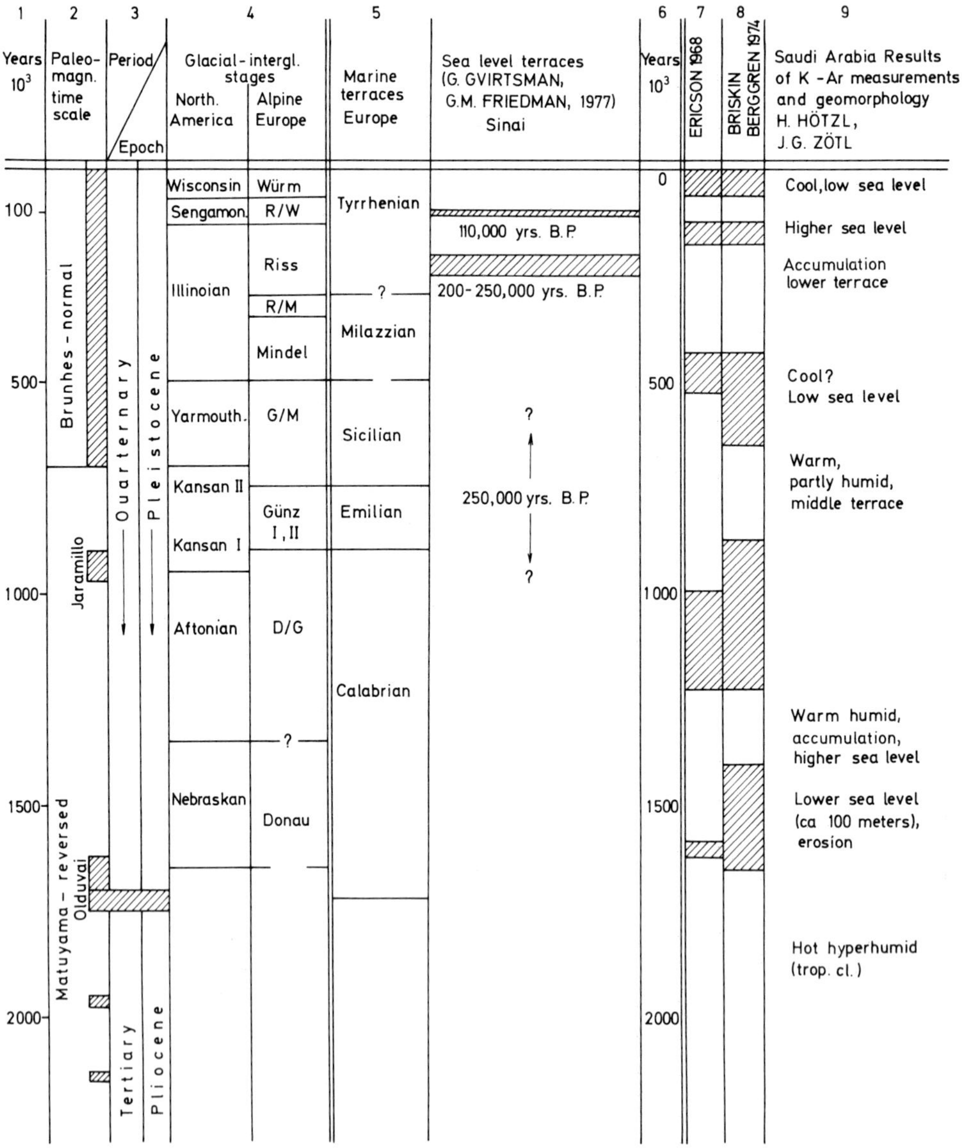

1	2	3	4		5		6	7	8	9
Years 10^3	Paleo-magn. time scale	Period / Epoch	Glacial-intergl. stages North America	Alpine Europe	Marine terraces Europe	Sea level terraces (G. GVIRTSMAN, G.M. FRIEDMAN, 1977) Sinai	Years 10^3	ERICSON 1968	BRISKIN BERGGREN 1974	Saudi Arabia Results of K-Ar measurements and geomorphology H. HÖTZL, J.G. ZÖTL
	Brunhes - normal	Quarternary / Pleistocene	Wisconsin	Würm	Tyrrhenian		0			Cool, low sea level
100			Sengamon	R/W		110,000 yrs. B.P.				Higher sea level
			Illinoian	Riss	?	200-250,000 yrs. B.P.				Accumulation lower terrace
				R/M	Milazzian					
				Mindel						
500			Yarmouth.	G/M	Sicilian	?	500			Cool? Low sea level
			Kansan II	Günz I, II	Emilian	250,000 yrs. B.P.				Warm, partly humid, middle terrace
	Jaramillo		Kansan I			?				
1000	Matuyama - reversed		Aftonian	D/G	Calabrian		1000			
				?						Warm humid, accumulation, higher sea level
1500			Nebraskan	Donau			1500			Lower sea level (ca 100 meters), erosion
	Olduvai									
		Tertiary / Pliocene								Hot hyperhumid (trop. cl.)
2000							2000			

Hatched parts in column 7 and 8 = cool epochs

The situation was different in the west, on the **Red-Sea Coast.** Here on the weather side there was considerably more rain in the humid periods than in the east, and in the foreland and the Shield escarpment, terrace systems have survived even arid climatic phases.

These terraces, their accumulations, dissection and inter-fingering with dated basalt flows, or the deposition of coral banks, are the basis of the climatic phases given in Table 58, and compared to the division of the Early and Middle Pleistocene as they have been understood up till now. As the K-Ar age determinations are rather indefinite and have large gray zones in their time limits (see Tables 5 and 19, ± 0.2 – ± 0.9 million years), it is difficult to compare and define the epochs chronologically.

It is nonetheless certain that these climatic changes in the pre-Würm took place, in the order given. This sheds new light on the matter, and establishes the fact that there were indeed significant changes in climate during Early and Middle Pleistocene.

That it is difficult to be more precise about the situation is understandable, considering that there were cold arid, hot moist and hot arid periods. The fact that there was also a north–south shift of climatic zones makes comparison all the more difficult. Universal evidence is to be found in worldwide changes in sea level with only slight chronological shifts, and the small morphological forms.

Numerous papers are written about the last glacial age and its equivalents in the arid zone. But then – except sea levels – it looks like a door is closed. Especially from arid regions absolute data for older climate epochs of the Quaternary are scarce. Every investigation is one step further. But how much is still to do that show the results of our Quaternary study in Saudi Arabia.

References

ADAMSON, D., WILLIAMS, F., 1980: Structural Geology, Tectonics and the Control of Drainage in the Nile Basin. The S.A.T.N., Rotterdam.

ALABOUVETTE, B., LE CHAPELAIN, J. C., PELLATON, C., 1975: Geology and mineral exploration of the Yanbu' Al Bahr quadrangle 24/38C: BRGM-Technical Record 75-JED-20, 42 p.

ALABOUVETTE, B., 1977: Geology and mineral exploration of the Jabal Al Buwanah quadrangle, sheet 24/37A, 1 : 100,000. Deputy Ministry for Mineral Resources. Jiddah.

ALABOUVETTE, B., MOTTI, E., REMOND, C. P., VILLEMUR, J. R., 1979: The Wadi Azlam prospect (27/35D), results of the 1974 program. Deputy Ministry for Mineral Resources, 38p., 28 ill. Jiddah.

ALABOUVETTE, B., PELLATON, C., 1979: Geology and mineral exploration of the Al Wajh quadrangle, 26/36C & D, 1 : 100,000. Deputy Ministry for Mineral Resources, Jiddah.

ALDRICH, L. T., BROWN, G. F., HEDGE, C., MARVIN, R., 1978: Geochronologic data for the Arabian Shield. U. S. Geol. Survey Saudi Arabian Proj. Rept. IR-240, 20p.

ALIMEN, H., 1963: Considérations sur la Chronologie du Quaternaire Saharien. Soc. Géol. France, Bull. Sér. 7, 5, 627–634, Paris.

ALIMEN, H., 1965: The Quaternary Era in the Northwest Sahara. Geol. Soc. of America, Spec. Paper **84,** 627–634, New York.

AL-KHATIB, EYAD, A. B., 1977: Hydrogeology of Usfan District. Inst. of Applied Geology, King Abdulaziz University, Jeddah.

ALLCHIN, B., et al., 1978: The Prehistory and Palaeogeography of the Great Indian Sand Desert. London: Academic Press.

AL NUJAIDI, H., 1978: Ground water studies in Khulais area. King Abdulaziz Univ., Inst. Appl. Geol. unpub. Master's Thesis.

AL-SAYARI, S. S., ZÖTL, J. G. (eds.), 1978: Quaternary Period in Saudi Arabia 1: Sedimentological, Hydrogeological, Hydrochemical, Geomorphological, and Climatological Investigations in Central and Eastern Saudi Arabia, 335 p. Wien–New York: Springer.

AL-SHANTI, A. M., 1966: Oolitic iron ore deposits in Wadi Fatimah between Jeddah and Mecca, Saudi Arabia: Saudi Arabian Dir. Gen. Mineral Resources Bull. **2,** 51 p.

AL-WATBAN, NASSER FAHD, 1976: Ground-Water Potentiality of Tabuk and Saq Aquifers in Qasim Region. M. Sc. Thesis, Institute of Applied Geology, King Abdulaziz University, Jeddah, 223 p.

ANTON, D., 1980: Climatic Influence in the Cenozoic Evolution of the Arabian Shield. 2ème Congrès Géologique Int., Paris.

AUFRERE, L., 1928: L'orientation des Dunes et la Direction des Vents. Acad. Sci. Comptes Rendus, Paris **187,** 833–835.

AUFRERE, L., 1930: L'orientation des Dunes Continentales. 12th Intl. Geog. Cong., Cambridge, England, 1928. Proc., pp. 220–231.

AUFRERE, L., 1932: Morphologie Dunaire et Météorologie Saharienne. Assoc. Géographes Français Bull. **56,** 34–47.

AUFRERE, L., 1934: Les Dunes du Sahara Algerien. Assoc. Géographes Français Bull. **83,** 130–142.

AWAD, H., 1963: Some aspects of the geomorphology of Morocco related to the Quaternary climates. Geol. J. **129,** 129–139, London.

BAGNOLD, R. A., 1933: A further journey through the Libyan Desert. Geogr. J. **82,** 103–129, London.

BAGNOLD, R. A., 1935: The movement of desert sand. Geogr. J. **85,** 342–369, London.

BAGNOLD, R. A., 1937: The transport of sand by wind. Geogr. J. **89,** 409–438, London.

BAGNOLD, R. A., 1941: The Physics of Blown Sand and Desert Dunes, p. 265. London: Methuen and Co., Lts.

BAGNOLD, R. A., 1951: Sand formations in Southern Arabia. Geogr. J. **117,** Pt. 1, 78–86, London.

BAKER, B. H., et al., 1971: Geology of the Eastern Rift System of Africa. Geol. Soc. Amer. Spec. Paper **136,** p. 67.

BAKKER, E. M., VAN ZINDEREN, 1962: A Late Glacial and Post-Glacial climate correlation between East Africa and Europe. Nature **194,** 201–203.

BALOUT, L., 1952: Pluviaux Interglaciaires et Préhistoire Saharienne. Inst. Rech. Sahar., Trav. **8,** 9–22, Alger.

BARBETTI, M., 1980: Geomagnetic strength over the last 50,000 years and changes in atmospheric ^{14}C concentrations: Emerging trends. Radiocarbon **22,** 192–199, New Haven.

BARTOV, Y., STEINITZ, G., EYAL, M., EYAL, Y., 1980: Sinistral movement along the Gulf of Aqaba – its age and relation to opening of the Red Sea. Nature **285,** 220–221.

BAUBRON, C. J., DELFOUR, J., VIALETTE, V., 1976: Geochronological measurements on rocks of the Arabian Shield, Kingdom of Saudi Arabia. French Bur. Recherches Géol. Min. Open-file Rept. 76-JED-22, 152 p. Jiddah.

BAUMGARTNER, A., REICHEL, E., 1975: Die Weltwasserbilanz, 179 p. München–Wien: Oldenbourg.

BAYER, H.-J., DULLO, W.-CHR., HÖTZL, H., JADO, A. R., QUIEL, F., STEPHAN, R., 1983: Zur Geologie und jungen Tektonik der Midyan-Region Nordwestarabien. Aus: Berichtsband des Sonderforschungsbereich 108, Spannung und Spannungsumwandlung in der Lithosphäre 1981–1983, pp. 509–540, Karlsruhe.

BEAUMONT, P., 1977: Water and development in Saudi Arabia. Geogr. J. **143,** 1, London.

BECKMAN, R. J., RAMSAY, J. B., 1978: Cluster Analyses of Water Wells of the Al Qatif and Al Hasa Areas. In: Quaternary Period in Saudi Arabia, 1 (AL-SAYARI, S. S., ZÖTL, J. G., eds.), pp. 135–151. Wien–New York: Springer.

BENASVILI, I. A., 1948: Der Wasserstand des Kaspischen Meeres (russ.). Leningrad.

BENDER, F., 1968: Geologie von Jordanien. Beitr. Region Geol. Erde **7,** p. 230. Berlin–Stuttgart: Borntraeger.

BENDER, F., 1975: Geology of the Arabian Peninsula; Jordan. U.S.G.S. Prof. Paper 560-I, p. 136. Washington.

BERGGREN, W. A., VAN COUVERING, J. A., 1974: The Late Neogene. S.P.C. Amsterdam: Elsevier.

BERRY, L., WHITEMAN, A. J., et al., 1966: Some Radiocarbon dates and their geomorphological significance: Emerged reef complex of Sudan. Z. Geomorph. **10,** 119–143, Berlin–Stuttgart.

BEYDON, Z. R., 1966: Geology of the Arabian Peninsula – Eastern Aden Protectorate and part of Dhufar. USGS, Prof. Paper 560-H, p. 49.

BEYDOUN, Z. R., GREENWOOD, J. E., 1968: Protectorate d'Aden et Dhufar. Lexique Stratigr. Intl. **3,** 10, b 2, p. 128, Paris.

BIGOT, M., ALABOUVETTE, B., 1976: Geology and mineralization of the Tertiary Red Sea coast of northern Saudi Arabia. DGMR Geoscience Map 1 : 100,000, with text; Djiddah.

BISHOP, W. W., 1971: The Late Cenozoic History of East Africa in Relation to Hominoid Evolution (TUREKIAN, K. K., ed.), Late Cenozoic Glacial Ages, pp. 493–527. New Haven and London: Yale University Press.

BISHOP, W. W., 1969: Pleistocene Stratigraphy in Uganda. Geol. Survey Uganda, Mem. 10, p. 128, Entebbe.

BISHOP, W. W., 1969: Background to Evolution in Africa. Proc. of the Symposium: Systematic Investigation of the African Later Tertiary and Quaternary, Burg Warthenstein, Austria 1965, p. 935. Chicago–London: Univ. Chicago Press.

BLANCHARD, R., 1926: Le Relief de l'Arabie Central. Rev. Géogr. Alpine **14,** 765–786, Grenoble.

BLANKENHORN, M., 1914: Syrien, Arabien und Mesopotamien. Handbuch der regionalen Geologie, **5,** 4, p. 159. Heidelberg: Winter.

BLISSENBACH, E., 1954: Geology of Alluvial Fans in Semi-arid Regions. Geol. Soc. Amer. Bull. **65,** 175–190, New York.

BLÜTHGEN, J., 1964: Allgemeine Klimageographie. 599 p. Berlin.

BOBEK, H., 1955: Klima und Landschaft Irans in vor- und frühgeschichtlicher Zeit. Geogr. Jber. Österreich **25,** 1–42, Wien.

BOBEK, H., 1963: Nature and Implications of Quaternary Climatic Changes in Iran. UNESCO, Arid Zone Res. **20,** 403–422, Paris.

BOGOCH, R., COOK, P., 1974: Calcite cementation of Quaternary conglomerate in southern Sinai. J. Sed. Petrology **44**, 3, 917–920, Tulsa, Oklahoma.

BOGUE, R. G., 1953: Geologic Reconnaissance in Northwestern Saudi Arabia. Saudi Arabia Ministry of Finance Open-File Rept. 30 p.

BOKHARI, M. M. A., 1981: Explanatory notes to the reconnaissance geologic map of the Maqna quadrangle, sheet 28/34 D, Kingdom of Saudi Arabia. Open-File-Report DGMR-OF-01-16: 32 pp., 3 figs., Djiddah.

BOTTGER, U., ERGENZINGER, P. J., et al., 1972: Quartäre Seebildungen und ihre Mollusken. Inhalte im Tibestigebirge und seinen Rahmenbereichen der Zentralen Ostsahara. Z. Geomorph. N. F. **16**, 2, 182–234, Berlin–Stuttgart.

BRAITHWAITE, C. J. R., 1982: Pattern of accretion of reefs in the Sudanese Red Sea. Marine Geology **46**, 297–325, Amsterdam.

BRAMKAMP, R. A., BROWN, G. F., 1948: Groundwater in the Najd, Saudi Arabia, New York Acad. Sci., Trans., Ser. 2, **10**, 236–237, New York.

BRAMKAMP, R. A., et al., 1956: Geologic Map of the Southern Tuwayq Quadrangle, Kingdom of Saudi Arabia. USGS, Misc. Geol. Investign. Map I-212A, 1 : 500,000, Washington.

BRAMKAMP, R. A., 1958: Geologic Map of the Northern Tuwayq Quadrangle, Kingdom of Saudi Arabia. USGS, Misc. Geol. Investign. Map I-207A, 1 : 500,000, Washington.

BRAMKAMP, R. A., RAMIREZ, L. F., 1959: Geologic Map of the Northwestern Rub'Al Khali Quadrangle, Kingdom of Saudi Arabia. USGS, Misc. Geol. Investign. Map 1-213A, 1 : 500,000, Washington.

BRAMKAMP, R. A., RAMIREZ, L. F., 1959: Geologic Map of the Wadi Al Batin Quadrangle, Kingdom of Saudi Arabia. USGS, Misc. Geol. Investign. Map I-203A, 1 : 500,000, Washington.

BRAMKAMP, R. A., et al., 1963: Geologic Map of the Western Rub'Al Khali Quadrangle, Kingdom of Saudi Arabia. USGS, Misc. Geol. Investign. Map I-218A, 1 : 500,000, Washington.

BRAMKAMP, R. A., RAMIREZ, L. F., 1961: Geologic Map of the Central Persian Gulf Quadrangle, Kingdom of Saudi Arabia. USGS, Misc. Geol. Investign. Map I-209A, 1 - 500,000, Washington.

BRAMKAMP, R. A., RAMIREZ, L. F., 1961: Geologic Map of the Northeastern Rub'Al Khali Quadrangle, Kingdom of Saudi Arabia. USGS, Misc. Geol. Investign. Map I-214A, 1 : 500,000, Washington.

BRAMKAMP, R. A., RAMIREZ, L. F., 1963: Geologic Map of the Darb Zubayadah Quadrangle, Kingdom of Saudi Arabia. USGS, Misc. Geol. Investign. Map I-202A, 1 : 500,000, Washington.

BRAMKAMP, R. A., et al., 1963: Geologic Map of the Wadi Ar Rimah Quadrangle, Kingdom of Saudi Arabia. USGS, Misc. Geol. Investign. Map I-206A, 1 : 500,000, Washington.

BRAMKAMP, R. A., et al., 1964: Geologic Map of the Jawf-Sakakah Quadrangle, Kingdom of Saudi Arabia. USGS, Misc. Geol. Investign. Map I-210A, 1 : 500,000, Washington.

BRAY, J. R., 1970: Temporal patterning of Post-Pleistocene glaciation. Nature **228.**

BRIEM, E., 1977: Beiträge zur Genese und Morphodynamik des ariden Formenschatzes unter besonderer Berücksichtigung des Problems der Flächenbildung. Berliner Geogr. Abh. **26**, 89 p.

BRIEM, E., 1976: Beiträge zur Talgenese im westlichen Tibesti Gebirge. Berliner Geogr. Abh. **24**, 45–54.

BROEKER, W. S., WALTON, A. F., 1959: The geochemistry of ^{14}C in freshwater systems. Geochim. Cosmochim. Acta **16**, 15–38, 200.

BROSCHE, K. U., MOLLE, H. G., 1976: Geomorphologische und klimageschichtliche Studien in Süd- und Zentraltunesien. Z. Geomorph. N. F. Suppl. **24**, 149–159, Berlin–Stuttgart.

BROWN, G. F., 1949: The Geology and Groundwater of Al-Kharj District, Najd, Saudi Arabia. New York Acad. Sci. Trans. Ser. 2, **10**, 370–375, New York.

BROWN, G. F., 1960: Geomorphology of Western and Central Saudi Arabia. Int. Geol. Congr., 21st Report **21**, 150–159, Copenhagen.

BROWN, G. F., 1970: Eastern Margin of the Red Sea and the Coastal Structures in Saudi Arabia. Royal Soc. London Phil. Trans. A267, 75–89, London.

BROWN, F. H., et al., 1970: Pliocene/Pleistocene Formations in the Lower Omo Basin, Southern Ethiopia. Quaternaria **13**, 247–268, Roma.

BROWN, G. F., 1972: Tectonic Map of the Arabian Peninsula, Kingdom of Saudi Arabia. Director Genl. Resources, Map AP-2, 1, 4,000,000, Jiddah.

BROWN, G. F., JACKSON, R. O., 1959: Geologic map of the Asir Quadrangle, Kingdom of Saudi Arabia: U.S. Geol. Survey Misc. Geol. Inv. Map I-217 A, scale 1 : 500,000.

BROWN, G. F., JACKSON, R. O., 1960: The Arabian Shield. Intl. Geol. Congr., 21st Report **9**, 69–77, Copenhagen.
BROWN, G. F., et al., 1963: Geologic Map of the Southern Hijaz Quadrangel. USGS, Misc. Geol. Investign. Map I-210A, 1 : 500,000, Washington.
BROWN, G. F., et al., 1963: Geologic Map of the Northeastern Hijaz Quadrangle, Kingdom of Saudi Arabia. USGS, Misc. Geol. Investign. Map I-205A, 1 : 500,000, Washington.
BROWN, G. F., JACKSON, R. O., BOGUE, R. G., 1973: Geographic map of the northwestern Hijaz quadrangle, Kingdom of Saudi Arabia: Saudi Arabian Dir. Gen. Mineral Resources Geo. Map I (GM)-204 B, scale 1 : 500,000.
BÜDEL, J., 1954: Sinai, die Wüste der Gesetzesbildung. Abhdl. Akad. f. Raumforschung **28,** Festschrift Mortensen, Bremen.
BÜDEL, J., 1955: Reliefgeneration und Plio-pleistozäner Klimawandel im Hoggar-Gebirge. Erdkde. **9,** 100–115, Bonn.
BÜDEL, J., 1963: Die Pliozänen und Quartären Pluvialzeiten der Sahara. Eiszeitalter und Gegenwart **14,** 161–187, Ohringen.
BUNKER, D. G., 1953: The Southwest Borderlands of the Rub'Al Khali. Geogr. J. **119,** pt. 4, 420–430, London.
BURDON, D. J., 1973: Groundwater resources of Saudi Arabia. Groundwater resources in Arab. countries, ALESCO, Science Monograph No. 2.
BURDON, D. J., ÖTKUN, G., 1967: The Groundwater Potential of Karst Aquifers in Saudi Arabia. Intl. Assoc. Hydrogeologists, Istanbul Meeting, Istanbul.
BURDON, D. J., ÖTKUN, G., 1968: Hydrogeological Control of Development in Saudi Arabia. Intl. Geol. Congr., 23rd, **12,** 145–153, Prag.
BUREK, P. J., 1974: Plattentektonische Probleme in der weiteren Umgebung Arabiens sowie der Danakil-Afar-Senke. Geotekton. Forsch. **47,** 1/2, 1–100, Stuttgart.
BUSH, P., 1973: Some Aspects of the Diagenetic History of the Sabkha in Abu Dhabi, Persian Gulf. In: The Persian Gulf (PURSER, B. H., ed.), pp. 395–406. Berlin–Heidelberg–New York: Springer.
BUTLER, G. P., 1969: Modern evaporite deposition and geochemistry of co-existing brines, the Sabkha, Trucial Coast, Arabian Gulf. J. Sed. Petrol. **39,** 7089, Tulsa, Oklahoma.
BUTLER, G. P., 1971: Origin and controls on distribution of Arid Supratidal (Sabkha) Dolomite, Abu Dhabi, Trucial Coast (abs.). AAPG, Bull. **55,** 332, Tulsa, Oklahoma.
BUTZER, K. W., 1958: Quaternary stratigraphy and climate in the Near East. Bonner Geogr. Abh. **24,** 1–57, Bonn.
BUTZER, K. W., 1961: Climatic change in Arid Regions since the Pliocene. UNESCO, Arid Zone Res. **17,** 13–56, Paris.
BUTZER, K. W., 1963: The last "Pluvial" phase of the Eurafrican Subtropics. UNESCO, Arid Zone Res. **20,** 211–221, Paris.
BUTZER, K. W., 1966: Climatic Changes in the Arid Zones of Africa During Early to Mid-Holocene Times. In: World Climate from 8000 to 0 B.C. (SAWYER, J. S., ed.), pp. 72–83. London: Royal Meteor. Soc.
BUTZER, K. W., 1975: Late Glacial and Postglacial climatic variation in the Near East. Erdkunde **11,** 21–35, Bonn.
BUTZER, K. W., 1975: The recent climatic fluctuations in lower latitudes and the general circulation of the Pleistocene. Geogr. Ann. **39,** 2/3, 105–113, Stockholm.
BUTZER, K. W., 1980: Pleistocene History of the Nile Valley in Egypt and Lower Nubia. The SATN, Rotterdam.
BUTZER, K. W., HANSEN, C. L., 1967: Upper Pleistocene Stratigraphy in Southern Egypt. In: Background to Evolution in Africa (BISHOP, W. W., CLARK, J. D., eds.), pp. 329–356. Chicago–London: Univ. Chicago Press.
BUTZER, K. W., HANSEN, C. L., 1968: Desert and River in Nubia, p. 562. Madison–London: University of Wisconsin Press.
BUTZER, K. W., HANSEN, C. L., 1968: Desert and River in the Nubia; Geomorphology and Prehistoric Environments of the Aswan Reservoir, p. 562. Madison, Milwaukee: University of Wisconsin Press.
BUTZER, K. W., et al., 1972: Radiocarbon dating of East African Lake Levels. Science **175,** 1069–1076.

CATON-THOMPSON, G., 1938: Geology and Archaeology of the Hadhramout, Southwest Arabia. Nature **142,** 3586, 139–142, London.

CHAPMAN, R. W., 1971: Climatic changes and the evolution of landforms in the Eastern Province of Saudi Arabia. Geol. Soc. Amer. Bull. **82,** 2713–2728, Boulder, Colorado.

CHAPMAN, R. W., 1974: Calcareous duricrust in Al-Hasa, Saudi Arabia. Geol. Soc. Amer. Bull. **85,** 119–130, Boulder, Colorado.

CHAPMAN, R. W., 1978: Geology. In: Quaternary Period in Saudi Arabia, 1 (AL-SAYARI, S. S., ZÖTL, J. G., eds.), pp. 4–18. Wien–New York: Springer.

CHAPMAN, R. W., 1978: Geomorphology. In: Quaternary Period in Saudi Arabia, 1 (AL-SAYARI, S. S., ZÖTL, J. G., eds.), pp. 19–30. Wien–New York: Springer.

CHAPMAN, R. W., 1978: Geomorphology of the Eastern Margin of the Shedgum Plateau. In: Quaternary Period in Saudi Arabia, 1 (AL-SAYARI, S. S., ZÖTL, J. G., eds.), pp. 77–84. Wien–New York: Springer.

CHAPPELL, J., POLACH, H. A., 1972: Some effects of partial recrystallisation on ^{14}C dating Late Pleistocene corals and molluscs. Quaternary Res. **2,** 244–252, New York.

CHAUDHRI, I. I., 1957: Succession of Vegetation in the Arid Regions of West Pakistan Plains. In: Food and Agricultural Council of Pakistan and UNESCO, Symposium on Soil Erosion and its Control in the Arid and Semiarid Zones, Karachi, Nov. 1957, pp. 141–156.

CHEPIL, W. S., 1945: The Transport Capacity of the Wind. Pt. 3 of Dynamics of Wind Erosion: Soil Sci. **60,** No. 6, 475–480.

CHEPIL, W. S., WOODRUFF, N. P., 1963: The Physics of Wind Erosion and its Control. In: Advances in Agronomy, Vol. 15 (NORMAN, A. G., ed.). New York–London: Academic Press.

CIMIOTTI, U. K., 1980: On the geomorphology of the Gulf of Elat-Aquaba and its borderlands. Berliner Geogr. Studien **7,** 155–176, Berlin.

CLARK, M. D., 1981: Geologic map of the Al Hamrah' Quadrangle, Sheet 23C, Jeddah, Saudi Arabia.

COLEMAN, R. G., 1973: Geologic background of the Red Sea: U. S. Geol. Survey Saudi Arabian Proj. openfile rept. **155,** 30 p.

COLEMAN, R. G., 1975: A Miocene ophiolote on the Red Sea coastal plain (abs.): EOS (Am. Geophys. Union Trans.) **56,** No. 12, 1080.

COLEMAN, R. G., 1977: Geologic background of the Red Sea. In: Red Sea Research 1970–1975 (HILPERT, L. S., ed.). Saudi Arabian Dir. Gen. Mineral Resources Bull. **22,** C1–C9.

CONRAD, G., 1969: L'évolution Continentale Post-Hercanienne du Sahara Algérien (Saoura, Erg, Chech-Tanezrouft, Ahnet-Mouydir). Centre de Recherches sur les Zones Arides, Paris, série géologique 10.

CONRAD, V., 1936: Die klimatologischen Elemente in ihrer Abhängigkeit von terrestrialen Einflüssen. In: Handbuch Klimat. I, B (KÖPPEN–GEIGER, ed.) Berlin, 5 Bde. 1930–1939 (unvollst.).

COQUE, R., JAUZEIN, R., 1967: The Geomorphology and Quaternary Geology of Tunisia (Transl. by D. COSTER, G. L.). In: Guidebook to the Geology and History of Tunisia (MARTIN, L., ed.). Petrol. Explor. Soc. Libya, Ann. Field, Conf., 9th, pp. 227–257, Tripoli.

COTECCHIA, V., GRASSI, D., ORSINI, A., 1970: Geological hydrogeological and geotechnical studies for water supplies in an arid region of the Asir (Jizan, Saudi Arabia) with special reference to the construction of Malaki Dam, Unpubl. Report.

COX, L. G., 1931: The geology of the Farasan Islands, Jizan and Kamaran Island, Red Sea, Part 2, Molluscan Paleontology: Geol. Mag. (Great Britain), Vol. 68, pp. 1–13.

CYPRUS MISSION: Reconnaissance Report of Northern Oman Water Resources and Development Prospects. Nicosia 1975, 53 p.

DADET, P., MARCHESSEAU, J., MÍLLON, R., MOTTI, E., 1970: Mineral occurrences related to stratigraphy and tectonics in Tertiary sediments near Umm Lajj, eastern Red Sea area, Saudi Arabia: Royal Soc. London Philos., Trans. **A 267,** 99–106.

DAGENS, E. T., HECKY, R. E., 1974: Paleoclimatic Reconstruction of Late Pleistocene and Holocene Based on Biogenic Sediments from the Black Sea and a Tropical African Lake. Colloques Internationaux du C.N.R.S., 219. Les méthodes quantitatives d'étude des variations du climat au cours du Pleistocene, pp. 13–24, Paris.

DAVIES, F. B., 1980: Reconnaissance geology of the Duba quadrangle, sheet 27/35D, Kingdom of Saudi Arabia. Directorate Gen. Mineral Resources, Geol. Map GM-57, 1 : 100,000. Jiddah.

DAVIES, F. B., 1981a: Reconnaissance geologic map of the Wadi Thalaba quadrangle, sheet 26/36A,

Kingdom of Saudi Arabia. Deputy Ministry for Mineral Resources, Geol. Map GM-42, 1 : 100,000. Jiddah.

DAVIES, F. B., 1981b: Geologic map of the Al Muwaylih quadrangle, sheet 27A, Kingdom of Saudi Arabia. Deputy Ministry for Mineral Resources, 1 : 250,000. Jiddah.

DEGENS, E. T., ROSS, D. A. (eds.), 1969: Hot Brines and Recent Heavy Metal Deposits in the Red Sea, 600 p. New York: Springer.

DELFOUR, J., 1966: Report on the Mineral Resources and Geology of the Hulayfah Musayna'h region (Sheet 78, Zone 1 North). Bur. Research, Geol. Minieres, Jeddah, SG 66 A 8, Saudi Arabian Dir. Gen. Mineral Resources, Open-file Rept., p. 64.

DELFOUR, J., 1970a: J'Balah Group. French Bur. Rech. Géol. Min. Open-file Rept. 70-JED-4, 31p.

DELFOUR, J., 1970b: Le groupe de J'Balah, une nouvelle unité du bouclier arabe. French. Bur. Rech. Géol. Min. Bull. (Ser. 2), Sec. 4, No. 4, 19–32.

DELFOUR, J., 1970c: Preliminary data on the Zarghat magnesite prospect. In: BRGM Annual Report 1969. French Bur. Rech. Géol. Min. Open-file Rept. 70-JED-1.

DELFOUR, J., 1970d: Results of an exploratory drillings at the As Safra copper prospect. In: BRGM Annual Report 1969. French Bur. Rech. Géol. Min. Open-file Rept. 70-JED-1.

DELFOUR, J., 1970e: Sulfide mineralization at Nuqrah and Jabal Sayid. French Bur. Rech. Géol. Min. Open-file Rept. 70-JED-22, 34 p.

DELFOUR, J., 1970f: Summary review on the Jabal Sayid deposit. French Bur. Rech. Géol. Min. Open-file Rept. 70-JED-19, 21p.

DELFOUR, J., 1977: Geology of the Nuqrah quadrangle, sheet 25E, Kingdom of Saudi Arabia. Saudi Arabian Directorate Gen. Mineral Resources Geol. Map. GM-28, 1 : 250,000.

DIESTER, L., 1972: Zur Spätpleistozänen und Holozänen Sedimentation im Zentralen und Östlichen Persischen Golf. Meteor. Forsch.-Ergebnisse, Reihe C, 8, pp. 37–83, Berlin, Stuttgart.

DIESTER, H. L., 1973: Holocene Climate in the Persian Gulf as Deduced from Grain-size and Pleropod Distribution. Marine Geol. **14,** 207–223, Amsterdam.

DINCER, T., 1978: Environmental Isotopes in the Dahna Sand Dune. Working Paper, IAEA Advisory Group Meeting on the Application of Isotope Techniques to Arid Zones Hydrology, Vienna, 6–9 November 1978.

DINCER, T., 1982: Estimating Aquifer Recharge due to Rainfall – A Comment. J. Hydrol. **58,** 179–182.

DINCER, T., et al., 1973: Study of Groundwater Recharge and Movement in Shallow and Deep Aquifers in Saudi Arabia with Stable Isotopes and Salinity Data, Manuscript SM 182/17, Ar Riyadh.

DINCER, T., AL-MUGRAIN, A., ZIMMERMANN, V., 1974: Study of the Infiltration and Recharge through the Sand Dunes in Arid Zones with Special References to the Stable Isotopes and Thermonuclear Tritium. J. Hydrol. **23,** 79–109, Amsterdam.

DOORNKAMP, J. C., et al. (eds.), 1980: Geology, Geomorphology and Pedology of Bahrain. Geoabstracts Lts. Univ. East Anglia, U. K.

DROST, W., MOSER, H., NEUMAIER, F., RAUERT, W., 1974: Isotope methods in groundwater hydrology. Eurisotop Office Information Booklet 61, Brussels.

DRUFFEL, E. M., 1980: Radiocarbon in annual coral rings of Belize and Florida. Radiocarbon **22,** 363–371, New Haven.

DUBERTRET, L., 1970: Review of structural geology of the Red Sea and surrounding areas: Royal Soc. London Philos. Trans. **A267,** 9–20.

DUBIEF, J., 1952: Le vent et le déplacement du sable au Sahara. Algier Univ. Inst. Recherches Sahariennes Travaux **8,** pp. 123–164.

DUBIEF, J., 1959: Le Climat du Sahara. Inst. Rech. Sahar., Mém. 1, Alger.

DUBIEF, J., 1963: Le Climat du Sahara. Inst. Rech. Sahar., Mém. 2, Alger.

DULLO, W.-CHR., 1983: Zur Diagenese aragonitischer Strukturen am Beispiel rezenter und pleistozäner Korallenriffe des Roten Meeres. Natur und Mensch, Jahresmitteilung, 109–115, Nürnberg.

DULLO, W.-CHR., HÖTZL, H., JADO, A. R., 1983: New stratigraphical results from the Tertiary sequence of the Midyan area, NW Saudi Arabia. Newsl. Stratigr. **12** (2), 75–83. Berlin, Stuttgart.

DUOZOY, G., 1972: Hydrogéologie des basaltes du Harrat Rahat. Bull. B.R.G.M., Sect. III, pp. 37–50.

EL RAMLY, I. M., 1971: Shorelines Changes during the Quaternary in the Western Desert Mediterranean Coastal Region (Alexandria-Salhum). UAR Quaternaria **15,** 285–292, Roma.

EMERY, K. O., 1956: Sediments and Water of the Persian Gulf. AAPG Bull. **40,** 2354–2383, Tulsa, Oklahoma.

EVANS, G., SCHMIDT, V., BUSH, P., NELSON, M., 1969: Stratigraphic and geologic history of the Sabkha, Abu Dhabi, Persian Gulf. Sedimentology **12,** 145–159, Amsterdam.

EVANS, P., 1972: The present status of age determination in the Quaternary (with special reference to the period between 70,000 and 1,000,000 years ago). Intern. Geol. Congr. 24th Section **12,** 16–21, Montreal.

EYAL, M., EYAL, Y., BARTOV, Y., STEINITZ, G., 1981: The tectonic development of the western margin of the Gulf of Elat (Aqaba) rift. Tectonophysics **80,** 39–66, Amsterdam.

FABER, E., SCHÖLL, M., 1978: Oxygen and hydrogen isotopic composition of Red Sea brines. Nature **275,** 436–437.

FAIRBRIDGE, R. W., 1963: Nile Sedimentation above Wadi Halfa during the last 20,000 years. Kush **11,** 96–107, Khartoum.

FAIRBRIDGE, R. W., 1964: African ice-age aridity. In: Problems in palaeoclimatology (NAIRN, A. E., ed.), Proc. NATO-Conf. on Palaeoclimate, 1963 Newcastle upon Tyne, pp. 356–360, London.

FAIRBRIDGE, R. W., 1972: Quaternary Sedimentation in the Mediterranean Region Controlled by Tectonics, Paleoclimates and Sea Level. In: The Mediterranean Sea; A Natural Sedimentation Laboratory (STANLEY, D. J., ed.), pp. 99–113. Stroudsburg, Pa., Dowden, Hutchinson and Ross.

FAIRBRIDGE, R. W., 1974: Eiszeitklima in Nordafrika. Geol. Rdsch. **54,** 399–414.

FAIRBRIDGE, R. W., 1977: Global climate chance during the 13,500 yr.b.p. Gothenburg geomagnetic excursion. Nature N 4–15, 1.

FAIRER, G. M., 1983: Reconnaissance geologic map of the Sabya quadrangle, sheet 17/42D, Kingdom of Saudi Arabia. Deputy Ministry for Mineral Resources Geol. Map. GM-68, 1 : 100,000, Jiddah.

FALCON, N. L., 1947: Raised Beaches and Terraces of the Iranian Makran Coast. Geogr. J. **109,** 149–151, London.

FALCON, N. L., 1967: The Geology of the Northern-East Margin of the Arabian Basement Shield. Advancement Sci. **24,** 31–42, London.

FALCON, N. L., GASS, I. G., GIRDLER, R. W., LAUGHTON, A. S., 1970: A discussion on the structure and evolution of the Red Sea and the nature of the Red Sea, Gulf of Aden, and Ethiopia rift junction. Royal Soc. London Philos Trans. **A267,** 417p.

FAURE, H., 1969: Lacs Quaternaires du Sahara. Mitt. Intl. Limnolog.

FAURE, H., HOANG, C., LALOU, C., 1974: Les Récifs Soulévés à l'ouest du Golfe d'Aden (T.F.A.I.) et les Chronologie et Paléoclimats Interglaciaires. Colloques Internationaux C.N.R.S. Les Méthodes quantitatives d'étude des variations du climat au cours du Pleistocene, pp. 103–114, Paris.

FELBER, H., et al., 1978: Sea Level Fluctuations During the Quaternary Period. In: Quaternary Period in Saudi Arabia, 1 (AL-SAYARI, S. S., J. G. ZÖTL, eds.), pp. 50–57, Wien–New York: Springer.

FIELD, H. C., 1958: Stone implements from the Rub'Al Khali, Southern Arabia. Man. **58,** No. 121, 93–94.

FIELD, H. C., 1960: Stone Implements from the Rub'Al Khali, Southern Arabia. Man. **60,** No. 30, 25–26.

FLECK, R. J., et al., 1973: Potassium-argon Geochronology of the Arabian Shields. USGS, Saudi Arabian Project Report **165,** 40, Jeddah.

FLECK, R. J., COLEMAN, R. G., CORNWALL, H. R., GREENWOOD, W. R., HADLEY, D. G., SCHMIDT, D. L., PRINZ, W. C., RATTÉ, J. C., 1976: Geochronology of the Arabian Shield, western Saudi Arabia: K-Ar results. Geol. Soc. Amer. Bull. **87,** 9–21, Boulder, Colorado.

FLINT, R. F., 1959: Pleistocene climates in eastern and southern Africa. Geol. Soc. Amer. Bull. **70,** New York.

FLINT, R. F., 1965: The Pliocene-Pleistocene Boundary. Geol. Soc. Amer. Inc. Special Paper **84** 497–533.

FLINT, R. F., 1971: Glacial and Quaternary Geology. New York.

FLINT, R. F., BRANDTNER, F., 1961: Climatic changes since the last Interglacial. Amer. J. Sci., New Haven.

FLOHN, H., 1952: Allgemeine atmosphärische Zirkulation und Paläoklimatologie. Geol. Rdsch. **40**, 153–178, Stuttgart.

FLOHN, H., 1957: Klimaschwankungen der letzten 1000 Jahre und ihre geophysischen Ursachen. Verh. Dtsch. Geogr.-Tag Würzburg.

FLOHN, H., 1960: Climatic fluctuations and their physical causes, especially in the Tropics. In: Bargman, Trop. Meteor. in Africa, Nairobi, 1960.

FLOHN, H., 1977: Modelle der Klimaentwicklung im 21. Jahrhundert. In: Das Klima (OESCHGER, H., et al., Hrsg.). Berlin–Heidelberg–New York: Springer.

FLOHN, H., 1977: Climate and Energy: a scenario to a 21st century problem. Clim. Change **1**, 5–20.

FLOHN, H., NICHOLSON, S., 1979: Climatic fluctuations in the arid belt of the "Old World" since the last glacial maximum; possible causes and future implications. Paleoecology of Africa, Vol. 12 (BAKKER, E. H., ZINDEREN, eds.).

FRANZ, H., 1967: On the Stratigraphy and Evolution of Climate in the Chad Basin During the Quaternary. In: Background to Evolution in Africa (BISHOP, W. W., CLARK, J. D., eds.), pp. 273–283. Chicago–London: Univ. Chicago Press.

FRENZEL, B., 1967: Die Klimaschwankungen des Eiszeitalters, 296 p. Braunschweig: Vieweg.

FRENZEL, B., 1980: Das Klima der letzten Eiszeit in Europa. In: Das Klima (OESCHGER, H., et al., Hrsg.). Berlin–Heidelberg–New York: Springer.

FREUND, R., GARFUNKEL, Z., ZAK, I., GOLDBERG, M., WEISSBROD, T., DERIN, B., 1970: The shear along the Dead Dea Rift: Royal Soc. London Philos. Trans. **A267**, 107–130.

FREUND, R., GARFUNKEL, Z. (eds.), 1981: The Dead Sea rift. Tectonophysics **80**, 303 p., Amsterdam.

FUJII, S., et al., 1971: Sea level changes in Asia during the past 11,000 years. Quaternaria **14**, 211–216, Roma.

FURRER, G., GAMPER-SCHOLLENBERGER, B., SUTER, J., 1980: Zur Geschichte unserer Gletscher in der Nacheiszeit in Methoden und Ergebnisse. In: Das Klima (OESCHGER, H., et al., Hrsg.). Berlin–Heidelberg–New York: Springer.

GABRIEL, E., 1968: Arabische Halbinsel. In: Westermann Lexikon der Geographie (TIETZE, W., Hrsg.), Bd. 1. Braunschweig: G. Westermann.

GABRIEL, E., 1977: Zum ökologischen Wandel im Neolithikum der östlichen Zentralsahara. Berliner Geogr. Abhandl. **27.**

GASS, I. G., 1970: The evolution of volcanism in the junction area of the Red Sea, Gulf of Aden and Ethiopian rifts. Royal Soc. London Philos. Trans. **A267**, 369–381.

GASS, I. G., 1970: Tectonic and magmatic evolution of the Afro-Arabian dome. In: African magnetism and tectonics (CLIFFORD, T. N., GASS, I. G., eds.) pp. 285–300. Edinburgh: Oliver and Boyd, Ltd.

GASSE, F., 1975: L'évolution des Lacs de l'Afar Central (Ethiopie et TFAI) du Pljo-Pleistocène à l'actuel; Reconstitution des Paléomilieux Lacustres à partir de l'étude des Diatomées, D. Sc. Thesis, University of Paris VI, p. 406, V. 3.

GASSE, F., 1977: Evolution of Lake Abhe (Ethiopia and TFAI) from 70,000 B. P. Nature **265**, 42–45.

GASSE, F., DELIBRIAS, J., 1977: Les Lacs de l'Afar Central (Ethiopie et TFAI) au Pleistocène Superieur. In: Paleolimnology of Lake Biwa and the Japanese Pleistocene **4**, 529–575.

GASSE, F., STIELTJES, L., 1973: Les Sédiments du Quaternaire Récent du Lac Asal (Afar Central, Territoire Français des Afars et des Issars). Bull. Bur. Rech. Géol. Min. (2ème sér.) **4 (4)**, 229–245.

GASSE, F., et al., 1974: Variations Hydrologiques et Extension des Lacs Holocènes du Désert Danakil. Palaeogeogr. Palaeoclimatol. Palaeoecol. **15**, 109–148.

GASSE, F., et al., 1980: Quaternary History of the Afar and Ethiopian Rift Lakes. The S.A.T.N., Rotterdam.

GERMAN CONSULT, 1978: Investigation and detailed studies for the agricultural development of South Tihama. Final reports, Frankfurt/Main.

GEUKENS, F., 1966: Geology of the Arabian Peninsula: Yemen. USGS Prof. Paper 560-B, p. 23, Washington.

GEYH, M. A., 1970: Isotopenphysikalische Untersuchungen an Kalksinter, ihre Bedeutung für die ^{14}C-Altersbestimmung von Grundwasser und die Erforschung des Paläoklimas. Geol. Jb. **88**, 149–158, Hannover.

GEYH, M. A., 1971: Die Anwendung der ^{14}C-Methode. Clausthaler Tektonische Hefte **11,** 118, Clausthal-Zellerfeld.

GEYH, M. A., JÄKEL, D., 1974a: Late glacial and holocene climatic history of the Sahara desert derived from a statistical assay of ^{14}C dates. Palaeogeogr. Palaeoclimatol. Palaeoecol. **15,** 205–208, Amsterdam.

GEYH, M. A., JÄKEL, D., 1974b: Spätpleistozene und holozäne Klimageschichte der Sahara auf Grund zugänglicher ^{14}C-Daten. Z. Geomorph. N. F. **18,** 82–98, Berlin–Stuttgart.

GEYH, M. A., JÄKEL, D., 1974: ^{14}C-Altersbestimmungen im Rahmen der Forschungsarbeiten der Außenstelle Bordai/Tibesti der Freien Universität Berlin. Freie Universität Berlin, Pressedienst Wissenschaft **5,** 107–117, Berlin.

GEYH, M. A., OBENAUF, K. P., 1974: Zur Frage der Neubildung von Grundwasser unter ariden Bedingungen. Ein Beitrag zur Hydrologie des Tibesti Gebirges. Freie Universität Berlin, Pressedienst Wissenschaft **5,** 70–91, Berlin.

GILLMAN, M., LETULLIER, A., RENOUARD, G., 1966: La Mer Rouge, géologie et problème pétrolier: Rev. l'Inst. Français Pétrole **21,** No. 10, 1467–1487.

GILLMAN, M., 1968: Primary results of a geological and geophysical reconnaissance of the Jizan coastal plain in Saudi Arabia: AIME Regional Tech. Symposium, Rept. 2d, Dhahran, pp. 189–208.

GIRDLER, R. W., STYLES, P., 1974: Two stage Red Sea floor spreading. Nature **247,** 7–11.

GLENNIE, K. W., 1970: Desert Sedimentary Environments. In: Developments, in Sedimentology **14,** p. 222. Amsterdam–London–New York: Elsevier Publishing Co.

GONFIANTINI, R., 1981: Investigating water resources of the desert: how isotopes can help. IAEA Bulletin **23,** (1), 3–10, Vienna.

GOUDIE, A. S., et al., 1973: The Former Extensions of the Great India Sand Desert. Geogr. J. **139,** pt. 2, 243–257.

GRANT-TAYLOR, T. L., 1972: Conditions for the use of calcium carbonate as a dating material. In: Proc. 8th Int. Conf. on Radiocarbon Dating (RAFTER, T. A., GRANT-TAYLOR, T. L., eds.). Lower Hutt, Wellington (New Zealand): Royal Soc. New Zealand.

GREENWOOD, W. R., 1973: The Hail Arch – A Key to Deformation of the Arabian Shield During Evolution of the Red Sea Rift. Dir. Gen. Min. Resources. Min. Res. Bull. **7,** 4, Jiddah.

GREENWOOD, W. R., 1975a: Geology of the Al Aqiq quadrangle, sheet 20/41D, Kingdom of Saudi Arabia. Saudi Arabian Directorate Gen. Mineral Resources Geol. Map GM-23, 1 : 100,000.

GREENWOOD, W. R., 1975b: Geology of the Biljurshi quadrangle, sheet 19/41B, Kingdom of Saudi Arabia, with sections on geophysical investigations by G. E. ANDREASEN and geochemical investigations and mineral resources by V. A. TRENT and T. H. KILSGAARD. Saudi Arabian Directorate Gen. Mineral Resources Geol. Map GM-25, 1 : 100,000.

GREENWOOD, W. R., 1975c: Geology of the Jabal Ibrahim quadrangle, sheet 20/41C, Kingdom of Saudi Arabia, with a section on economic geology, by R. G. WORL and W. R. GREENWOOD. Saudi Arabian Directorate Gen. Mineral Resources Geol. Map GM-22, 1 : 100,000.

GREENWOOD, W. R., 1975d: Reconnaissance geology of the Jabal Shada quadrangle, sheet 19/41A, Kingdom of Saudi Arabia. Saudi Arabian Directorate Gen. Mineral Resources Geol. Map GM-20, 1 : 100,000.

GREENWOOD, W. R., 1980: Reconnaissance geology of the Wadi Wassat quadrangle, sheet 18/44C, Kingdom of Saudi Arabia. Saudi Arabian Directorate Gen. Mineral Resources Geol. Map GM-40, 1 : 100,000.

GREENWOOD, J. E., BLEAKLEY, D., 1976: Geology of the Arabian Peninsula: Aden Protectorate. USGS, Prof. Paper 560 C, p. 96, Washington.

GREENWOOD, W. R., BROWN, G. F., 1972: Petrology and Chemical Analysis of Selected Plutonic Rocks from the Arabian Shield. USGS, Saudi Arabian Proj. Rep. **147,** 21, Jiddah.

GREENWOOD, W. R., ANDERSON, R. E., FLECK, R. J., ROBERTS, R. J., 1980: Precambrium geologic history and plate tectonic evolution of the Arabian Shield. Saudi Arabia Directorate General of Mineral Resources. Min. Res. Bull. **24,** 35p., Jiddah.

GROVE, A. T., 1959: The Former Extent of Lake Chad. Geogr. J. **125,** 465–467, London.

GROVE, A. T., 1958: The Ancient Erg of Hausaland and Similar Formations on the South Side of the Sahara. Geogr. J. **134,** London.

GROVE, A. T., 1960: Geomorphology of the Tibesti Region with Special Reference to Western Tibesti. Geogr. J. **126,** pt. 1, 18–31, London.

GROVE, A. T., 1980: Geomorphic Evolution of the Sahara and the Nile. The S.A.T.N. Rotterdam.
GROVE, A. T., GOUDIE, A. S., 1971: Late Quaternary Lake Levels in the Rift Valley of Southern Ethiopia and Elsewhere in Tropical Africa. Nature **234**, 403–405, London.
GROVE, A. T., PULLAN, R. A., 1964: Some Aspects of the Pleistocene Palaeogeography of the Chad Basin. In: Africa Ecology and Human Evolution (HOWELL, F. C., BOURLIERE, F., eds.). Viking Fund Publ. Anthrop. **36**, 230–245, London.
GROVE, A. T., STREET, F. A., et al., 1975: Former Lake Levels and Climatic Change in the Rift Valley of Southern Ethiopia. Geogr. J. **141**, 177–202, London.
GROVE, A. T., WARREN, A., 1968: Quaternary Landforms and Climate in the South Side of the Sahara. Geogr. J. **134**, 194–208, London.
GUILCHER, A., 1969: Pleistocene and Holocene Sea Level Changes. Earth-Science Reviews **5**, 69–97.
GVIRTZMAN, G., BUCHBINDER, B., SNEH, A., NIR, Y., FRIEDMAN, G. M., 1977: Morphology of the Red Sea fringing reefs: A result of the erosional pattern of the last glacial low stand sea level and the following Holocene recolonisation. 2ème Symp. Int. sur les coreaux récifs coralliens fossils, Paris Sept. 1975. Mem. B.R.G.M. **89**, 480–491, Paris.
GVIRTZMAN, G., FRIEDMAN, G. M., 1977b: Sequence of progressive diagenesis in coral reefs. Amer. Ass. Petrol Geol., Studies in Geology **4**, 357–380, 25 Figs., 7 Tab., Tulsa.

HADLEY, D. G., FLECK, R. J., 1980: Reconnaissance geologic map of the Al Lith quadrangle, sheet 20/40 C, Kingdom of Saudi Arabia. Saudi Arabia Directorate General of Mineral Resources Geologic Map GM-32, scale 1 : 100,000, Jiddah.
HADLEY, D. G., SCHMIDT, D. L., 1979: Proterozoic sedimentary rocks and basins of the Arabian Shield and their evolution. Saudi Arabia Directorate General of Mineral Resources, USGS Project Report 242.
HALL, S. A., ANDREASEN, G. F., GIRDLER, R. W., 1977: Total intensity magnetic anomaly map of the Red Sea and adjacent coastal areas, a description and preliminary interpretation. In: Red Sea Research 1970–1975, Saudi Arabia, Dir. Gen. Mineral Resour. Bull. **22**, F1–F15, Jeddah.
HAMZA, W., 1982: Zur Hydrogeologie und Geologie der Nord-Tihama (Saudi Arabien). Giessener Geol. Schriften **33**, 172 p., Giessen.
HAQ, B. U., BERGGREN, W. A., VAN COUVERING, J. A., 1977: Corrected age of the Pliocene/Pleistocene boundary. Nature **269**, 483–488.
HAYS, J. D., BERGGREN, W. A., 1971: Quaternary boundaries and correlations. In: Micropaleontology of the Oceans (FUNNEL, B. M., RIEDEL, W. R., eds.), pp. 669–691. Cambridge University Press.
HEIM, A., 1928: Die Artesischen Quellen der Bahrein-Inseln im Persischen Golf. Eclogae Geol. Helv. **21**, 1–6, Basel.
HEINZELEIN, J. D. E., 1967: Pleistocene Sediments and Events in Sudanese Nubia. In: Background to Evolution in Africa (BISHOP, W. W., CLARK, J. D., eds.), pp. 313–328. Chicago–London: Univ. Chicago Press.
HÖLL, K., 1980: Wasser (Untersuchung, Beurteilung, Aufbereitung, Chemie, Bakteriologie, Biologie), 423 p. Berlin.
HOLM, D. A., 1953: Dome-shaped dunes of Central Nejd, Saudi Arabia. Int. Congr., 19th Alger 1952. Compt. Rend., Section 7, 107–112, Alger.
HOLM, D. A., 1957: Sigmoidal Dunes – A Transitional Form (abs.). Geol. Soc. Amer. Bull. **68**, No. 12, pt. 2, 1746.
HOLM, D. A., 1960: Desert Geomorphology in the Arabian Peninsula. Science **132**, 1369–1379, Washington.
HOLM, D. A., 1968: Sand Dunes. In: The Encyclopedia of Geomorphology (FAIRBRIDGE, R. W., ed.), pp. 973–979. New York: Reinhold Book Corp.
HÖLTING, B., 1980: Hydrogeologie. (Einführung in die Allgemeine und Angewandte Hydrogeologie.), 340 p. Stuttgart.
HOOKE, R. L., 1967: Processes on Arid Region Alluvial Fans. J. Geol. **75**, 438–460, Chicago.
HÖTZL, H., 1978: Wadi Ad Dawasir and its Hinterland – General Geology. In: Quaternary Period in Saudi Arabia, 1 (AL-SAYARI, S. S., ZÖTL, J. G., eds.), pp. 228–230. Wien–New York: Springer.
HÖTZL, H., 1977: Herkunft der Grundwässer im Osten der Arabischen Halbinsel. Die Umschau.
HÖTZL, H., FELBER, H., MAURIN, V., ZÖTL, J. G., 1978: Accumulation Terraces of Wadi Hanifah

and Wadi Al Luhy. In: Quaternary Period in Saudi Arabia, 1 (AL-SAYARI, S. S., ZÖTL, J. G., eds.), pp. 202–209. Wien–New York: Springer.

HÖTZL, H., FELBER, J., ZÖTL, J. G., 1978: The Quaternary Development of the Upper Part of Wadi Ar Rimah. In: Quaternary Period in Saudi Arabia, 1 (AL-SAYARI, S. S., ZÖTL, J. G., eds.), pp. 173–182. Wien–New York: Springer.

HÖTZL, H., JADO, A. R., 1983: Geologische Vorerkundung an der Roten Meer Küste zwischen Yanbu und Umm Lajj, Saudi Arabien. In: Berichtsband des Sonderforschungsbereich 108: Spannung und Spannungsumwandlung in der Lithosphäre 1981–1983. Karlsruhe.

HÖTZL, H., JOB, C., MOSER, H., RAUERT, W., STICHLER, W., 1978: Hydrogeological and Hydrochemical Investigations in the Upper Part of the Wadi Ar Rimah. In: Quaternary Period in Saudi Arabia, 1 (AL-SAYARI, S. S., ZÖTL, J. G., eds.), pp. 182–194. Wien–New York: Springer.

HÖTZL, H., KRÄMER, F., MAURIN, V., 1978: Quaternary Sediments. In: Quaternary Period in Saudi Arabia, 1 (AL-SAYARI, S. S., ZÖTL, J. G., eds.), pp. 264–301. Wien–New York: Springer.

HÖTZL, H., LIPPOLT, H. J., MAURIN, V., MOSER, H., RAUERT, W., 1978: Quaternary Studies on the Recharge Area Situated in Crystalline Rock Regions. In: Quaternary Period in Saudi Arabia, 1 (AL-SAYARI, S. S., ZÖTL, J. G., eds.), pp. 230–239. Wien–New York: Springer.

HÖTZL, H., MAURIN, V., 1978: Wadi Birk. In: Quaternary Period in Saudi Arabia, 1 (AL-SAYARI, S. S., ZÖTL, J. G., eds.), pp. 209–214. Wien–New York: Springer.

HÖTZL, H., MAURIN, V., ZÖTL, J. G., 1978: Studies of Quaternary Development of the Eastern Part of the Recharge Area of Wadi Ad Dawasir. In: Quaternary Period in Saudi Arabia, 1 (AL-SAYARI, S. S., ZÖTL, J. G., eds.), pp. 239–246. Wien–New York: Springer.

HÖTZL, H., MAURIN, V., ZÖTL, J. G., 1978: Gulf Coastal Region and its Hinterland. Geologic History of the Al Hasa Area Since the Pliocene. In: Quaternary Period in Saudi Arabia, 1 (AL-SAYARI, S. S., ZÖTL, J. G., eds.), pp. 58–77. Wien–New York: Springer.

HÖTZL, H., et al., 1980: Isotope methods as a tool for Quaternary studies in Saudi Arabia. Arid-Zone Hydrology: Investigations with Isotope Techniques. IAEA, Vienna, pp. 215–235.

HOUBOLT, J. J., 1957: Surface Sediments of the Persian Gulf near the Qatar Peninsula. Doctoral Thesis, University of Utrecht, Den Haag.

HUME, W. F., CRAIG, J. I., 1911: The Glacial Period and Climatic Changes in North-East Africa. Report, 80th Meeting of the British Assoc. for the Advancement of Science, pp. 382–383.

Hydrogeology Department, Centre for Applied Geology, Jeddah, The Kingdom of Saudi Arabia, 1975: Major groundwater problems in developing countries of the arid zone. Microfiche in: HEINDL, L. A.: Hidden waters in arid lands, 18 p. Int. Developm. Research Centre Ottawa.

IAEA (International Atomic Energy Agency), 1968: Guidebook on nuclear techniques in hydrology. Vienna.

ILLIES, J. H., 1970: Graben tectonics as related to crust-mantle interaction. In: Graben problems (ILLIES, J. H., MUELLER, S., eds.), pp. 4–27. Stuttgart: Schweizerbartsche Verlagsbuchhandlung.

Inst. Mar. Fish and Oceanogr. (Hrsg.), 1959: Vieljährige Änderungen des Haushalts des Kaspischen Meeres. Transact., Bd. 38 (russ.). Moskau.

Italconsult, 1965: Land and water surveys on the Wadi Jizan. United Nations Special fund project. FAO, Rome.

Italconsult, 1967: Final report on water supply survey for Jeddah-Mecca-Taif area. Italconsult, unpubl. Rept. to Saudi Arabian Ministry of Agriculture and Water.

Italconsult, 1969: Water and agricultural development surveys for areas II and III. Final Report, Wadi Bishah selected area. Min. of Agricult. and Water (Kingdom of Saudi Arabia), Rome, 52 p.

Italconsult, 1969: Water and agricultural development studies for area IV. Final Report, Drilling Investigations, Discharge Tests, Water Point Inventory. Unpublished Report, Rome.

JACKSON, R. O., et al., 1963: Geologic Map of the Southern Najd Quadrangle, Kingdom of Saudi Arabia. USGS, Misc. Geol. Investign. Map I-211A, 1 : 500,000, Washington.

JÄKEL, D., 1971: Erosion und Akkumulation in Enneri Bardague – Araye des Tibestigebirges (Zentrale Sahara) während des Pleistozäns und Holozäns. Berliner Geogr. Abh. **10**, 1–55, Berlin.

JÄKEL, D., 1977: Abfluß und fluviatile Formungsvorgänge im Tibestigebirge als Indikatoren zur Rekonstruktion einer Klimageschichte der Zentralsahara im Spätpleistozän und Holozän. Birmingham: X. INQUA.

JÄKEL, D., 1978: Eine Klimakurve für die Zentralsahara. In: Sahara 10.000 Jahre zwischen Weide und Wüste. Museen der Stadt Köln, pp. 382–396.

JÄKEL, D., 1983: Runoff and Erosion Processes in the Tibesti Mountains as Indicators of Climatic History in the Central Sahara During the Late Pleistocene and Holocene. Palaeoecology of Africa **11.**

JOB, C., 1978: Hydrochemical Investigations in the Areas of Al Qatif and Al Hasa with Some Remarks on Water Samples from Wadi Al Miyah and Wadi As Sah'ba near Haradh. In: Quaternary Period in Saudi Arabia, 1 (AL-SAYARI, S. S., ZÖTL, J. G., eds.), pp. 93–135. Wien–New York: Springer.

JOB, C., MOSER, H., RAUERT, W., STICHLER, W., 1978: Hydrochemical Investigations and Isotope Measurements in the Areas of Riyadh Al Khabra, Wadi Ar Rimah and Wadi Maraghan. In: Quaternary Period in Saudi Arabia, 1 (AL-SAYARI, S. S., ZÖTL, J. G., eds.), pp. 187–194. Wien–New York: Springer.

JOB, C., MOSER, H., PAK, E., RAUERT, W., STICHLER, W., 1978: Hydrochemical Investigations and Measurements of the Content of Isotopes of Wells in Wadi Ad Dawasir. In: Quaternary Period in Saudi Arabia, 1 (AL-SAYARI, S. S., ZÖTL, J. G., eds.), pp. 246–252. Wien–New York: Springer.

JOB, C., MOSER, H., RAUERT, W., STICHLER, W., 1978: Chemistry and Isotope Content of Some Wadi Groundwaters in the Central Parts of the Tuwayq Mountains. In: Quaternary Period in Saudi Arabia, 1 (AL-SAYARI, S. S., ZÖTL, J. G., eds.), pp. 216–228. Wien–New York: Springer.

JOHANSON, D. C., et al., 1972: Geological Framework of the Pliocene Hadar Formation (Afar, Ethiopia) with Notes on Palaeontology Including Hominids. In: Geological Background to Early Man (BISHOP, W. W., ed.), pp. 549–564. Edinburgh, Scotland: Academic Press.

JOHNSON, D. H., 1978: Gulf Coastal Region and Its Hinterland. In: Quaternary Period in Saudi Arabia, 1 (AL-SAYARI, S. S., ZÖTL, J. G., eds.), pp. 45–50. Wien–New York: Springer.

JOHNSON, D. H., KAMAL, M. R., PIERSON, G. O., RAMSAY, J. B., 1978: Sabkhahs of Eastern Saudi Arabia. In: Quaternary Period in Saudi Arabia, 1 (AL-SAYARI, S. S., ZÖTL, J. G., eds.), pp. 84–93. Wien–New York: Springer.

KABBANI, F. K., 1970: Geophysical and Structural aspects of the central Red Sea rift valley: Royal Soc. London Philos. Trans. **A267,** 89–97.

KAISER, K., KEMP, E. K., LEROI-GOURHAN, A., SCHULT, H., 1973: Quartärstratigraphische Untersuchungen aus dem Damaskus-Becken und seiner Umgebung. Z. Geomorph. N. F. **17,** 3, 263–353, Berlin, Stuttgart.

KARPOFF, R., 1957a: Sur quelques failles et „grabens" de l'Arabie centrale et septentrionale: Soc. géol. France, Comptes rendus, pp. 294–296.

KARPOFF, R., 1957b: L'Antécambrien de la Péninsule Arabique: Internat. Geol. Cong., 21st, Copenhagen 1960, rept. 21, pt. 9, 79–94.

KARPOFF, R., 1957c: Sur l'existence du Maestichtien au nord de Djeddah (Arabie Séoudite): Paris, Acad. Sci., Comptes rendus **245,** No. 16, 1322–1324.

KARPOFF, R., 1957d: Esquisse géologique de l'Arabie Séoudite. Soc. géol. Franc. Bull. **7,** 653–697.

KARPOFF, R., 1960: L'Antécambrian de la Péninsule Arabique. 21. Internat. Geol. Cong. Copenhagen Rept. **21,** 78–94.

KASSLER, P., The Structural and Geomorphic Evolution of the Persian Gulf. In: The Persian Gulf (PURSER, B. H., ed.), pp. 11–32. Berlin–Heidelberg–New York: Springer.

KELLOGG, W. W., 1977: Effects of human activities on global climate. WMD Tech. Note 165, No. 486, 47.

KELLOGG, W. W., 1980: Review of Human Impact on Climate. In: Das Klima (OESCHGER, H., et al., Hrsg.). Berlin–Heidelberg–New York: Springer.

KENDALL, C. G., SKIPWITH, S. P., 1968: Recent Algal Mats of a Persian Gulf Lagoon. J. Sed. Petrology **38,** 1040–1058, Tulsa, Oklahoma.

KERDANY, M. T., 1968: Note on the planktonic zonation of the Miocene in the Gulf of Suez region, U.A.R.: Internat. Union Geol. Sci., Comm. Mediterranean Neogene Stratigr., Sess. 4, Proc., pt. 3, 167–178.

KNETSCH, G., SHATA, A., DEGENS, E., MÜNNICH, K. O., VOGEL, J. C., SHAZLY, M. M., 1962: Untersuchungen an Grundwässern der Ost-Sahara. Geol. Rdsch. **52,** 587.

KÖPPEN, W., 1931: Grundriss der Klimakunde. Berlin–Leipzig, 1931.

KÖPPEN, W., GEIGER, R., 1930–1939: Handbuch der Klimatologie. 5 Bde. (unvollst.) Berlin.

LABEYRIE, J., DUPLESSY, J. C., DELIBRIAS, G., LETOLLE, R., 1967: Etude des températures des climats ancien par la mesure de l'oxygène-18, du carbone-13 et du carbone-14 dans les concretions des cavernes. Radioactive Dating and Low-Level Counting, IAEA, 153–160, Wien.

LAMB, H. H., 1977: Climate: Present, Past and Future, Vol. II. London: Methuen.

LAUER, W., FRANKENBERG, P., 1979: Klima- und Vegetationsgeschichte der westlichen Sahara. Ak. Wiss. Lit. Mainz **1**, 1–61.

LAURENT, D., 1970: Raw materials for a cement factory in the Yanbu'Al Bahr area: French Bureau de Recherches Géologiques et Minières Technical Record 70-JED-23, 27p., 1 fig., 3 maps. Jeddah.

LAURENT, D., DAESLE, M., BERTON, Y., DEHLAVI, M., 1973: Engineering geology map of ground conditions in Jiddah (21/39C). DGMR/BRGM, 2 pl. (1 : 100,000).

LE PICHON, X., FRANCHETEAU, J., 1978: A plate tectonic analysis of the Red Sea-Gulf of Aden area: Tectonophysics **46**, 369–406.

LEES, G. M., 1928: The Geology and Tectonics of Oman and Parts of South Eastern Arabia. Quat. J. Geol. Soc. London **84**, 4, 585–670, London.

LEES, G. M., 1928: The Physical Geography of South Eastern Arabia. Geogr. J. **71**, 441–452, London.

LEES, G. M., FALCON, N. L., 1952: The Geographical History of the Mesopotamian Plains. Geogr. J. **118**, 24–39, London.

LEIDLMAIR, A., 1962: Klimamorphologische Probleme in Hadramaut. Hermann V. Wissmann-Festschr., pp. 162–179, Tübingen.

LIND, E. M., MORRISON, M. S., 1974: East African Vegetation, p. 257. London: Longmans.

LIVINGSTONE, D. A., 1980: Environmental Changes in the Nile Headwater. The S.A.T.N.

LLOYD, J. W., FRITZ, P., CHARLESWORTH, D., 1980: A conceptual hydrochemical model for alluvial aquifers on the Saudi Arabian Basement Shield. Arid-Zone Hydrology, IAEA, Vienna.

LOWELL, J. D., GENIK, G. J. 1972: Sea-floor spreading and structural evolution of the southern Red Sea: Am. Assoc. Petroleum Geologists Bull. **56**, no. 2, 247–259.

MACFADYEN, W. A., 1930: The Undercutting of Coral Reef Limestone on the Coasts of Some Islands in the Red Sea. Geogr. J. **75**, 27–34, London.

MACFADYEN, W. A., COX, L. R., BRIGHTON, A. G., 1930: The Geology of the Farasan Islands, Jizan and Kamaran Island, Red Sea. Geol. Mag. (Great Britain) **68**, 1–13, 323–333.

MCCLURE, H. A., 1976: Radiocarbon Chronology of Late Quaternary Lakes in the Arabian Desert. Nature **263**, 755–756, London.

MCCLURE, H. A., 1978: Ar Rub'Al Khali. In: Quaternary Period in Saudi Arabia, 1 (AL-SAYARI, S. S., ZÖTL, J. G., eds.), pp. 252–263. Wien–New York: Springer.

MCCLURE, H. A., SWAIN, F. M., 1974: The Fresh water and Brackish Water Fossil Quaternary Ostracoda from the Rub'Al Khali, Saudi Arabia. Paper read at the 6th African Micropalaeontological Colloquium, Tunis.

MCKEE, E. D., (ed.), 1979: A Study of Global Sand Seas. Geol. Survey, Prof. Paper 1052, USGPO, Washington.

MCKEE, E. D., BREED, C. S., 1974: An Investigation of Major Sand Seas in Desert Areas throughout the World. In: Third Earth Resources Technology Satellite-1 Symposium, Washington, D. C., Dec. 10–14, 1973. Natl. Aeronautics and Space. Spec. Pub. NASA SP-351, pp. 665–579.

MCLAREN, Intl. Ltd., 1978: Agricultural Investigations in the Southern Arabian Shield. Unpublished Report, Taif, Toronto.

MAINGUET, M., et al., 1980: Le Sahara: Géomorphologie et Paléomorphologie Eoliennes. The S.A.T.N., Rotterdam.

MALEY, J., 1973: Mécanisme des changements climatiques aux basses latitudes. Palaeogeogr. Palaeoclimatol. Palaeoecol. **14**, 193–227.

MALEY, J., 1977a: Paleoclimates of Central Sahara during the early Holocene. Nature **269**, 573–577.

MALEY, J., 1977b: Analyses polliniques et paléoclimatologie des douze dernières millémaires du Bassin du Tschad. Rech. fr. Quat. Suppl. Bull. FEQ, INQUA, pp. 187–197.

MALEY, J., 1980: Les Changements Climatiques de la Fin du Tertiaire en Afrique: leur consequence sur l'Apparition du Sahara et de sa Végétation. The S.A.T.N., Rotterdam.

MARTIN, G., KRAPP, L., 1977: Die Hydrogeologie des Umm Er Radhuma Dolomit-Wasserträgers im „King Faisal Bedouin Settlement"-Project bei Haradh, Saudi Arabien. Geol. Jb. **C17**, 37–57, Hannover.

MASON, J. F., MOORE, Q. M., 1970: Petroleum developments in Middle East countries in 1969. AAPG Bull. **54**, 1524–1547.

MATTHES, F. E., 1939: Report of Committee on Glaciers. Am. Geophys. Union Trans. **20.**

MATTHESS, G., 1973: Die Beschaffenheit des Grundwassers. Lehrb. d. Hydrogeologie, Bd. 2, 324 p. Berlin–Stuttgart.

MECKELEIN, W., 1959: Forschungen in der zentralen Sahara. I. Klimamorphologie. Braunschweig.

MEHRINGER, P. J., et al., 1979: A Pollen Record from Birket Qarun and the Recent History of the Fayum, Egypt. Quaternary Research **11**, No. 2.

MENSCHING, H., 1958: Glacis, Fußfläche, Pediment. Z. Geomorph. N. F. **2**, 165–186, Berlin–Stuttgart.

MENSCHING, H., 1970: Flächenbildung in der Sudan- und Sahel Zone (Ober-Volta und Niger). Z. Geomorph. N. F. Suppl. **10**, 1–29, Berlin–Stuttgart.

MERGNER, H., SCHUHMACHER, H.: Morphologie, Ökologie und Zonierung von Korallenriffen bei Aqaba (Golf von Aqaba, Rotes Meer). Helgoländer Meeresuntersuchungen **26**, 238–358, 13 Abb., 20 Tab., Hamburg 1974.

MESSERLI, B., 1980: Die afrikanischen Hochgebirge und die Klimageschichte Afrikas in den letzten 20.000 Jahren. In: Das Klima (OESCHGER, H., et al., Hrsg.). Berlin–Heidelberg–New York: Springer.

MICHAEL, H. N., RALPH, E. K., 1971: Dating Techniques for the Archaeologist, 227 p. Cambridge (Mass.)–London: MIT Press.

MILRON, D. I., 1967: Geology of the Arabian Peninsula, Kuwait. USGS, Prof. Paper 560-F, p. 7, Washington.

MITCHELL, R. C., 1957: Notes on the Geology of Western Iraq and Northern Saudi Arabia. Geol. Rdsch. **46**, 476–493, Stuttgart.

MORTON, D. M., 1959: The Geology of Oman. World Petroleum Cong., 5th, New York 1959, Proc. sec. 1, 277–294.

MOSER, H., et al., 1978: Isotopic Composition of Waters of Al Qatif and Al Hasa Areas. In: Quaternary Period in Saudi Arabia, 1 (AL-SAYARI, S. S., ZÖTL, J. G., eds.), pp. 153–163. Wien–New York: Springer.

MOTTI, E., TEIXIDO, L., VAZQUEZ LOPEZ, R., VIAL, A., 1982: Le massif de Maqna; résultats des campagnes 1977–1978 et 1978–1979. French Bureau de Recherches Géologiques et Minières Open-File Report.

MÜNNICH, K. O., VOGEL, J. C., 1962: Untersuchungen an pluvialen Wässern der Ost-Sahara. Geol. Rdsch. **52**, 611–624, Stuttgart.

NAIMI, A. I., 1965: The Groundwater of Northeastern Saudi Arabia. Fifth Arab Petroleum Congress, March 16–23, Cairo.

NEBERT, K., 1970: Geology of Jabal Sauron and Jabal Farsan region. Direc. Gen. Min. Resour., Bull. 4, Jeddah.

NEBERT, K., 1970: Geology of the area north of Wadi Fatimah. I.A.G. Bull. 1, Jeddah.

NEEV, D., HALL, J. K., 1977: Climatic Fluctuations during the Holocene as Reflected by the Dead Sea Levels. In: Desertic Terminal Lakes (GREER, D. C., ed.), pp. 53–60. Proc. of the Intl. Conf. on Desertic Terminal Lakes. Utah Water Research Laboratory, Logan, Utah.

NEUMANN, G., 1944: Das Schwarze Meer. Z. Ges. Edkd., Berlin.

NICHOLSON, S. E., 1976: A climatic chronology for Africa: Synthesis of Geological, Historical and Meteorological Information and Data. Thesis Dept. Meteorol., Univ. Wisconsin.

NICHOLSON, S. E., 1980: Saharan Climates in Historic Times. The S.A.T.N., Rotterdam.

NORRIS, R. M., 1969: Dune Reddening and Time. J. Sed. Petrology **39**, No. 1, 7–11.

NOY, D. J., 1978: A comparison of magnetic anomalies in the Red Sea and the Gulf of Aden. In: Tectonophysics and Geophysics of Continental Rifts (RAMBERG, I. B., NEUMANN, E. R., eds.), pp. 279–286. Dordrecht.

NOZAKI, Y., RYE, D. M., TUREKIAN, K. K., DIDGE, R. E., 1978: A 200-year record of carbon-13 and carbon-14 variations in a Bermuda coral. Geophys. Research Letters **5**, 825–828, Washington.

OBENAUF, P., 1967: Beobachtungen zur Spätpleistozänen und Holozänen Talformung im Nord-west Tibesti. Berliner Geogr. Abh. **5**, 27–38, Berlin.

OESCHGER, H., MESSERLI, B., SVILAR, M. (Hrsg.), 1980: Das Klima. Analysen und Modelle, Geschichte und Zukunft. Berlin–Heidelberg–New York: Springer.

OLSSON, I. U., GÖKSU, Y., STENBERG, A., 1968: Further investigations of storing and treatment of Foraminifera and mollusks for C^{14}dating. Geol. Fören. Stockholm Förh. **90**, 417–426, Stockholm.

OLSSON, I. U., 1974: Some problems in connection with the evaluation of C^{14} dates. Geol. Fören. Stockholm Förh. **96**, 311–320, Stockholm.

ÖTKUN, G., 1969: Outlines of Ground Water Resources of Saudi Arabia. Int. Conf. on Arid. Lands, Tucson, Arizona, June 3–13, 16 p.

ÖTKUN, G., 1972: Observations on Mesozoic Sandstone Aquifers in Saudi Arabia. 24th Int. Geol. Congr., Section 11, Riyadh, pp. 28–35.

ÖTKUN, G., 1973: General aspects of Palaeocene carbonate aquifer in Saudi Arabia. Int. Symp. Devel. Groundwater Resources, Madras, pp. 1–13.

PACHUR, H.-J., 1974: Geomorphologische Untersuchungen im Raum des Tibesti (Zentralsahara). Berliner Geogr. Abh. **17**, 62 p.

PATTERSON, R. J., KINSMAN, D. J. J., 1981: Hydrological Framework of a Sabkha along the Arabian Gulf. AAPG **65**, No. 8, 1457–1475.

PELLATON, C., 1979: Geologic map of the Yanbu'Al Bahr quadrangle, GM-48 sheet 24 C Kingdom of Saudi Arabia (with explanatory notes). Saudi Arabian Directorate General of Mineral Resources, Jeddah.

PELLATON, C., 1982a: Geologic map of the Umm Lajj quadrangle, sheet 25B, Kingdom of Saudi Arabia. Deputy Ministry for Mineral Resources Geol. Map GM-61A, 1 : 250,000, Jiddah.

PELLATON, C., 1982b: Geologic map of the Jabal Al Buwanah quadrangle, sheet 248, Kingdom of Saudi Arabia. Deputy Ministry for Mineral Resources. Geol. Map GM-62A, 1 : 250,000, Jiddah.

PENCK, A., 1913: Die Formen der Landoberflächen und die Verschiebungen der Klimagürtel. Sitz.-Ber. Kgl. Preuss. Akad. Wiss. Halbbd. I, 4, 77–97.

PENCK, A., BRÜCKNER, E.: Die Alpen im Eiszeitalter. 3 Bde., Leipzig 1901–1909.

PESCE, A., 1968: Gemini Space Photographs of Libya and Tibesti. A Geological and Geographical Analysis: Tripoli, Petroleum Exploration Soc. Libya, p. 81.

PHILIPP HOLZMANN A. G.: 1967: Al Hassa – Oase Hofuf. Sonderdruck aus Z. Kulturtechnik und Flurbereinigung **8**, H. 6, 321–344.

PILGER, A., RÖSLER, A. (eds.), 1975: Afar Depression of Ethiopia. Inter Union Comm. on Geodynamics, Scientific Report **14**, 416 p. Stuttgart: Schweizerbart.

PILGER, A., RÖSLER, A. (eds.), 1976: Afar between continental and oceanic rifting. Inter Union Commission on Geodynamics, Scientific Report **16**, 216 p. Stuttgart: Schweizerbart.

POWERS, R. W., 1968: Arabia Séoudite. Lexique Stratigr. Intl. **3**, 10b, 1, p. 177, Paris.

POWERS, R. W., RAMIREZ, L. F., REDMOND, C. D., ELBERG, E. L., JR., 1966: Geology of the Arabian Peninsula. Sedimentary geology of Saudi Arabia. U. S. Geol. Survey Prof. Paper 560-D, 147 p.

PRICE, W. A., 1950: Saharan Sand Dunes and the Origin of the Longitudinal Dunes. A Review. Geogr. Rev. **40**, No. 3, 462–465.

QUEZEL, P., MARTINEZ, G., 1959: Le Dernier Interpluvial au Sahara Central. Libya Antrop. Prehist. Ethnogr. **5**, 211–255, Alger.

RAIKES, R. L., DYSON, R. H., 1961: The Prehistoric Climate of Baluchistan and the Undus Valley. Amer. Anthropol. **63**, 265–281, Lancaster.

RICHARDSON, J. L., RICHARDSON, A. E., 1972: History of an African Rift Lake and its Climatic Implications. Ecological Monographs **42**, 499–534.

RICHTER, W., LILLICH, W., 1975: Abriss der Hydrogeologie, 281 p. Berlin–Stuttgart: Borntraeger.

RIDLEY, A. P., SEELEY, M. W., 1979: Evidence of recent coastal uplift near Al Jubail, Saudi Arabia. Tectonophysics **52**, 319–327.

ROGNON, P., FASSE, F., 1973: Depots Lacustres Quaternaires de la Basse Vallée de l'Awash (Efar, Ethiopie); Leurs rapports aves la tectonique et la Volcanisme Sous-aquatique. Rev. Geogr. Phys. Geol. Dyn. **15 (2)**, 295–316.

ROGNON, P., 1975: Précisions Chronologiques et Paléogéographiques sur le Façonnement du Glacis principal des Basins de l'Afar Central, Colloque Géomorphologie des Glacis. Univ. of Tours p. 81–84.

ROGNON, P., 1976: Essai d'Interprétation des Variations Climatiques au Sahara Depuis 40,000 ans. Revue de Géographie Physique et Géologie Dynamique **2,** 18, 251–282.

ROSS, D. A., 1974: The Red and Black Seas. In: Man's finite earth, Part two, Oceanography. Minneapolis, Burgess Publ. Co., pp. 77–88.

ROSS, D. A., SCHLEE, J., 1973: Shallow structure and geologic development of the southern Red Sea. Geol. Soc. America Bull. **84,** 3827–3848.

RUTTNER, A. W., RUTTNER-KOLISKO, A. E., 1972: Some Data on the Hydrology of the Tabas-Shirgesht – Ozbak – Kuh Area (East Iran). Jb. Geol. B.-A. **115,** 1–48, Wien.

SAID, R., 1962: The geology of Egypt, 377 p. Amsterdam–New York: Elsevier Publishing Co.

SAID, R., 1969: General stratigraphy of the adjacent land areas of the Red Sea. In: Hot brines and recent heavy metal deposits in the Red Sea (DEGENS, E. T., ROSS, D. A., eds.), pp. 71–81. Berlin–Heidelberg–New York: Springer.

SANDER, N. J., 1962: Aperçu Paléontologique et Stratigraphique de Paléogène on Arabie Séoudite Orientale. Rev. de Micropaléontologie **6,** No. 1, 3–40.

SARNTHEIN, M., 1972: Sediments and history of the postglacial transgression in the Persian Gulf and northwestern Gulf of Oman. Marine Geol. **12,** 245–266, Amsterdam.

SARNTHEIN, M., 1978: Sand Desert During Glacial Maximum and Climatic Optimum. Nature **272,** 24–45, London.

SARNTHEIN, M., 1979: Nature **272,** 355–358, London.

SAXEN, A., 1967: Situation der bewässerten Landwirtschaft in der Ostprovinz Saudi Arabiens. Z. Kulturtechnik Flurbereinigung **8,** 6 321–344, Berlin–Hamburg.

SCHARLAU, K., 1958: Zum Problem der Pluvialzeiten in Nordost-Iran. Z. Geomorph. N. F. **2,** 258–277, Berlin–Stuttgart.

SCHMIDT, D. L., HADLEY, D. G., GREENWOOD, W. R., GONZALES, L., COLEMAN, R. G., BROWN, G. F., 1973: Stratigraphy and Tectonism of the southern part of the Precambrian Shield of Saudi Arabia. Dir. Gen. Mineral Resources, Min. Res. Bull. **8,** 13, Jiddah.

SCHMIDT, D. L., HADLEY, D. G., STOESSER, D. G., 1978: Late Proterozoic crustal history of the Arabian Shield southern Najd Province. Saudi Arabia Directorate General of Mineral Resources, USGS Project Report 251. Jiddah.

SCHOELL, M., FABER, E., 1976: Survey on the isotopic composition of waters from NE Africa. Geol. Jb. **D17,** 197–213.

SCHULZ, E., 1974: Pollenanalytische Untersuchungen quartärer Sedimente des Nordwest-Tibesti. Forschungsstation Bardai, FU-Berlin **5,** 59–69.

SCHWARZBACH, M., 1974: Das Klima der Vorzeit. Stuttgart: F. Enke.

SCHYFSMA, E., 1978a: As Sub Plateau. General Geology. In: Quaternary Period in Saudi Arabia, 1 (AL-SAYARI, S. S., ZÖTL, J. G., eds.), pp. 163–166. Wien–New York: Springer.

SCHYFSMA, E., 1978b: Climate of the Arabian Peninsula. In Quaternary Period in Saudi Arabia, 1 (AL-SAYARI, S. S., ZÖTL, J. G., eds.), pp. 31–44. Wien–New York: Springer.

SCHYFSMA, E., 1978c: Cuesta Region of the Tuwayq Mountains. General Geology and Stratigraphy. In: Quaternary Period in Saudi Arabia, 1 (AL-SAYARI, S. S., ZÖTL, J. G., eds.), pp. 194–202. Wien–New York: Springer.

SERVANT, M., SERVANT, S., 1969: Chronologie du Quaternaire Récent des Bases du Tchad. Acad. Sci. J. Paris. Compt. Rendus, Serie D **269,** 1603–1606, Paris.

SHAMPINE, W. J., DINCER, T., NOORY, M., 1979: An Evaluation of Isotope Concentrations in the Groundwater of Saudi Arabia. IAEA-SM-228/23, Vienna, pp. 443–463.

SHINN, E. A., 1969: Submarine Lithification of Holocene Carbonate Sediments in the Persian Gulf. Sedimentology **12,** 109–144, Amsterdam.

SKIPWITH, P., 1973: The Red Sea and Coastal Plain of the Kingdom of Saudi Arabia. Dir. Gen. Min. Resources, Techn. rec. TR-1973-1, p. 149, Jiddah.

SKOCEK, V., SAADALLAH, A. A., 1972: Heavy Minerals in Eolian Sands, Southern Desert, Iraq. Sedimentary Geology **8,** 29–45, Amsterdam.

SMALLEY, I. J., VITA, F. C., 1968: The Formation of Fine Particles in Sandy Desert and the Nature of Desert Loess. J. Sed. Petrology **38,** 766–774, Tulsa, Oklahoma.

SMIRTH, G. I., 1974: Quaternary Deposits in Southwestern Afghanistan. Quaternary Research **4**, 39–52.

SMITH, J. W., 1979: Geology of the Wadi Azlam quadrangle sheet 27/36 C Kingdom of Saudi Arabia. Saudi Arabia Directorate General of Mineral Resources Geologic Map GM 36, scale 1 : 200,000 Jeddah.

SOGREA, 1970: Final report of area VI, water resources. Saudi Arabian Ministry of Agriculture and Water. Unpublished Report.

SONNTAG, CH., 1979: Palaeoclimatic Information from Deuterium and Oxygen-18 in Carbon-14 dated North Saharian Groundwaters. IAEA-SM-228/28, Vienna, pp. 569–581.

SONNTAG, CH., KLITZSCH, E., EL SHAZLY, E. M., KALINKE, CHR., MÜNNICH, K. O., 1978: Paläoklimatische Information im Isotopengehalt ^{14}C-datierter Saharawässer: Kontinentaleffekt in δ^2H and $\delta^{18}O$. Geol. Rdsch. **67**, 413–424.

SONNTAG, CH., et al., 1978: Paleoclimatic Information from D and O-18 in C-14 dated North Saharian Groundwaters. Groundwater Formation in the Past. Isotope Hydrology, IAEA, Vienna.

SRDOČ, D., OBELIC, B., HORVATINČIC, N., 1980: Radiocarbon dating of calcareous tufa: How reliable data can we expect? Radiocarbon **22**, 858–862, New Haven.

STEIGER, R. H., JÄGER, E., 1977: Subcommission on Geochronology: Convention on the use of decay constants in geo- and cosmochronology. Earth Planet. Sci. Letters **36**, 359–362, Amsterdam.

STEINEKE, et al., 1958: Geologic map of the Western Persian Gulf Quadrangle: Kingdom of Saudi Arabia. USGS, Misc. Geol. Investign., Map I-208A, 1 : 500,000, Washington.

STIFFERS, P., ROSS, D. A., 1974: Sedimentary history of the Red Sea. In: Initial reports of the Deep Sea Drilling Project, Vol. 23 (WHITMARSH, R. B., WESER, O. E., ROSS, D. A., et al., eds.), pp. 849–865. Washington, U. S. Govt. Printing Office.

STREET, F. A., GROVE, A. T., 1976: Environmental and Climatic Implications of Late Quaternary Lake-level Fluctuations in Africa. Nature **261**, 385–390, London.

STREET, F. A., 1979: Chronology of Late Pleistocene and Holocene Lake-level Fluctuations. Ziway-Shala Basin, Ethiopia. Proc. VIII Pan Afr. Congr. Prehist. Quat. Studs., Nairobi.

STREET, F. A., GROVE, A. T., 1979: Global Maps of Lake-level Fluctuations since 30,000 yr. B. P. Quaternary Research **12**, No. 1, July.

STYLES, P., HALL, S. A., 1980: A comparison of the seafloor spreading histories of the western Gulf of Aden and the central Red Sea. In: Geodynamic evolution of the Afro-Arabian rift system. Accad. Naz. Lincei, Rome, pp. 587–606.

SUESS, H. E., 1980: The radiocarbon record in tree rings of the last 8000 years. Radiocarbon **22**, 200–209, New Haven.

SUGDEN, W., 1963: The Hydrology of the Persian Gulf and its Significance with Respect to Evaporite Deposition. Amer. J. Sci. **261**, 741–755, New Haven.

SUGDEN, W., STANDRING, A. J., 1975: Qatar Peninsula. Lexiq. Stratigr. Intern. **3**, 10b, 3, 120, Paris.

TATCHER, L., RUBIN, M., BROWN, G. F., 1961: Dating desert groundwater. Science **134**, 105–106, Washington.

TAYLOR, J. C. M., ILLING, L. V., 1969: Holocene intertidal calcium carbonate cementation Qatar, Persian-Gulf. Sedimentology **12**, 69–107, Amsterdam.

THRALLS, H. W., HASSON, R. C., 1956: Geology and Oil Resources of Eastern Saudi Arabia. Internat. Geol. Cong. 20th, Mexico, Symposium Sobre Yacimentos de Petroleo, Gas **2**, 9–32.

TORRENT, H., SAUVEPLANE, S., 1977: Orientation Map for Groundwater Exploration Related to Mineral Investigations in the Arabian Shield. BRGM, Jeddah, 6 p.

TRICART, J., 1956: Tentative de Corrélation des Périodes Pluviales Africaines et des Périodes Glaciaires. Comptes Rendus Sommaires de la Société Géologique de France 9–10, pp. 164–167.

TURNER, W. M., 1971: Quaternary Sea Levels of Western Cyprus. Quaternaria **15**, 197–202, Roma.

UHLIG, D., 1971: Al Hassa Irrigation and Drainage Project Kingdom of Saudi Arabia. WAKUTI, Siegen.

UHLIG, D., 1971: Anwendung von Stahlbetonfertigteilen für das Bewässerungssystem des Al-Hassa-Projektes. Die Bautechnik **48**, 109–113, Berlin–München–Düsseldorf.

U.S.G.S.: Arabian-American Oil Company, 1963: Geologic Map of the Arabian Peninsula. USGS, Misc. Geol. Investign. Map I-270A.

U.S. Geological Survey Saudi Arabian Project, 1972: Topographic Map of the Arabian Peninsula. Arabian Peninsula Series Map Ap-1, 1 : 4,000,000, Washington.

VAN DAALHOFF, H., 1974: Mineral Locality Map of the Arabian Shield. Metalliferous Minerals. Saudi Arabian Dir. Gen. Min. Res. Geol. Map GM-15, Plate 1, 2,000,000, Jiddah.

VAN DAALHOFF, H., 1974: Mineral Locality Map of the Arabian Shield. Saudi Arabian Dir. Gen. Min. Res. Geol. Map GM-15, Plate 2, 1 : 2,000,000, Jiddah.

VAQUEZ-LOPEZ, R., 1981: Prospecting in the sedimentary Formations of the Red Sea coast between Yanbu Al Bahr and Maqna 1968–1979. Technical record BRGM-TR-01-1 Deputy Ministry for Mineral Resources, Jiddah.

VITA-FINZI, C., 1979: Rates of Holocene folding in the coastal Zagros near Bandar Abbas. Nature **278,** 632–634, London.

VITA-FINZI, C., 1980: ^{14}C dating of recent coastal movements in the Persian Gulf and Iranian Makran. Radiocarbon **22,** 763–773.

VITA-FINZI, C., 1982: Recent coastal deformation near the Strait of Hormuz. Proc. R. Soc. Lond. **A382,** 441–457, London.

VOSS, B., 1979: Ein Beitrag zur Hydrogeologie und Wasserwirtschaft der Oase Al Hassa in Saudi Arabien. Fortschritt – Berichte der VDI Zeitschriften, Reihe 4, **48,** 1–165, Düsseldorf.

VOUTE, C., 1960: Climate and Landscape in the Zagros Mountains (Iran). Intern. Geol. Cong. 21st, Copenhagen 1960, Rep. 4, pp. 81–87, Copenhagen.

VOUTE, C., WEDMAN, E. J., 1963: The Quaternary Climate as a Morphological Agent in Iraq. UNESCO, Arid Zone Research **20,** 395–402, Paris.

WADIA, D. N., 1939: Geology of India, p. 460. London: Macmillan and Co.

WALKER, B. M., 1979: Shell dissolution: Destructive diagenesis in a meteoric environment. Scanning Electron Microscopy **2,** 463–468, AMF O'Hare (I1).

WALTHER, J. K., 1888: Die Korallenriffe der Sinai Halbinsel. Geologische und biologische Betrachtungen. Akad. Wiss. Leipzig Abh., Math.-Naturwiss. Kl., **14.**

WHITEMAN, A. J., 1971: The Geology of the Sudan Republic, p. 290. Oxford: Oxford Univ. Press, Clarenden.

WHITEMAN, M. A. J., 1975: Late Pleistocene Tropical Aridity Synchronous in both Hemispheres. Nature **253,** 617–618, London.

WHITMARSH, R. B., ROSS, D. A., ALI, S. A., BOUDREAUX, J. E., COLEMAN, R. G., FLEISHER, R. L., GIRDLER, R. W., MANHEIM, F. T., MATTER, A., NIGRINI, C., STOFFERS, P., SUPKO, R., 1974: Sites 225–230. In: Initial reports of the Deep Sea Drilling Project, Vol. 23 (WHITMARSH, R. B., WESER, O. E., ROSS, D. A., et al., eds.), pp. 539–812. Washington, U. S. Govt. Printing Office.

WHITMARSH, R. B., WESER, O. E., ROSS, D. A., et al., 1974: Initial reports of the Deep Sea Drilling Project, Vol. 23, 1180 p. Washington, U. S. Govt. Printing Office, 1180 p.

WILCOX, L. V., 1948: The quality of water for irrigation use. U. S. Dept. Agr. Techn. Bull., **962,** 40 p., Washington.

WILLIAMS, G. E., POLACH, H. A., 1971: Radiocarbon dating of arid-zone calcareous paleosols. Geol. Soc. Amer. Bull. 82, 3069–3086, Boulder, Colorado.

WILLIAMS, M. A., 1976: Radiocarbon Dating and Late Quaternary Sahara Climates. A discussion. Z. Geomorph. N. F. **20,** 3, 361–362, Berlin, Stuttgart.

WILLIAMS, M. A. J., 1980: Late Quaternary Depositional History of the Blue and White Nile Rivers in Central Sudan. The S.A.T.N., Rotterdam.

WILLIAMS, M. A. J., et al., 1977: Late Quaternary Lake Levels in Southern Afar and the Adjacent Ethiopian Rift. Nature **267,** 690–693, London.

WILLIAMS, M. A. J., WILLIAMS, F. M., 1980: Evolution of the Nile Basin. The S.A.T.N., Rotterdam.

WILLIAMS, M. A. J., FAURE, 1980: The Sahara and the Nile. A. A. Balkema, Rotterdam.

WILSON, I. G., 1971: Desert Sandflow Basins and a Model for the Development of Ergs. Geogr. J. **137,** pt. 2, 180–199, London.

WIRTH, K., 1974: Spurenelementgehalte in Quellwässern und ihre Beziehungen zum durchflossenen Gestein. Geol. Mitt. **12,** 367–388, Aachen.

WOLDSTEDT, P., 1958–1965: Das Eiszeitalter, Bd. 1, 3. Aufl. 1961, 374 p., Bd. 2, 2. Aufl. 1958, 438 p., Bd. 3, 2. Aufl. 1965, 328 p. Stuttgart: Enke.

WOLFART, R., 1961: Geologic Hydrogeologic Research for the Utilization of Groundwater in the

Wadi Nisah Area. A supplement to the water supply of Ar-Riyadh, Saudi Arabia. Unpubl. Report. Bundesanstalt für Bodenforschung, Hannover.

WOLFART, R., 1961: Hydrogeology of the Central Tuwayq Mountains and Adjoining Regions (Saudi Arabia) Intern. Assoc. Sci. Hydrology, Publ. **57**, 98–112, Gentbrugge.

WOLFART, R., 1965: Geologie und Hydrogeologie des Mittleren Tuwayq-Gebirges und der angrenzenden Gebiete (Saudi Arabien). Geol. J. **83**, 149–190, Hannover.

WOLFART, R., 1966: Zur Geologie und Hydrogeologie von Syrien. Geol. Jb., Beihefte 68, 129, Hannover.

WOLFART, R., 1967: Geologie von Syrien und dem Libanon. Beitr. Reg. Geol. Erde **6**, Berlin: Borntraeger.

WRIGHT, H. E., 1961: Pleistocene Glaciation in Kurdistan. Eiszeitalter und Gegenwart **13**, 131–164, Ohringen.

WRIGHT, H. E., JR., FREY, D. G., eds., 1965: The Quaternary of the United States. Princeton.

WRIGHT, H. E., JR., 1976: The Environmental Setting for Plant Domestication in the Near East. Science **194**, 385–389.

YAALON, D. H., GANOR, E., 1973: The Influence of Dust on Soils During the Quaternary. Science **116**, 146–155.

ZAKIR, F. A. R., 1982: Preliminary study of the geology and tectonics of the Raghama Formation Maqna area, Wadi As' Sirhan quadrangle Kingdom of Saudi Arabia. Unpublished Ph. D. thesis, 240 p. Rapid City: South Dakota School of Mines and Technology.

ZEIST, W. V., 1967: Late Quaternary Vegetation History of Western Iran. Rev. Palaeobot. Palynol. **2**, 303–311, Amsterdam.

ZEUNER, F. E., 1952: Pleistocene Shore-lines. Geol. Rdsch. **40**, 39–50, Stuttgart.

ZEUNER, F. E., 1953: Das Problem der Pluvialzeiten. Geol. Rdsch. **41**, 242–253, Stuttgart.

ZEUNER, F. E., 1954: Neolithic Sites from the Rub'Al Khali. Man **209**, 133–136, London.

Index of Geographical Names

As there is no universally agreed on scheme for the transliteration of Arabic words into English, we have found some Arabic names spelled in many different ways in a series of various maps and reports. It was therefore decided to use a different approach when assigning place names.

The spelling of Arabic names in this volume aims at producing correct or nearly correct pronunciation by a person unfamiliar with Arabic but used to common Latin spelling. As it is not possible to transliterate directly from Arabic into English, the names were spelled phonetically according to their Latin pronunciation. The results may differ in some cases from other versions but it is believed that the spelling adopted here will result in pronunciation closer to the original Arabic than is usual with other schemes.

Transcription of Names Used in Figures and Plates

p. 7, Fig. 2:	Read Sea, correct: Red Sea
p. 7, Fig. 2:	Riyadh, correct: Ar Riyadh
p. 29, Fig. 8:	Jabal ar Raghama, correct: Jabal Ar Raghamah
p. 41, Fig. 15:	Ra's Ash Shaykh Humayd, correct: Ra's Ash Sheik Humayd
p. 55, Fig. 24:	Sha'ib an Nakhlah, correct: Wadi Sha'ib An Nakhla
p. 61, Fig. 28:	Aynunah, correct: El Aynunah
p. 79, Fig. 35:	Wadi al Hamd, correct: Wadi Al Hamd
p. 83, Fig. 38:	Jabal al Jarra, correct: Jabal Al Jarra
	Jabal al Buwana, correct: Jabal Al Buwana
	Jabal an Nabah, correct: Jabal An Nabah
	Ra's al Lakk, correct: Ra's Al Lakk
	Sharm al Kawr, correct: Sharm Al Kawr
p. 103, Fig. 47:	Wadi Hamd, correct: Wadi Al Hamdh
p. 125, Fig. 54:	Wadi Faydah, correct: Wadi Fayidah
	Wadi Ash Shamiah, correct: Wadi Ash Shamiyah
	Wadi Na'am, correct: Wadi Na'man
p. 138, Fig. 59:	Khanat, correct: kanat
p. 139, Fig. 61:	Khanat, correct: kanat
p. 146, Fig. 62:	Seismic date, correct: seismic data
p. 160, Fig. 64:	Malaki Bridge, correct: Maloki bridge
p. 179, Fig. 73:	Wadi Ta'ashar, correct: Wadi Ta'ashshar
p. 184, Fig. 75:	Wadi Ta'ashar, correct: Wadi Ta'ashshar
p. 189, Fig. 77:	Wadi Ta'ashar, correct: Wadi Ta'ashshar
p. 191, Fig. 79:	Wadi Ta'ashar, correct: Wadi Ta'ashshar
p. 197, Fig. 82:	Malaki Reservoir, correct: Maloki Reservoir
p. 271, Fig. 114:	Bir al Dumariyah, correct: Bir Al Dumariyah
	Bir at Harradiyah, correct: Bir At Harradiyah
Plate I:	Al Bad, correct: Al Bad'
	Aynunah, correct: El Aynunah
	Legend: Quarternary, correct: Quaternary
Plate II:	Wadi as Schamiyah, correct: Wadi Ash Shamiyah
	Wadi as Suqah, correct: Wadi As Suqah
	Wadi Na'am, correct: Wadi Na'man
Plate III:	Wadi Magab, correct: Wadi Maqab
	Wadi Harad, correct: Wadi Haradh
	Wadi Ta'ashar, correct: Wadi Ta'ashshar
Plate V:	Ar Rub' al Khali, correct: Ar Rub' Al Khali
Plate VI:	Duba, correct: Dhuba

Filmsatz und Offsetdruck: Ferdinand Berger & Söhne Gesellschaft m.b.H., 3580 Horn